…
GULF WAR and HEALTH

VOLUME 4

HEALTH EFFECTS OF SERVING IN THE GULF WAR

Committee on Gulf War and Health: A Review of the Medical Literature Relative to the Gulf War Veterans' Health

Board on Population Health and Public Health Practice

INSTITUTE OF MEDICINE
OF THE NATIONAL ACADEMIES

THE NATIONAL ACADEMIES PRESS
Washington, D.C.
www.nap.edu

THE NATIONAL ACADEMIES PRESS • 500 Fifth Street, NW • Washington, DC 20001

NOTICE: The project that is the subject of this report was approved by the Governing Board of the National Research Council, whose members are drawn from the councils of the National Academy of Sciences, the National Academy of Engineering, and the Institute of Medicine. The members of the committee responsible for the report were chosen for their special competences and with regard for appropriate balance.

This study was supported by Contract V101(93)P-2155 between the National Academy of Sciences and the Department of Veterans Affairs. Any opinions, findings, conclusions, or recommendations expressed in this publication are those of the author(s) and do not necessarily reflect the view of the organizations or agencies that provided support for this project.

International Standard Book Number-10: 0-309-10176-X
International Standard Book Number-13: 978-0-309-10176-9

Library of Congress Control Number: 2006934960

Additional copies of this report are available from the National Academies Press, 500 Fifth Street, NW, Lockbox 285, Washington, DC 20055; (800) 624-6242 or (202) 334-3313 (in the Washington metropolitan area); Internet, http://www.nap.edu.

For more information about the Institute of Medicine, visit the IOM home page at **www.iom.edu.**

Copyright 2006 by the National Academy of Sciences. All rights reserved.

Printed in the United States of America.

The serpent has been a symbol of long life, healing, and knowledge among almost all cultures and religions since the beginning of recorded history. The serpent adopted as a logotype by the Institute of Medicine is a relief carving from ancient Greece, now held by the Staatliche Museen in Berlin.

*"Knowing is not enough; we must apply.
Willing is not enough; we must do."*
—Goethe

INSTITUTE OF MEDICINE
OF THE NATIONAL ACADEMIES

Advising the Nation. Improving Health.

THE NATIONAL ACADEMIES
Advisers to the Nation on Science, Engineering, and Medicine

The **National Academy of Sciences** is a private, nonprofit, self-perpetuating society of distinguished scholars engaged in scientific and engineering research, dedicated to the furtherance of science and technology and to their use for the general welfare. Upon the authority of the charter granted to it by the Congress in 1863, the Academy has a mandate that requires it to advise the federal government on scientific and technical matters. Dr. Ralph J. Cicerone is president of the National Academy of Sciences.

The **National Academy of Engineering** was established in 1964, under the charter of the National Academy of Sciences, as a parallel organization of outstanding engineers. It is autonomous in its administration and in the selection of its members, sharing with the National Academy of Sciences the responsibility for advising the federal government. The National Academy of Engineering also sponsors engineering programs aimed at meeting national needs, encourages education and research, and recognizes the superior achievements of engineers. Dr. Wm. A. Wulf is president of the National Academy of Engineering.

The **Institute of Medicine** was established in 1970 by the National Academy of Sciences to secure the services of eminent members of appropriate professions in the examination of policy matters pertaining to the health of the public. The Institute acts under the responsibility given to the National Academy of Sciences by its congressional charter to be an adviser to the federal government and, upon its own initiative, to identify issues of medical care, research, and education. Dr. Harvey V. Fineberg is president of the Institute of Medicine.

The **National Research Council** was organized by the National Academy of Sciences in 1916 to associate the broad community of science and technology with the Academy's purposes of furthering knowledge and advising the federal government. Functioning in accordance with general policies determined by the Academy, the Council has become the principal operating agency of both the National Academy of Sciences and the National Academy of Engineering in providing services to the government, the public, and the scientific and engineering communities. The Council is administered jointly by both Academies and the Institute of Medicine. Dr. Ralph J. Cicerone and Dr. Wm. A. Wulf are chair and vice chair, respectively, of the National Research Council.

www.national-academies.org

COMMITTEE ON GULF WAR AND HEALTH: A REVIEW OF THE MEDICAL LITERATURE RELATIVE TO GULF WAR VETERANS' HEALTH

LYNN R. GOLDMAN, MD, MPH, *(chair)* Professor, Bloomberg School of Public Health, Johns Hopkins University, Baltimore, MD
MARCIA ANGELL, MD, Senior Lecturer on Social Medicine, Department of Social Medicine, Harvard Medical School, Boston, MA
W. KENT ANGER, PhD, Associate Director for Occupational Research, Center for Research on Occupational and Environmental Toxicology, Oregon Health and Science University, Portland, OR
MICHAEL BRAUER, ScD, Professor, School of Occupational and Environmental Hygiene, University of British Columbia, Vancouver, British Columbia
DEDRA S. BUCHWALD, MD, Director, Harborview Medical Center, University of Washington, Seattle, WA
FRANCESCA DOMINICI, PhD, Associate Professor, Bloomberg School of Public Health, Johns Hopkins University, Baltimore, MD
ARTHUR L. FRANK, MD, PhD, Professor, Chair, Department of Environmental and Occupational Health, Drexel University School of Public Health, Philadelphia, PA
FRANCINE LADEN, ScD, Assistant Professor of Medicine, Channing Laboratory, Harvard Medical School, Boston, MA
DAVID MATCHAR, MD, Director, Center for Clinical Health Policy Research, Duke University Medical Center, Durham, NC
SAMUEL J. POTOLICCHIO, MD, Professor, Department of Neurology, George Washington University Medical Center, Washington, DC
THOMAS G. ROBINS, MD, MPH, Professor, Department of Environmental Health Sciences, University of Michigan School of Public Health, Ann Arbor, MI
GEORGE W. RUTHERFORD, MD, Professor, Vice-Chair, Department of Epidemiology and Biostatistics, Division of Preventive Medicine and Public Health, School of Medicine, University of California, San Francisco, CA
CAROL A. TAMMINGA, M.D., Professor, Department of Psychiatry, University of Texas, Southwestern Medical Center, Dallas, TX

STAFF

CAROLYN FULCO, Senior Program Officer
ABIGAIL MITCHELL, Senior Program Officer
DEEPALI PATEL, Senior Program Associate
MICHAEL SCHNEIDER, Senior Program Associate
JUDITH URBANCZYK, Senior Program Associate
HOPE HARE, Administrative Assistant
PETER JAMES, Research Associate
DAMIKA WEBB, Research Assistant
RENEE WLODARCZYK, Intern
NORMAN GROSSBLATT, Senior Editor
ROSE MARIE MARTINEZ, Director, Board on Population Health and Public Health Practice

CONSULTANTS

MIRIAM DAVIS, Independent Medical Writer, Silver Spring, MD
ANNE STANGL, Tulane School of Public Health and Tropical Medicine, New Orleans, LA

REVIEWERS

This report has been reviewed in draft form by persons chosen for their diverse perspectives and technical expertise in accordance with procedures approved by the National Research Council's Report Review Committee. The purpose of this independent review is to provide candid and critical comments that will assist the institution in making its published report as sound as possible and to ensure that the report meets institutional standards for objectivity, evidence, and responsiveness to the study charge. The review comments and draft manuscript remain confidential to protect the integrity of the deliberative process. We wish to thank the following for their review of this report:

ARTHUR K. ASBURY, MD, Department of Neurology, University of Pennsylvania, Philadelphia, PA

SHARON COOPER, PhD, Professor and Chair, Department of Epidemiology and Biostatistics, Texas A & M University School of Rural Public Health, College Station, TX

PETER J. DYCK, MD, Director, Peripheral Nerve Research Laboratory, Mayo Clinic College of Medicine, Rochester, MN

DAVID GAYLOR, PhD, MS, President, Gaylor & Associates, LLC, Eureka Springs, AR

JACK M. GORMAN, MD, President and Psychiatrist in Chief, McLean Hospital, Belmont, MA

PHILIP GREENLAND, MD, Executive Associate Dean for Clinical and Translational Research, Northwestern University Feinberg School of Medicine, Chicago, IL

HOWARD KIPEN, MD, MPH, Director, Clinical Research and Occupational Medicine Division, Environmental & Occupational Health Sciences Institute, UMDNJ-Robert Wood Johnson Medical School, Piscataway, NJ

JOSEPH LADOU, MD, Editor, *International Journal of Occupational and Environmental Health*, Professor, Division of Occupational and Environmental Medicine, University of California, San Francisco, CA

ELLEN REMENCHIK, MD, MPH, Assistant Professor, Occupational and Environmental Medicine, The University of Texas Health Center, Tyler, TX

KATHERINE S. SQUIBB, PhD, Associate Professor & Head, Division of Environmental Epidemiology & Toxicology, University of Maryland School of Medicine, Baltimore, MD

Although the reviewers listed above have provided many constructive comments and suggestions, they were not asked to endorse the conclusions or recommendations nor did they see the final draft of the report before its release. The review of this report was overseen by **David J. Tollerud,** Professor and Chair, Department of Environmental and Occupational Health Sciences, University of Louisville and by **Harold Sox,** editor, *Annals of Internal Medicine*, American College of Physicians of Internal Medicine. Appointed by the National Research Council, Dr. Sox was responsible for making certain that an independent examination of this report was carried out in accordance with institutional procedures and that all review comments were carefully considered. Responsibility for the final content of this report rests entirely with the authoring committee and the institution.

PREFACE

The 1990-1991 Persian Gulf War was brief and entailed few US casualties in comparison with other wars, and yet it had a profound impact on the lives of many of the troops. Among the 700,000 US military personnel deployed in the battle theater, many veterans have reported chronic symptoms and illnesses that they have attributed to their service in the gulf. Numerous studies have been conducted to characterize the long-term adverse health consequences of deployment to the Persian Gulf.

Potential exposures to numerous hazardous substances have been identified in association with the Gulf War. Most alarming are the smoke from oil-well fires that were set by Iraqis as they retreated at the end of the war and the potential exposures arising from the US military bombing of a poison-gas munitions dump at a location called Khamisiyah. Military personnel have also been reported to have had other exposures, such as to fuels, vaccines, pharmaceuticals, and pathogens. Most recently, the Department of Defense published a report documenting a large amount of pesticide use in the war theater. For most of those exposures, it is difficult or impossible to reconstruct doses because of lack of exposure measurements on either the individual or group level. The situation is compounded by the stress experienced by many veterans during deployment and in some cases after deployment. Stress is known to have serious acute and chronic health effects, but at the time of the Gulf War relatively little attention was given to reduction of stress and its consequences.

The Department of Veterans Affairs (VA) and the US Congress have secured the assistance of the Institute of Medicine (IOM) in evaluating the scientific literature regarding possible health outcomes associated with exposures that might have occurred in the Gulf War, IOM has published several volumes that review the clinical diseases that might be associated with exposures, such as exposure to sarin gas, depleted uranium, pesticides, solvents, rocket propellants, fuels, and combustion products. Such reviews continue and will provide information about illnesses related to exposure to pathogens, stress, and chemical agents. The congressional request regarding the possible association between illness and exposures in the gulf is similar to the approach Congress took after the Vietnam War to address the potential adverse health effects of exposure to Agent Orange.

The current report, however, takes a different approach, which is to identify the adverse health effects, if any, that are occurring among Gulf War veterans and thus might warrant further attention, either on the individual level or for the Gulf War veterans as a whole. Many of the relevant studies are limited by the lack of objective exposure information. Although there is a blood test that can provide an indication of exposure to Agent Orange and dioxin that occurred many years ago, there is not biological measure that can be employed today to assess exposures during the Gulf War. Another limitation is that most studies have relied on self-reports of symptoms and symptom-based case definitions to determine whether rates of diseases were increased among Gulf War veterans. Nonetheless, some studies do point to psychiatric disorders and neurologic end points that might be associated with Gulf War service and for which it might be possible to develop new approaches to prevention and clinical treatment that could benefit not only Gulf War veterans but also veterans of later conflicts. Our committee does not recommend that more such studies be undertaken for the Gulf War veterans, but, there would be value in continuing to monitor the veterans for some health end points, specifically, cancer, especially brain and testicular cancers, neurologic diseases including Amyotrophic Lateral Sclerosis (ALS), and causes of death. Therefore, despite the serious limitations of the available studies as a group, they do point the way to actions that might benefit Gulf War and other combat veterans.

I am deeply appreciative of the expert work of our committee members: Marcia Angell, W. Kent Anger, Michael Brauer, Dedra S. Buchwald, Francesca Dominici, Arthur L. Frank, Francine Laden, David Matchar, Samuel J. Potolicchio, Thomas G. Robins, George W. Rutherford, and Carol Tamminga. Although our committee developed conclusions independently of input from IOM and its staff, we deeply appreciate their hard work and attention to detail and the extensive research that they conducted to ensure that we had all the information that we needed from the outset. It has been a privilege and a pleasure to work with the IOM staff directed by Carolyn Fulco and with our consultant, Miriam Davis. Without them, this report would not have been possible. Most of all, our committee appreciates the veterans who served in the Gulf War and who have volunteered again and again to participate in the health studies that we reviewed. It is for them that we do this work. We hope this report will inform those who have given so much to our nation about what researchers have been able to learn about their health.

LYNN R. GOLDMAN, MD, MPH
PROFESSOR
JOHNS HOPKINS UNIVERSITY

CONTENTS

Summary 1
   Charge to the Committee 1
   Committee's Approach to Its Charge...1
   Limitations of the Gulf War Studies 2
   Overview of Health Outcomes...2
      Outcomes Based Primarily on Symptoms or Self-Reports...3
      Outcomes with Objective Measures or Diagnostic Medical Tests 5
   Recommendations..7
      Predeployment and Postdeployment Screening..7
      Exposure Assessment 7
      Surveillance for Adverse Outcomes 8
   Brief Summary of Findings and Recommendations 9

1   Introduction..11
   Background 11
   The Gulf War Setting 12
      Deployment...12
      Living Conditions...13
      Environmental and Chemical Exposures 13
      Threat of Chemical and Biologic Warfare 14
   Charge to the Committee 15
   Committee's Approach to Its Charge...15
   Inclusion Criteria..15
   Complexities in Resolving Gulf War and Health Issues...16
      Multiple Exposures and Chemical Interactions 16
      Limitations of Exposure Information 16
      Individual Variability..17
      Unexplained Symptoms 17
   Organization of the Report...18
   References..18

2   Exposures in the Persian Gulf..21
   Exposure Assessment in Epidemiologic Studies 21
   Studies Assessing Exposures with Questionnaires 21
      Exposure to Oil-Well Fire Smoke 22
      Exposure to Vaccination...22

Exposure to Pyridostigmine Bromide...23
Exposure to Depleted Uranium 24
General Cohort Studies (Prevalence Studies) 25
Studies Using Simulation to Assess the Potential Magnitude of Exposures 26
Tent Heaters..26
Khamisiyah Demolition and Potential Exposure to Sarin and Cyclosarin.......................26
Epidemiologic Studies Using Fate and Transport Models
    to Assess Exposure to Sarin and Cyclosarin..35
Studies Using Environmental Fate and Transport Models for Specific Exposures............37
Studies Using Biologic Monitoring for Specific Exposures..39
Depleted Uranium..39
Oil-Well Fire Smoke..40
Summary and Conclusions...41
References...41

3 Considerations in Identifying and Evaluating the Literature...45
Types of Epidemiologic Studies 45
Cohort Studies 45
Case-Control Studies 47
Cross-Sectional Studies 47
General Remarks 48
Defining a New Syndrome..48
Statistical Techniques Used to Develop a Case Definition..49
Inclusion Criteria..51
Additional Considerations..51
Bias 52
Confounding 52
Chance 52
Multiple Comparisons 52
Assignment of Causality..53
Limitations of Gulf War Veteran Studies 53
Summary 53
References...54

4 Major Cohort Studies..55
General Limitations of Gulf War Cohort Studies and Derivative Studies........................56
Organization of This Chapter..58
Population-Based Studies 58
The Iowa Study...58
Department of Veterans Affairs Study 60
Oregon and Washington Veteran Studies..63
Kansas Veteran Study..64
Canadian Veteran Study 65
United Kingdom Veteran Studies..65

          Danish Peacekeeper Studies 68
          Australian Veteran Studies 69
     Military-Unit-Based Studies 70
          Ft. Devens and New Orleans Cohort Studies 70
          Seabee Reserve Battalion Studies...71
          Larger Seabee Cohort Studies 73
          Pennsylvania Air National Guard Study...74
     Other Cohort Studies...75
          Hawaii and Pennsylvania Active Duty and Reserve Study 76
          New Orleans Reservist Studies...76
          Air Force Women Study...76
          Connecticut National Guard 77
     References..105

5 Health Outcomes..115
     Cancer (ICD-10 C00-D48)...115
          Primary and Secondary Studies..116
          Summary and Conclusion...118
     Mental and Behavioral Disorders (ICD-10 F00-F99) 122
          Primary Studies...123
          Secondary Studies...127
          Summary and Conclusion...127
     Neurobehavioral and Neurocognitive Outcomes (ICD-10 F00-F99) 131
          Neurobehavioral Tests and Confounding Factors 131
          Studies That Respond to Question 1 (Outcomes in Gulf War-Deployed Veterans
               vs Veterans Deployed Elsewhere or Not Deployed) 132
          Studies That Respond to Question 2 (Symptomatic vs Nonsymptomatic Veterans) 135
          Related Findings: Malingering and Association of Symptoms
               with Objective Test Results 140
          Summary and Conclusion...140
     Diseases of the Nervous System (ICD-10 G00-G99) 153
          Amyotrophic Lateral Sclerosis 153
          Summary and Conclusion...155
          Peripheral Neuropathy and Other Neurologic Outcomes.....................157
          Summary and Conclusion...159
     Chronic Fatigue Syndrome 161
          Primary Studies...162
          Secondary Studies...162
          Summary and Conclusion...163
     Diseases of the Circulatory System (ICD-10 I00-I99)...............................166
          Primary Studies...166
          Secondary Studies...167
          Summary and Conclusion...168
     Diseases of the Respiratory System (ICD-10 J00-J99)..............................170
          Associations of Respiratory Outcomes with Deployment in the Gulf War Theater 170

Associations of Respiratory Outcomes with Specific Exposures
    Experienced by Gulf War Veterans During Their Deployment 172
  Summary and Conclusion..174
Diseases of the Digestive System (ICD-10 K00-K93) 180
  Primary Studies..180
  Secondary Studies...181
  Summary and Conclusion..181
Diseases of the Skin and Subcutaneous Tissue (ICD-10 L00-L99)....................................183
  Primary Studies..183
  Secondary Studies...183
  Summary and Conclusion..183
Diseases of the Musculoskeletal System and Connective Tissue (ICD-10 M00-M99)....185
  Arthritis and Arthralgia..185
  Summary and Conclusion..186
Fibromyalgia..188
  Primary Studies..188
  Secondary Studies...189
  Summary and Conclusion..190
Birth Defects and Adverse Pregnancy Outcomes (ICD-10 O00-Q99)..............................192
  Birth Defects..192
  Summary and Conclusion..194
  Adverse Pregnancy Outcomes..195
  Summary and Conclusion..195
  Male Fertility Problems and Infertility 196
Symptoms, Signs, and Abnormal Clinical
    and Laboratory Findings (ICD-10 R00-R99)...202
  Unexplained Illness 202
  Hospitalizations for Unexplained Illness...202
  Factor-Analysis Derived Syndromes 203
  Cluster Analysis...212
  Summary and Conclusion..213
Injury and External Causes of Morbidity and Mortality (ICD-10 S00-Y98) 219
  Primary Studies..219
  Secondary Studies...220
  Summary and Conclusion..220
All-Cause Hospitalization Studies 223
  Primary Studies..223
  Summary and Conclusion..224
Multiple Chemical Sensitivity..227
  Primary Studies..227
  Secondary Studies...228
  Summary and Conclusion..229
References...232

6	Conclusions and Recommendations 247

    Quality of the Studies...247
    Overview of Health Outcomes...247
        Outcomes Based Primarily on Symptoms and Self-Reports 248
        Outcomes with Objective Measures or Diagnostic Medical Tests................251
    Recommendations..254
        Predeployment and Postdeployment Screening...254
        Exposure Assessment 254
        Surveillance for Adverse Outcomes 254
    References..255

Index 261

SUMMARY

Although the 1990-1991 Persian Gulf War was considered a brief and successful military operation with few injuries and deaths among coalition forces, many returning veterans soon began reporting numerous health problems that they believed to be associated with their service in the Persian Gulf.

In 1998, in response to the growing concerns of the ill Gulf War veterans, Congress passed two laws: PL 105-277, the Persian Gulf War Veterans Act, and PL 105-368, the Veterans Programs Enhancement Act. Those laws directed the secretary of veterans affairs to enter into a contract with the National Academy of Sciences (NAS) to review and evaluate the scientific and medical literature regarding associations between illness and exposure to toxic agents, environmental or wartime hazards, and preventive medicines or vaccines associated with Gulf War service and to consider the NAS conclusions when making decisions about compensation. Those studies were assigned to the Institute of Medicine (IOM).

This study, conducted at the request of the Department of Veterans Affairs (VA), differs from the previous work of IOM in that it summarizes in one place the current status of health effects in veterans deployed to the Persian Gulf irrespective of exposure information. One can confidently assess health responses associated only with deployment in the Gulf War Theater. Estimating the veterans' health risks associated with particular environmental exposures is challenged by the lack of exposure monitoring and of biomarkers to quantify individual exposures of veterans during the deployment retrospectively.

CHARGE TO THE COMMITTEE

The charge to this IOM committee was to review, evaluate, and summarize peer-reviewed scientific and medical literature addressing the health status of Gulf War veterans. The study was to help to inform the VA of illnesses among Gulf War veterans that might not be immediately evident.

COMMITTEE'S APPROACH TO ITS CHARGE

The committee began its evaluation by presuming neither the existence nor the absence of illnesses associated with deployment. It sought to characterize and weigh the strengths and limitations of the available evidence. The committee did not concern itself with policy issues, such as decisions regarding disability, potential costs of compensation, or any broad policy implications of its findings.

Extensive searches of the scientific and medical literature were conducted, and over 4,000 potentially relevant references were retrieved. After assessment of the titles and abstracts references found in of the initial searches, the committee focused on 850 potentially relevant epidemiologic studies for its review and evaluation.

The committee limited its review of the literature primarily to epidemiologic studies of Gulf War veterans to determine the prevalence of diseases and symptoms in that population. Those studies typically examine veterans' health outcomes in comparison with outcomes in their nondeployed counterparts.

The committee decided to use only peer-reviewed published literature on which to base its conclusions. The process of peer review by fellow professionals increases the likelihood of a high-quality study but does not guarantee its validity or the generalizability of its findings to the entire group of subjects under review. Accordingly, committee members read each study critically and considered its relevance and quality. The committee did not collect original data, nor did it perform any secondary data analysis (exception to calculate response rates for consistency among studies).

After securing the full text of the peer-reviewed epidemiologic studies it would review, the committee determined which studies would be considered primary or secondary studies. Primary studies provide the basis of the committee's findings. To be included in the committee's review as a primary study, a study had to meet specified criteria. The criteria include studies that provide information about specific health outcomes, demonstrate rigorous methods, describe its methods in sufficient detail, include a control or reference group, have the statistical power to detect effects, and include reasonable adjustments for confounders. Other studies were considered secondary for the purpose of this review and provided background information or "context" for the report. Another step that the committee took in organizing its literature was to determine how all the studies were related to one another. Numerous Gulf War cohorts have been assembled, from several different countries; from those original cohorts many derivative studies have been conducted. The committee organized the literature into the major cohorts and derivative studies because they didn't want to interpret the findings of the same cohorts as though they were results from unique groups (Chapter 4).

LIMITATIONS OF THE GULF WAR STUDIES

Overall, the studies of Gulf War veterans' health are of varied quality. Although, they have provided valuable information, many of them have limitations that hinder accurate assessment of the veterans' health status. Common study limitations include use of a population that was not representative of the entire Gulf War population, reliance on self-reports rather than objective measures of symptoms, low participation rates, and a period of investigation that was too brief to detect health outcomes with long latency such as, cancer. In addition, many of the US studies are cross-sectional, and this limits the opportunity to learn about symptom duration, long-term health effects, latency of onset, and prognosis.

OVERVIEW OF HEALTH OUTCOMES

While examining health outcomes in Gulf War-deployed veterans, numerous researchers have attempted to determine whether a set of symptoms reported by veterans could be defined as a unique syndrome or illness. Investigators have attempted, by using factor or cluster analysis, to define a unique health outcome, but none has been identified. Every study reviewed by this committee found that veterans of the Gulf War report higher rates of nearly all symptoms examined than their nondeployed counterparts. That finding was applied not only to Gulf War veterans from the United States but also to the Gulf War veterans deployed from the UK, Canada, Australia, and Denmark. Some studies examined performance on neurocognitive tests in association with symptoms that were considered possibly indicative of neurological or cognitive impairment (such as headache, confusion, and memory problems). Those few studies seemed to indicate that Gulf War veterans with such symptoms demonstrated neurobehavioral deficits, but, most of the studies did not include control groups (or, in some cases, valid control groups).

In many studies, investigators found a higher prevalence not only of individual symptoms but also of chronic multisymptom illnesses among Gulf War-deployed veterans than among the nondeployed. Multisymptom-based medical conditions reported to occur more frequently among deployed Gulf War veterans include fibromyalgia, chronic fatigue syndrome (CFS), and multiple chemical sensitivity (MCS). However, the case definitions for those conditions are based on symptom reports, and there are no objective diagnostic criteria that can be used to validate the findings, so, it is not clear whether the literature supports a true excess of the conditions or whether the associations are spurious and result from the increased reporting of symptoms across the board. The literature also demonstrates that deployment places veterans at increased risk for symptoms that meet diagnostic criteria for a number of psychiatric illnesses, particularly posttraumatic stress disorder (PTSD), anxiety, depression, and substance abuse. In addition, comorbidities have been reported, for example, symptoms of both PTSD and depression. The committee felt confident that several studies validated the increased risk of psychiatric disorders.

Some studies indicate that Gulf War veterans are at increased risk for amyotrophic lateral sclerosis (ALS). With regard to birth defects, there is weaker evidence that Gulf War veterans' offspring might be at risk for some birth defects; the findings are inconsistent. There were increased rates of transportation-related injuries and mortality among deployed Gulf War veterans, however, that increase appears to have been restricted to the first several years after the war. Finally, long-term exacerbation of asthma appeared to be associated with oil-well fire smoke, but there were no objective measures of pulmonary function in the studies.

The health outcomes presented above are discussed in some detail in the following pages. They are grouped according to whether the findings were based on objective measures and diagnostic medical tests.

Outcomes Based Primarily on Symptoms or Self-Reports

The largest and most nationally representative survey of US veterans found that nearly 29% of deployed veterans met a case definition of "multisymptom illness", compared with 16% of nondeployed veterans. Those figures indicate that unexplained illnesses are the most prevalent health outcome of service in the Gulf War. Several researchers have tried to determine whether the symptoms that have been reported by Gulf War veterans cluster in such a way as to make up a unique syndrome, such as "Gulf War illness". The results of that research indicate that although deployed veterans report more symptoms and more severe symptoms than their nondeployed counterparts, there is not a unique symptom complex (or syndrome) in deployed Gulf War veterans.

Among the many symptoms reported by Gulf War veterans are deficits in neurocognitive ability. Obviously such reports are of concern because of the potential for those deficits to have adverse effects on the lives of the veterans. Primary studies of deployed Gulf War veterans and non-Gulf War-deployed veterans, however, have not demonstrated differences in cognitive and motor measures as determined with neurobehavioral testing. But studies of returning Gulf War veterans with at least one commonly reported symptom (fatigue, memory loss, confusion, inability to concentrate, mood swings, somnolence, gastrointestinal distress, muscle and joint pain or skin or mucous-membrane complaints) demonstrated poorer performance on cognitive tests than by returning Gulf War veterans who did not report such symptoms. Most of those studies did not include control groups (or in some cases valid control groups) so it is not possible to determine whether the combination of symptoms and neurocognitive-test decrements is uniquely associated with Gulf War service.

Several studies focused on multisymptom-based medical conditions: fibromyalgia, CFS, and MCS. Those conditions have several features in common: they do not fit a precise diagnostic category; case definitions are symptom-based (supplemented, in the case of fibromyalgia, by report of pain on digital palpation of tender points in a physical examination); there are no objective criteria independent of patient reports, such as laboratory test results, for validating the case definitions; and the symptoms among those syndromes are to some extent overlapping. Gulf War-deployed veterans report higher rates of symptoms that are consistent with the case definitions of MCS, CFS, and fibromyalgia.

Several large or population-based studies of Gulf War veterans found, by questionnaire, that the prevalence of MCS-like symptoms ranged from 2% to 6%. However, no two of the primary studies used the same definition of MCS, so it is difficult to compare them, and none performed medical evaluations to exclude other explanations, as would be required by the case definition of MCS.

The prevalence of CFS among Gulf War veterans is highly variable from study to study; most studies used the Centers for Disease Control and Prevention case definition. One primary study demonstrated a higher prevalence of CFS in deployed than in nondeployed veterans (1.6% vs 0.1%). Secondary studies also showed a higher prevalence of CFS and CFS-like illnesses among veterans deployed to the Persian Gulf than in to their counterparts who were not deployed or who were deployed elsewhere.

The diagnosis of fibromyalgia is based on symptoms and a very limited physical examination that consists of determining whether pain is elicited by pressing on several points on the body; there are no laboratory tests with which to confirm the diagnosis. Only one of the available cross-sectional studies included both Gulf War-deployed and -nondeployed veterans and used the full American College of Rheumatology case definition of fibromyalgia, including the physical-examination criteria. It found a statistically significant difference in prevalence of fibromyalgia between deployed and nondeployed veterans (2.0% vs 1.2%). Other studies using a case definition based on symptoms alone reported inconsistent results.

Other symptoms that are self-reported more often by deployed veterans are gastrointestinal symptoms, particularly dyspepsia; dermatologic conditions, particularly atopic dermatitis and warts; and joint pains.

There were many reports of gastrointestinal symptoms in Gulf War-deployed veterans. Those symptoms seem to be linked to reports of exposures to contaminated water and burning of animal waste in the war theater. The committee notes that several studies reported a higher rate of self-reported dyspepsia in deployed Gulf War veterans than in nondeployed veterans. In the context of nearly all symptoms being reported more frequently for Gulf War veterans, it is difficult to interpret those findings.

For dermatologic conditions, a few studies have included an examination of the skin and thus would be more reliable than self-reports. Those studies have reported that a few unrelated skin conditions occurred more frequently among Gulf War-deployed veterans; however, the findings are not consistent. From one study that did conduct a skin examination, there is some evidence of a higher prevalence of two distinct dermatologic conditions, atopic dermatitis and verruca vulgaris (warts), in Gulf War-deployed veterans.

Arthralgias (joint pains) were more frequently reported among Gulf War veterans. Likewise, self-reports of arthritis were more common among those deployed to the gulf. Again, in the context of global reporting increases, such data are difficult to interpret. Moreover, studies that included a physical examination did not find evidence of an increase in arthritis.

Finally, Gulf War veterans consistently have been found to suffer from a variety of psychiatric conditions. Two well-designed studies using validated interview-based assessments reported that several psychiatric disorders, most notably PTSD and depression, are 2-3 times more likely in Gulf War -deployed than in nondeployed veterans. Moreover, comorbidities were reported among a number of veterans, with co-occurrence of PTSD, depression, anxiety, or substance abuse. Most of the additional studies administered well-validated symptom questionnaires, and the findings were remarkably similar: an overall increase by a factor of 2-3 in the prevalence of psychiatric disorders.

Outcomes with Objective Measures or Diagnostic Medical Tests

A number of studies examined rates of injuries in Gulf War veterans. Those studies provide evidence of a modest increase in transportation-related injuries and deaths among deployed than among non-deployed Gulf War veterans in the decade immediately after deployment. However, studies with longer followup indicate that the increased injury rate was restricted to the first several years after the war.

With regard to all causes of hospitalization, studies provide some reassurance that excess hospitalizations did not occur among veterans of the Gulf War who remained on active duty through 1994, inasmuch as it has been noted that Gulf War veterans who left the military reported worse health outcomes than those who remained. Those studies, however, are limited by their inability to capture hospitalizations from illnesses that might have longer latency, such as some cancers. In addition, hospitalization data on people separated from the military and admitted to nonmilitary (Department of Veterans Affairs [VA] and civilian) hospitals or those who used outpatient facilities might be incomplete.

Veterans are understandably concerned about increases in cancer, and the studies reviewed did not demonstrate consistent evidence of increased overall cancer in the Gulf War veterans compared with nondeployed veterans. However, many veterans are young for cancer diagnoses, and, for most cancers, the time since the Gulf War is probably too short to expect the onset of cancer. Incidence of and mortality from cancer in general and brain and testicular cancer in particular have been assessed in cohort studies. An association of brain-cancer mortality with possible nerve-agent exposure was observed in one study, but however, there were many uncertainties in the exposure model used. Results for testicular cancer were mixed: one study concluded that there was no evidence of an excess risk, and another, small registry-based study suggested that there might be an increased risk.

Another concern for veterans has been whether ALS is increased in Gulf War veterans. Two primary studies and one secondary study found that deployed veterans appear to be at increased risk of for ALS. One primary study that had the possibility of underascertainment of cases in the nondeployed population was confirmed by a secondary analysis that documented a nearly 2-fold increase in risk. A secondary study that used general population estimates as the comparison group found a slightly higher relative risk.

Peripheral neuropathy has been studied in Gulf War veterans. One large, well-designed study conducted by VA which used a thorough and objective evaluation and a stringent case definition, did not find evidence of excess peripheral neuropathy. Several other secondary studies supported no excess risk. Thus, there does not appear to be an increase in the prevalence of peripheral neuropathy in deployed vs nondeployed veterans, as defined by history, physical examination, and electrophysiologic studies.

With regard to cardiovascular disease, primary studies found no significant differences between deployed and nondeployed veterans in rates of hypertension. One study did report a small but significant increase in hospitalizations due to cardiovascular disease among a subset of deployed veterans who were possibly exposed to the Khamisiyah plume compared with Gulf War-deployed veterans who were not in the suspected exposure area. The increased hospitalizations were due entirely to an increase in cardiac dysrhythmias. In secondary studies, deployed veterans were generally more likely to report hypertension and palpitations, but those reports were not confirmed with medical evaluations. Thus, it does not appear that there is a difference in the prevalence of cardiovascular disease or diabetes between deployed Gulf War veterans and nondeployed.

Many veterans are understandably concerned about the possibility of birth defects in their offspring. Two primary studies yielded some evidence of increased risk of birth defects among offspring of Gulf War veterans. However, the specific defects with increased prevalence (cardiac, kidney, urinary tract, and musculoskeletal abnormalities) in the two studies were not consistent. Overall, the studies are difficult to interpret because of the relative rarity of specific birth defects, use of small sample, timing of exposure (before or after conception), and whether the mother or the father was exposed. There was no consistent pattern of one of more birth defects with a higher prevalence in the offspring of male or female Gulf War veterans. Only one set of defects (that is, urinary tract abnormalities) has been found to be increased in more than one well-designed study. With regard to other adverse reproductive outcomes, the results of one primary study, which had hospital discharge data available, were suggestive of an increased risk of spontaneous abortions and ectopic pregnancies in Gulf War veterans.

Numerous studies in several countries examined respiratory outcomes related to deployment to the Gulf War Theater. The overwhelming majority of studies conducted among Gulf War veterans, whether from the United States, the UK, Canada, Australia, or Denmark, have found that several years after deployment, those deployed report higher rates of respiratory symptoms and respiratory illnesses than nondeployed troops. However, in all five studies, representing four distinct cohorts from three countries (the United States, Australia, and Denmark) that examined associations of Gulf War deployment with pulmonary-function measures or respiratory disease diagnoses based in part on such measures, such associations were not found. The uniformity of the findings is striking, especially given that the same five studies found that Gulf War deployment status was significantly associated with self-reports of respiratory symptoms among three of the four cohorts.

Whereas the studies discussed above examined respiratory outcomes associated simply with deployment vs nondeployment, other studies examined respiratory outcomes associated with specific environmental exposures experienced by Gulf War veterans, including exposure to oil-well fires and nerve agents. The methodologically strongest such study used objective exposure measures and methods and found significant associations between exposure to oil-well

SUMMARY

fire smoke and a doctor-assigned diagnosis of asthma in veterans. However, the strongest study was limited by the self-selection of participants. A second study, which had the advantage of being population-based, had the key limitation that case definitions were purely symptom-based, and it did not find associations between the same objective measures of exposure to oil-well fire smoke and asthma symptoms. A third study found no significant associations between the same objective measures of exposure to smoke from oil-well fires and later hospitalization for asthma, acute bronchitis, chronic bronchitis, or emphysema; however, the participants were all active-duty veterans, and young adults are seldom hospitalized for those diagnoses, so most cases would not be expected to be captured.

With regard to modeled exposure to nerve agents at Khamisiyah, one study found a small increase in postwar hospitalization for respiratory system disease. However, limitations of that study include probable substantial exposure misclassification based on Department of Defense (DOD) exposure estimates that were later revised, lack of control for tobacco-smoking, lack of a clear dose-response pattern, and low biologic plausibility for this target organ system in a setting in which no effect on nervous system diseases was seen. A second study using revised DOD exposure estimates found no associations between pulmonary-function measures and exposure to nerve agents at Khamisiyah.

RECOMMENDATIONS

The adequacy of the government's response has been both praised and criticized, VA and DOD have expended enormous effort and resources in attempts to address the numerous health issues related to the Gulf War veterans. The information obtained from those efforts, however, has not been sufficient to determine conclusively the origins, extent, and potential long-term implications of health problems potentially associated with veterans' participation in the Gulf War. The difficulty in obtaining meaningful answers, as noted by numerous past Institute of Medicine committees and the present committee agrees, is due largely to inadequate predeployment and postdeployment screening and medical examinations, and lack of monitoring of possible exposures of deployed personnel.

Predeployment and Postdeployment Screening

Predeployment and postdeployment data-gathering needs to include physician verification of data obtained from questionnaires so that one could have confidence in baseline and postdeployment health data. Collection and archiving of biologic samples might enable the diagnosis of specific medical conditions and provide a basis of later comparison. Meticulous records of all medications, whether used for treatment or prophylactically, would have improved the data and their interpretation in many of the studies reviewed.

Exposure Assessment

Environmental exposures were usually not assessed directly, and that critically hampers the assessment of the effects of specific exposures on specific health outcomes. There have been detailed and laudable efforts to simulate and model exposures, but those efforts have been hampered by lack of the input data required to link the exposure scenarios to specific people or even to specific units or job categories. Moving beyond the current state requires that more detailed information be gathered during future military deployments. Specifically, working toward the development of a job-task-unit-exposure matrix in which information on people with specific jobs or tasks or attached to specific units (according to routinely available records) is linked to exposures by expert assessment or simulation studies would enable quantitative assessment of the effects of specific exposures.

Surveillance for Adverse Outcomes

The committee noted that several health outcomes seemed to be appearing with higher incidence or prevalence in the Gulf War-deployed veterans. For those outcomes, the committee recommends continued surveillance to determine whether there is actually a higher risk in Gulf War veterans. Those outcomes are cancer (particularly brain and testicular), ALS, birth defects (including Goldenhar syndrome and urinary tract abnormalities) and other adverse pregnancy outcomes (such as, spontaneous abortion and ectopic pregnancy), and postdeployment psychiatric conditions. The committee also recommends that cause-specific mortality in Gulf War veterans continue to be monitored. Although there was an increase in mortality in the first few years after the Gulf War, the deaths appear to have been related to transportation injuries.

BRIEF SUMMARY OF FINDINGS AND RECOMMENDATIONS

Outcomes Based Primarily on Symptoms or Self-Reports

- No unique syndrome, unique illness, or unique symptom complex in deployed Gulf War veterans. Veterans of the Gulf War report higher rates of nearly all symptoms or sets of symptoms than their nondeployed counterparts; 29% of veterans meet a case definition of "multisymptom illness", as compared with 16% of nondeployed veterans.
- Multisymptom-based medical conditions reported to occur more frequently among deployed Gulf War veterans include fibromyalgia, chronic fatigue syndrome, and multiple chemical sensitivity (MCS).
- Deployment places veterans at increased risk for symptoms that meet diagnostic criteria for a number of psychiatric illnesses, particularly post traumatic stress disorder (PTSD), anxiety, depression, and substance abuse. In addition, co-morbidities were reported among a number of veterans, with PTSD, depression, anxiety, and/or substance abuse.
- Studies of deployed Gulf war veterans vs non-Gulf War deployed have not demonstrated differences in cognitive and motor measures as determined through neurobehavioral testing.
- Studies of returning Gulf War veterans with at least one of the symptoms most commonly reported by Gulf War veterans (i.e., fatigue, memory loss, confusion, inability to concentrate, mood swings, somnolence, GI distress, muscle and joint pain, skin/mucous membrane complaints) found poorer performance on cognitive tests when compared to returning Gulf War veterans who did not report such symptoms.
- Other symptoms that appear to be self-reported more often by deployed veterans are gastrointestinal symptoms, particularly dyspepsia; dermatologic conditions, particularly atopic dermatitis and warts; and joint pains (arthralgias).

Outcomes with Objective Measures or Diagnostic Medical Tests

- Studies of mortality provide evidence for a modest increase in transportation-related injuries and mortalities among deployed compared to non-deployed Gulf War veterans in the decade immediately following deployment. However, studies with longer followup indicate that the increased injury rate was likely to have been restricted to the first several years after the war.
- With regard to all-causes of hospitalization, excess hospitalizations did not occur among veterans of the Gulf War who remained on active duty through 1994. However, Gulf War veterans who left the military reported worse health outcomes than those who remained.
- The studies do not demonstrate consistent evidence of increased overall cancer in the Gulf War veterans compared to nondeployed veterans. Studies of testicular cancer produced inconsistent results, but the latency period for many cancers may not have been reached among Gulf War veterans.
- Studies indicate that Gulf War veterans might be at increased risk for amyotrophic lateral sclerosis (ALS).
- There does not appear to be an increase in the prevalence of peripheral neuropathy in deployed vs non-deployed veterans, as defined by history, physical examination, and electrophysiologic studies.
- It does not appear that there is a difference in the prevalence of cardiovascular disease or diabetes between deployed Gulf War veterans and nondeployed.
- Overall there is no consistent pattern of one of more birth defects with a higher prevalence for the offspring of male or female Gulf War veterans. Only one set of defects, urinary tract abnormalities, has been found to be increased in more than one well-designed study.
- Respiratory symptoms are strongly associated with Gulf War deployment when using comparison groups of non-deployed veterans in most studies addressing this question. However, studies with objective pulmonary function measures find no association between respiratory illnesses with Gulf War deployment across the four cohorts in which this has been investigated.

Specific Gulf War Exposures

Outcomes with Objective Measures or Diagnostic Medical Tests

- Among studies that examined pulmonary outcomes in associations with specific exposures in the Gulf War Theater, exacerbation of asthma associated with oil-well fire smoke has been indicated.
- With respect to nerve agents at Khamisiyah, no study using objective estimates of exposure has found associations with pulmonary function measures or physician-diagnosed respiratory disease. Another study indicated that there might be an increase in brain cancer among such veterans, however, the exposure models are highly uncertain.

RECOMMENDATIONS

- Pre- and post-deployment screening of health status
- Assessment of exposures

Surveillance for adverse health outcomes, specifically: cancer, ALS, birth defects, adverse pregnancy outcomes, post-deployment psychiatric outcomes, and mortality.

1

INTRODUCTION

More than 15 years have passed since the Iraqi invasion of Kuwait in August 1990 and the offensive by coalition troops in January 1991. Oil-well fires became visible in satellite images as early as February 9, 1991; the ground war began on February 24; and by February 28, 1991, the war was over. The military operation in the gulf was brief: an official cease-fire was signed in April 1991, and the last troops to participate in the ground war returned home on June 13, 1991. In all, about 697,000 US troops had been deployed to the Persian Gulf during the conflict.

Although the Persian Gulf War was considered a successful military operation with few injuries and deaths among coalition forces, many returning veterans soon began to report numerous health problems that they believed were associated with their service in the gulf. Although most Gulf War veterans returned to normal activities, some have had a wide array of symptoms and unexplained illnesses. This volume summarizes the overall health effects in veterans and notes which health outcomes are more evident in Gulf War veterans than in their nondeployed counterparts.

An impressive body of literature details the veterans' symptoms and illnesses. At the request of the Department of Veterans Affairs (VA), the Institute of Medicine (IOM) appointed a committee (the Committee on Gulf War and Health: A Review of the Medical Literature Relative to Gulf War Veterans' Health) to review that body of literature and to summarize what is known about the current status of the veterans' health.

Previous IOM committees and their reports focused on associations between biologic or chemical agents to which veterans might have been exposed in the gulf and health outcomes. Those committees typically relied on studies of occupational groups exposed to the putative agents and, when available, included studies of veterans. The present committee, however, did not use occupational groups as surrogates of exposure to the putative agents that might have been found in the gulf, but rather reviewed the research on Gulf War veterans themselves that details their symptoms and illnesses. The numerous studies that have been conducted in the intervening years since the war have typically compared Gulf War veterans with their nondeployed counterparts. Thus, within the limitations of each study, it is possible to determine which symptoms and illnesses are associated with deployment to the Persian Gulf.

BACKGROUND

In 1998, in response to the growing concerns of ill Gulf War veterans, Congress passed two laws: PL 105-277, the Persian Gulf War Veterans Act, and PL 105-368, the Veterans Programs Enhancement Act. Those laws directed the secretary of veterans affairs to enter into a contract with the National Academy of Sciences (NAS) to review and evaluate the scientific and medical literature regarding associations between illness and exposure to toxic agents, environmental or wartime hazards, or preventive medicines or vaccines associated with Gulf War service and to consider the NAS conclusions when making decisions about compensation. The study was assigned to the IOM.

The Persian Gulf War legislation directs IOM to study diverse biologic, chemical, and physical agents. Exposures to most of the Gulf War agents have been extensively studied and characterized, primarily in occupational settings (for example, exposure to pesticides, solvents, and fuels), but exposures to others have not been as well studied and characterized in human populations (for example, exposure to nerve agents and vaccines).

Given the large number of agents to study, IOM divided the task into several reviews, which are now complete: Gulf War and Health, Volume 1: Depleted Uranium, Pyridostigmine Bromide, Sarin, Vaccines (IOM 2000); Gulf War and Health, Volume 2: Insecticides and Solvents (IOM 2003); Gulf War and Health, Volume 3: Fuels, Combustion Products, and Propellants (IOM 2005); and Gulf War and Health: Updated Literature Review of Sarin (IOM 2004). Three other studies are underway: one examining the long-term sequelae of infectious diseases that are endemic to the Persian Gulf, another reviewing the long-term health effects that might be associated with deployment-related stress, and a third reviewing whether there is an increased risk of amyotrophic lateral sclerosis in all veteran populations. The present report summarizes health effects in veterans deployed to the Persian Gulf irrespective of specific exposures.

THE GULF WAR SETTING[1]

Although the committee's charge was not to review the scientific evidence on the possible health effects of various agents to which Gulf War veterans were potentially exposed, the committee recognized that it needed to have as complete an understanding of the Gulf War experience as possible. Furthermore, information on the likelihood or magnitude of specific exposures might be helpful in interpreting epidemiologic studies that are reviewed in detail in Chapters 4 and 5. It should be noted, that in addition to reviewing studies from the United States, the committee reviewed studies from Australia, Canada, Denmark, and the United Kingdom.

The information in this section provides a context for the many scientific articles that the committee reviewed and an appreciation (albeit limited) of the collective experience of Gulf War veterans. It is compiled from many sources (Gunby 1991) and from presentations by veterans and other speakers at the committee's public meeting (Hyams et al. 1995; IOM 1995; IOM 1996; IOM 1999; Joellenbeck et al. 1998; Lawler et al. 1997; NIH Technology Assessment Workshop Panel 1994; PAC 1996; PAC 1997; Persian Gulf Veterans Coordinating Board 1995; U.S. Department of Veterans Affairs 1998; Ursano and Norwood 1996).

Deployment

The pace of the buildup for the Gulf War was unprecedented. Within 5 days after Iraq invaded Kuwait, the United States began moving troops into the region as part of Operation Desert Shield. By September 15, 1990, the number of American service members reached 150,000 and included nearly 50,000 reservists. Within the next month, another 60,000 troops

---

[1] This section is adapted from *Gulf War and Health, Volume 1* (IOM, 2000).

INTRODUCTION

arrived in Southwest Asia; in November, an additional 135,000 reservists and National Guard members were called up. By February 24, 1991, more than 500,000 US troops had been deployed to the Persian Gulf region. In addition to the US troops, a coalition force of 34 member countries was eventually assembled.

The Gulf War reflected many changes from previous wars, particularly in the demographic composition of military personnel and the uncertainty of conditions for many reservists. Of the nearly 700,000 US troops who fought in Operation Desert Shield and Operation Desert Storm, almost 7% were women and about 17% were from National Guard and reserve units. Military personnel were, overall, older than those who had participated in previous wars with a mean age of 28 years. Seventy percent of the troops were non-Hispanic/White; 23% were black, and 5% were Hispanic (Joseph 1997). Rapid mobilization exerted substantial pressure on those who were deployed, disrupting lives, separating families, and, for reserve and National Guard units, creating uncertainty about whether jobs would be available when they returned to civilian life.

Living Conditions

Combat troops were crowded into warehouses and tents on arrival and then often moved to isolated desert locations. Most troops lived in tents and slept on cots lined up side by side, affording virtually no privacy or quiet. Sanitation was often primitive, with strains on latrines and communal washing facilities. Hot showers were infrequent, the interval between laundering uniforms was sometimes long, and desert flies were a constant nuisance, as were scorpions and snakes. Military personnel worked long hours and had narrowly restricted outlets for relaxation. Troops were ordered not to fraternize with local people, and alcoholic drinks were prohibited in deference to religious beliefs in the host countries. A mild, traveler's type of diarrhea affected more than half of the troops in some units. Fresh fruits and vegetables from neighboring countries were identified as the cause and were removed from the diet. Thereafter, the diet consisted mostly of packaged foods and bottled water.

For the first 2 months of troop deployment (August and September 1990) the weather was extremely hot and humid, with air temperatures as high as 115°F and sand temperatures reaching 150°F. Except for coastal regions, the relative humidity was less than 40%. Troops had to drink large quantities of water to prevent dehydration. Although the summers were hot and dry, temperatures in winter (December-March) were low, with wind-chill temperatures at night dropping to well below freezing. Wind and blowing sand made protection of skin and eyes imperative. Goggles and sunglasses helped somewhat, but visibility was often poor.

Environmental and Chemical Exposures

The most visually dramatic environmental event of the Gulf War was the smoke from more than 750 oil-well fires. Smoke plumes from individual fires rose and combined to form giant plumes that could be seen for hundreds of kilometers. There were additional potential sources of exposure to petroleum-based combustion products. Kerosene, diesel, and leaded gasoline were used in unvented tent heaters, cooking stoves, and portable generators. Exposures to tent-heater emissions were not specifically documented, but a simulation study was conducted after the war to determine exposure (see Chapter 2). Petroleum products, including diesel fuels, were also used to suppress sand and dust, and petroleum fuels were used to aid in the burning of waste and trash.

Pesticides, including dog flea collars, were widely used by troops in the Persian Gulf to combat the region's ubiquitous insect and rodent populations; and although guidelines for use were strict, there were many reports of misuse. The pesticides used included methyl carbamates, organophosphates, pyrethroids, and chlorinated hydrocarbons. The use of those pesticides is reported in numerous reports (e.g., RAND 2000), however objective information regarding individual levels of pesticide exposure is generally not available.

Many exposures could have been related to particular occupational activities in the Gulf War. The majority of occupational chemical exposures appear to have been related to repair and maintenance activities, including battery repair (corrosive liquids), cleaning and degreasing (solvents, including chlorinated hydrocarbons), sandblasting (abrasive particles), vehicle repair (asbestos, carbon monoxide, and organic solvents), weapon repair (lead particles), and welding and cutting (chromates, nitrogen dioxide, and heated metal fumes). In addition, troops painted vehicles and other equipment used in the gulf with a chemical-agent-resistant coating either before being shipped to the gulf or at ports in Saudi Arabia. Working conditions in the field were not ideal and recommended occupational-hygiene standards might not have been followed at all times.

Exposure of US personnel to depleted uranium (DU) occurred as the result of "friendly-fire" incidents, cleanup operations, and accidents (including fires). Others might have inhaled DU dust through contact with DU-contaminated tanks or munitions. DU exposure is discussed in more detail in Chapter 2. Assessment of DU exposure, especially high exposure, is considered to be more accurate than assessment of exposure to most other agents because of the availability of biologic monitoring information.

Threat of Chemical and Biologic Warfare

When US troops arrived in the gulf, they had no way of knowing whether they would be exposed to biologic and chemical weapons. Iraq previously had used such weapons in fighting Iran and in attacks on the Kurdish minority in Iraq. Military leaders feared that the use of such weapons in the gulf could result in the deaths of tens of thousands of Americans. Therefore, in addition to the standard vaccinations before military deployment, about 150,000 troops received anthrax vaccine and about 8,000 botulinum toxoid vaccine. In some cases, vaccination records were kept, and they provide an objective measure of exposure in addition to self-reporting by troops.

Troops were also given blister packs of 21 tablets of pyridostigmine bromide (PB) to protect against agents of chemical warfare, specifically nerve gas. Troops were to take PB on the orders of a commanding officer when a chemical-warfare attack was believed to be imminent. Chemical sensors and alarms were distributed throughout the region to warn of such attacks. The alarms were extremely sensitive and could be triggered by many substances, including some organic solvents, vehicle-exhaust fumes, and insecticides. Alarms sounded often and troops responded by donning the confining protective gear and ingesting PB as an antidote to nerve gas. In addition to the alarms, there were widespread reports of dead sheep, goats, and camels, which troops were taught could be indication of the use of chemical or biologic weapons. The sounding of the alarms, the reports of dead animals, and rumors that other units had been hit by chemical-warfare agents caused the troops to be concerned that they would be or had been exposed to such agents.

Despite the small numbers of US personnel injured or killed during combat in the Gulf War, the troops, as in any war, faced the fear of death, injury, or capture by the enemy. After the war, there was the potential for other exposures, including US demolition of a munitions storage complex at Khamisiyah, Iraq, which—unbeknownst to demolition troops at the time—contained stores of sarin and cyclosarin. The potential exposures to sarin and cyclosarin from the Khamisiyah incident are discussed in Chapter 2 and have been the subject of specific studies.

It has been documented from the Civil War to the Gulf War that a variety of physical and psychologic stressors have placed military personnel at high risk for adverse health effects (Engel et al. 2004; Hyams et al. 1996; Jones et al. 2002; Soetekouw et al. 2000). In addition to the threat or experience of combat, the Gulf War involved rapid and unexpected deployment, harsh living conditions, and anticipation of exposure to chemical and biologic agents, environmental pollution from burning oil fires, and family disruption and financial strain.

CHARGE TO THE COMMITTEE

The charge to this IOM committee is different from charges to previous IOM Gulf War committees in that this one does not associate health outcomes with specific biologic or chemical agents believed to have been present in the gulf, but rather it examines health outcomes related to deployment. For that reason, the committee did not review toxicologic or experimental studies. Thus, the committee has limited its review to epidemiologic studies of health outcomes in Gulf War veterans to determine their health status. The specific charge to the committee, as requested by the VA, was to review, evaluate, and summarize peer-reviewed scientific and medical literature addressing the health status of Gulf War veterans.

COMMITTEE'S APPROACH TO ITS CHARGE

The committee began its evaluation by presuming neither the existence nor the absence of illnesses associated with deployment. It sought to characterize and weigh the strengths and limitations of the available evidence. The committee did not address policy issues, such as decisions regarding compensation, potential costs of compensation, or any broader policy implications of its findings.

Extensive searches of the epidemiologic literature were conducted and over 4000 potentially relevant references were retrieved. After an assessment of the titles and abstracts of the initial searches, the committee focused on some 850 potentially relevant epidemiologic studies for review and evaluation.

The committee adopted a policy of using only peer-reviewed published literature as the basis for its conclusions. The process of peer review by fellow professionals increases the likelihood of high quality but does not guarantee the validity of a study or the ability to generalize its findings. Accordingly, committee members read each study critically and considered its relevance and quality. The committee did not collect original data, nor did it perform any secondary data analysis (except to calculate response rates for consistency among studies).

INCLUSION CRITERIA

The committee's next step, after securing the full text of the epidemiologic studies it would review, was to determine which studies would be considered primary or secondary studies. The committee developed inclusion criteria for studies (Chapter 3). Primary studies provide the basis of the committee's findings. For a study to be included in the committee's review as a primary study it had to meet specified criteria. For example, it would have to include information about specific health outcomes; demonstrate rigorous methods, such as being published in a peer-reviewed journal; include details of its methods; include a control or reference group; have the statistical power to detect effects; and include reasonable adjustments for confounders. A secondary study provides background information or context for this report. Secondary studies, although mentioned, are not written-up in detail and typically are not included in tables.

COMPLEXITIES IN RESOLVING GULF WAR AND HEALTH ISSUES

Investigations of the health effects of past wars have often focused on narrowly defined hazards or health outcomes, such as infectious diseases (for example, typhoid and malaria) during the Civil War, specific chemical hazards (for example, mustard gas in World War I and Agent Orange and other herbicides in Vietnam), and combat injuries. A discussion of the possible health effects of Gulf War service, however, involves many complex issues, some of which are explored below. They include exposure to multiple biologic and chemical agents, limitations of exposure information, individual variability factors, and illnesses that are often nonspecific and lack defined medical diagnoses or treatment protocols. The committee was not tasked with addressing those issues, but it presents them in this introductory chapter to acknowledge the difficulties faced by veterans, researchers, policymakers, and others in reaching an understanding about the veterans' ill health.

Multiple Exposures and Chemical Interactions

Although Operation Desert Shield and Operation Desert Storm were relatively brief, military personnel were potentially exposed to numerous harmful agents. They include agents administered as preventive measures (such as, PB, vaccines, pesticides, and insecticides), hazards of the natural environment (such as, sand and endemic diseases), job-specific agents (such as, paints, solvents, and diesel fumes), war-related agents (such as, smoke from oil-well fires and DU), and hazards from cleanup operations (such as, sarin and cyclosarin). Thus, military personnel might have been exposed to various agents at various doses for various periods. Many of the exposures are not specific to the Gulf War, but the number and combination of agents to which the veterans might have been exposed make it difficult to determine whether any agent or combination of agents is the cause of Gulf War veterans' illnesses.

Limitations of Exposure Information

Determining whether Gulf War veterans face an increased risk of illness because of their exposures during the war requires extensive information about each exposure (such as the actual agents, the duration of exposure, the route of entry and the internal dose) and documentation of adverse reactions. But very little is known about most Gulf War exposures. After the ground war, an environmental-monitoring effort was initiated primarily because of concerns related to smoke from oil-well fires and exposure to sarin and cyclosarin rather than for the other agents to which the troops might have been exposed. Consequently, exposure data on other agents are lacking or are severely limited.

Various exposure assessment tools are being used in research to fill gaps in exposure information, but there are limitations in reconstruction of past exposure events. For example, veterans are surveyed to obtain recollections about agents to which they might have been exposed, although survey results might be limited by recall bias (see Chapter 3). Models have been refined to estimate exposures to sarin and cyclosarin, but it is difficult to incorporate intelligence information, meteorologic data, transport and dispersion data, and troop-unit location information accurately (see Chapter 2). Extensive efforts have been made to model and obtain information on potential exposures to DU, smoke from oil-well fires, and other agents. Although modeling efforts are important for discerning the details of exposures of Gulf War veterans, they require external review and validation. Furthermore, even if there were accurate troop location data, the location of individual soldiers would be very uncertain. Because of the limitations in the exposure data, it is difficult to determine the likelihood of increased risk for disease or other adverse health effects in Gulf War veterans that are due specifically to biologic and chemical agents.

Individual Variability

Differences among people in their genetic, biologic, psychologic, and social vulnerabilities add to the complexities in determining health outcomes related to specific agents. Sensitive people will exhibit different responses to the same agents than people without the susceptibility. For example, a person who is a poor metabolizer of a particular substance, depending on his or her genetic makeup, might be at higher or lower risk for specific health effects due to exposure to the substance. Researchers are investigating the genotypes that code for two forms of an enzyme that differ in the rate at which they hydrolyze particular organophosphates (including sarin). Lower hydrolyzing activity would mean that despite identical exposure to sarin, more sarin would be bioavailable in people who are poor metabolizers and could result in increased anticholinesterase effects.

Unexplained Symptoms

Many Gulf War veterans suffer from an array of health problems and symptoms (for example, fatigue, muscle and joint pain, memory loss, gastrointestinal disorders, and rashes) that are not specific to any disease and are not easily classified with standard diagnostic coding systems. Population-based studies have found a higher prevalence of self-reported symptoms in Gulf War veterans than in nondeployed Gulf War-era-veterans or other control groups (see Chapters 4 and 5; Goss Gilroy Inc. 1998; Iowa Persian Gulf Study Group 1997; Unwin et al. 1999). That Gulf War veterans do not all experience the same array of symptoms has complicated efforts to determine whether there is a "Gulf War syndrome" or overlap with other symptom-based disorders. The nature of the symptoms suffered by many Gulf War veterans does not point to an obvious diagnosis, etiology, or standard treatment.

ORGANIZATION OF THE REPORT

Chapter 2 is a background chapter that details many of the specific biologic and chemical agents in the gulf and provides a context for the rest of the committee's report. Chapter 3 provides a brief background in epidemiology and describes the committee's methods for choosing the epidemiologic studies that are reviewed in later chapters. Chapter 4 describes the major Gulf War cohorts and provides information about the numerous studies that have been derived from them; the chapter includes a summary table that lists all the original cohorts and their derivative studies. Chapter 5 describes and analyzes the studies of health outcomes in Gulf War veterans; it also provides the basis for the committee's conclusions and recommendations, which are presented in Chapter 6.

REFERENCES

Engel CC, Jaffer A, Adkins J, Riddle JR, Gibson R. 2004. Can we prevent a second 'Gulf War syndrome'? Population-based healthcare for chronic idiopathic pain and fatigue after war. *Advances in Psychosomatic Medicine* 25:102-122.

Goss Gilroy Inc. 1998. *Health Study of Canadian Forces Personnel Involved in the 1991 Conflict in the Persian Gulf.* Ottawa, Canada: Goss Gilroy Inc. Department of National Defence.

Gunby P. 1991. Physicians provide continuum of care for Desert Storm fighting forces. *Journal of the American Medical Association* 265(5):557, 559.

Hyams KC, Hanson K, Wignall FS, Escamilla J, Oldfield EC, 3rd. 1995. The impact of infectious diseases on the health of U.S. troops deployed to the Persian Gulf during operations Desert Shield and Desert Storm. *Clinical Infectious Diseases* 20(6):1497-1504.

Hyams KC, Wignall FS, Roswell R. 1996. War syndromes and their evaluation: From the U.S. Civil War to the Persian Gulf War. *Annals of Internal Medicine* 125(5):398-405.

IOM (Institute of Medicine). 1995. *Health Consequences of Service During the Persian Gulf War: Initial Findings and Recommendations for Immediate Action.* Washington, DC: National Academy Press.

IOM. 1996. *Health Consequences of Service During the Persian Gulf War: Recommendations for Research and Information Systems.* Washington, DC: National Academy Press.

IOM. 1999. Gulf War Veterans: Measuring Health. Washington, DC: National Academy Press.

IOM. 2000. *Gulf War and Health, Volume 1: Depleted Uranium, Sarin, Pyridostigmine Bromide, Vaccines.* Washington, DC: National Academy Press.

IOM. 2003. *Gulf War and Health, Volume 2: Insecticides and Solvents.* Washington, DC: The National Academies Press.

IOM. 2004. *Gulf War and Health: Updated Literature Review of Sarin.* Washington, DC: The National Academies Press.

IOM. 2005. *Gulf War and Health, Volume 3: Fuels, Combustion Products, and Propellants.* Washington, DC: The National Academies Press.

Iowa Persian Gulf Study Group. 1997. Self-reported illness and health status among Gulf War veterans: A population-based study. *Journal of the American Medical Association* 277(3):238-245.

Joellenbeck LM, Landrigan PJ, Larson EL. 1998. Gulf War veterans' illnesses: A case study in causal inference. *Environmental Research* 79(2):71-81.

Jones E, Hodgins-Vermaas R, McCartney H, Everitt B, Beech C, Poynter D, Palmer I, Hyams K, Wessely S. 2002. Post-combat syndromes from the Boer war to the Gulf war: A cluster analysis of their nature and attribution. *British Medical Journal* 324(7333):321-324.

Joseph SC. 1997. A comprehensive clinical evaluation of 20,000 Persian Gulf War veterans. *Military Medicine* 162(3):149-155.

Lawler MK, Flori DE, Volk RJ, Davis AB. 1997. Family health status of National Guard personnel deployed during the Persian Gulf War. *Families, Systems and Health* 15(1):65-73.

NIH Technology Assessment Workshop Panel. 1994. The Persian Gulf experience and health. *Journal of the American Medical Association* 272(5):391-395.

PAC (Presidential Advisory Committee). 1996. *Presidential Advisory Committee on Gulf War Veterans' Illnesses: Final Report*. Washington, DC: US Government Printing Office.

PAC. 1997. *Presidential Advisory Committee on Gulf War Veterans' Illnesses: Special Report*. Washington, DC: Presidential Advisory Committee on Gulf War Veterans' Illnesses.

Persian Gulf Veterans Coordinating Board. 1995. Unexplained illnesses among Desert Storm veterans: A search for causes, treatment, and cooperation. *Archives of Internal Medicine*. 155(3):262-268.

RAND. 2000. *Review of the Scientific Literature As It Pertains to Gulf War Illnesses. Volume 8: Pesticides*. Santa Monica, CA: RAND Corporation.

Soetekouw PM, de Vries M, van Bergen L, Galama JM, Keyser A, Bleijenberg G, van der Meer JW. 2000. Somatic hypotheses of war syndromes. *European Journal of Clinical Investigation* 30(7):630-641.

U.S. Department of Veterans Affairs. 1998. *Consolidation and Combined Analysis of the Databases of the Department of Veterans Affairs Persian Gulf Health Registry and the Department of Defense Comprehensive Clinical Evaluation Program*. Washington, DC: Environmental Epidemiology Service, Department of Veterans Affairs.

Unwin C, Blatchley N, Coker W, Ferry S, Hotopf M, Hull L, Ismail K, Palmer I, David A, Wessely S. 1999. Health of UK servicemen who served in Persian Gulf War. *Lancet* 353(9148):169-178.

Ursano RJ, Norwood AE. 1996. *Emotional Aftermath of the Persian Gulf War: Veterans, Families, Communities, and Nations*. Washington, DC: American Psychiatric Publishing.

2

EXPOSURES IN THE PERSIAN GULF

The purpose of this chapter is to summarize in a general way what is known about the many exposures that might have been present in the Gulf War Theater and to discuss the effect of exposure information on the interpretation of human health outcomes in the available studies of Gulf War veterans. The committee does not draw conclusions about the association between specific exposures in the gulf and health outcomes. In general, three main types of studies are reviewed in this chapter: major cohort studies that typically assessed exposures with questionnaires administered to study subjects, simulation studies to assess the potential magnitude of exposures encountered under specific circumstances, and environmental fate and transport models that were then used in epidemiologic analyses (in some cases, incorporating information from simulation studies). In addition, some exposures—specifically, exposures to depleted uranium compounds—were assessed primarily with biologic monitoring.

The rarity of direct assessment of exposure critically hinders evaluation of the potential health effects of specific exposures. There have been detailed and laudable efforts to simulate and model exposures, but they are hampered by the lack of input data required to link exposure scenarios to specific people or even specific units or job categories. One can confidently compare health responses only between deployment in the theater and nondeployment or deployment elsewhere. To move beyond the current state requires that more-detailed information be gathered during future military deployments. Specifically, a job-task-unit-exposure matrix in which information on people with specific jobs or tasks or attached to specific units (typically available from routinely collected records) is linked to exposures through expert assessment or simulation studies would enable quantitative assessment of the effects of specific exposures.

EXPOSURE ASSESSMENT IN EPIDEMIOLOGIC STUDIES

As described in more detail in later chapters, most of the Gulf War literature is based on veterans' reports of their own exposures. In addition, many studies report the prevalence of specific health outcomes or clusters of outcomes among Gulf War veterans; those studies might or might not compare prevalence with that of control groups.

STUDIES ASSESSING EXPOSURES WITH QUESTIONNAIRES

Gulf War epidemiologic studies that assess exposures with questionnaires administered to study subjects typically rely on self-reports of exposure and are generally considered to be subject to recall or reporting bias (see Chapter 3) even when objective measures of health status are collected. The potential for bias is increased in studies in which both exposure and health-outcome information is based on self-reports. The use of self-reported exposure information was unavoidable in most Gulf War literature, but a number of attempts have been made to compare self-reported exposures with other estimates of exposure. Although such alternative estimates might appear to be more objective, for most comparisons between self-reported exposure and other measures there is no "gold standard" of exposure. Accordingly, the studies simply report on comparisons of different estimates of exposure rather than provide an objective assessment of the validity of self-reports. Exceptions might include exposure to vaccinations, for which records are available, and exposure to depleted uranium (DU), which can be verified with biologic monitoring. The following sections describe comparisons between self-reported and other measures of exposure.

Exposure to Oil-Well Fire Smoke

A number of studies have, at least indirectly, examined the validity of self-reported exposures. With respect to the oil-well fires, Lange et al. (2002) reported moderate correlations ($r = 0.4$ and 0.5) between self-reports of low and high exposures to oil-fire smoke as assessed with a dispersion model linked to troop-unit location information. At each level of self-reported exposure (based on the number of days exposed), the modeled exposures were highly variable. Cowan et al. (2002) reported a low interclass correlation coefficient (kappa) of 0.13 for self-reported exposure to oil-fire smoke vs cumulative modeled exposure (according to the model used by Lange et al.) to oil-fire smoke or days with high modeled exposure. Wolfe et al. (2002) reported that responses to a yes-no question regarding oil-fire exposure in the Fort Devens cohort did not correlate well with modeled particle exposures. Higher correlations were found when information regarding the self-reported frequency, duration, and intensity of exposure was considered.

Exposure to Vaccination

The strongest analysis of reporting bias with regard to vaccine exposure was conducted by Mahan et al. (2004) in their study of anthrax vaccination. Veterans were asked whether they received anthrax vaccination or were uncertain about receiving it. In a cohort of 11,441 Gulf War veterans who completed a health and exposure survey, 352 respondents also were on a Department of Defense (DOD) list of 7,691 people who were vaccinated at least once. The list was compiled from several sources and is the largest compilation of Gulf War veterans identified as receiving anthrax vaccination. In the full cohort, 4,601 (40%) reported receiving the vaccine, 2,979 (26%) reported not receiving it, and 3,861 (34%) were uncertain. Of the subset of 352 who were on the DOD vaccination list, 260 (74%) reported receiving the vaccine, 34 (10%) reported not having received it, and 58 (16%) reported that they were uncertain. This comparison indicates a 26% false-negative rate, but the lack of a documented "nonvaccinated" group makes it impossible to determine the false-positive rate. The study also provides some evidence of reporting bias.

Although immunization history was self-reported in most studies, Unwin et al. (1999) asked survey respondents to refer to their own vaccination records, if available, in a study of UK veterans deployed to the gulf compared with those deployed to Bosnia or other Gulf-War-era veterans. Some 32% of the Gulf War veterans in the survey reported that they had vaccination records, and confirmation by the investigators suggested that those with records had used them when completing the questionnaire. Only 2.8% of veterans without records reported receiving pertussis vaccination despite the fact that the anthrax and pertussis vaccines were always administered simultaneously. Of those with records, 36% reported receiving pertussis vaccination. Reporting of biologic-warfare vaccinations (for example, anthrax, plague, and pertussis) was associated with "CDC syndrome" (that is, Centers for Disease Control and Prevention multisymptom syndrome), irrespective of the use of records. However, an association of routine vaccinations (for example, hepatitis, typhoid, and cholera) with CDC syndrome was present only in those who did not use their records. The analysis, therefore, provides some evidence of bias with regard to self-reporting of vaccinations.

The investigators limited a later analysis focused on the same cohort to the subset of personnel who had vaccination records (Hotopf et al. 2000). That analysis concluded that multiple vaccinations received during deployment (but not multiple vaccinations received before deployment) were associated with symptom clusters.

Kelsall et al. (2004a) specifically asked Australian Gulf War veterans to refer to their own immunization booklets for information regarding the number and timing of immunizations relative to their Gulf War deployment. Although data were not provided in the paper, the authors report that the 52% of the 1,418 survey respondents who had immunization booklets reported higher total numbers of immunizations than those without booklets and were less likely to report not having received any immunizations. That suggests a general pattern of underreporting of exposures among veterans who provided self-reported vaccination information. The paper does not provide specific information on the types of immunizations reported, nor does it evaluate any potential bias regarding the source of vaccination records and the reporting of health outcomes.

Exposure to Pyridostigmine Bromide

Pyridostigmine bromide (PB) is a drug that was used during the Gulf War as a pretreatment to prevent the harmful effects of nerve agents because of its ability to reversibly bind to acetylcholinesterase (AChE).[1] The bound fraction is thereby protected from exposure to nerve agents that would irreversibly bind to AChE. PB is not an antidote (it has no value when administered after nerve-agent exposure) and is not a substitute for atropine or 2-pralidoxime chloride; rather, it enhances their efficacy (Madsen 1998).

DOD reported that 5,328,710 doses were fielded and estimated that about 250,000 personnel took at least some PB during the Gulf War.[2] It was supplied in a 21-tablet blister pack; the dosage prescribed was one 30-mg tablet every 8 hours. Each pack provided a 1-week supply of PB for one person, and military personnel were issued two blister packs each. Recommended long-term storage was at 2-80°C, and blister packs removed from refrigeration were to be used

---

[1] AChE is an enzyme necessary to remove acetylcholine (ACh). ACh transmits nerve signals at the cholinergic neuromuscular junction or synapses in the central nervous system. Anticholinesterase agents inhibit (inactivate) AChE, and this results in an accumulation of ACh. The accumulation repetitively activates the ACh receptors, resulting in exaggerated responses of organ (such as excess salivation).
[2] The number of doses fielded was learned through a search of archived Defense Personnel Support Center logistic records for Operations Desert Shield and Desert Storm and reflects the amount of product ordered and sent through supply channels. In most cases, only a review of people's own medical-treatment records would report the actual number of doses administered, and few records were maintained by them (Office of the Secretary of Defense 1998).

within 6 months (Madsen 1998). Variation in use occurred, however, because it was self-administered and was to be taken only on orders of a unit commander (PAC 1996b).

Keeler and colleagues (1991) conducted an uncontrolled retrospective survey of the medical officers of the XVIII Airborne Corps. The unit's 41,650 soldiers were instructed to take PB at the onset of Operation Desert Storm in January 1991. Use varied from one to 21 tablets taken over 1-7 days; 34,000 soldiers reported taking the medication for 6-7 days. Reported side effects of PB were estimated to have been present in half the troops; they were not incapacitating, however, and were primarily gastrointestinal. An estimated 1% of the soldiers believed that they had symptoms that warranted medical attention, but less than 0.1% had effects sufficient to warrant discontinuation of the drug (Keeler et al. 1991; Schumm et al. 2002). Although the prescribed PB dosage was three tablets per day, the veterans reported varied compliance with this regime: only 24% took the tablets in accordance with the directions that were supposed to have been given, 6% of those who reported the number of tablets per day indicated that they took six or more per day (twice the recommended number), and 61% took only one or two pills per day (below the recommended number).

Other papers reviewed included only general self-reports of PB consumption (for example, yes or no for PB consumption) or total numbers of tablets consumed during service; therefore, it is not possible to assess dosage or compliance with directions. As with many of the exposures, there were no data available for comparison between self-reported exposure and an objective measure of PB exposure. The Schumm et al. (2002) report described above, however, does suggest a strong possibility of dosages that were higher and lower than the recommended amounts.

Exposure to Depleted Uranium

The validity of self-reported exposure to DU can be evaluated on the basis of measurement of urinary uranium. Uranium activity decreases over time, but it has a very long radiologic half-life (4.5×10^9 years for U-238), so those with high exposures can be identified many years later. High urinary uranium is presumed to result from continuing mobilization of DU from metal fragments that oxidize in situ (McDiarmid et al. 2004a). As discussed in detail later in this chapter, although many veterans reported a potential for DU exposure, only those with retained DU-containing shrapnel had high urinary uranium. For example, 9-11 years after "exposure", high DU in urine was present in only six of 446 people in the Department of Veterans Affairs DU surveillance program, and all six individuals had embedded shrapnel (McDiarmid et al. 2004b). It should be recognized, however, that at the time of many of the large Gulf War cohort surveys, the potential for DU exposure (for example, involvement in a friendly-fire incident, rescue where DU-containing munitions or vehicles were used, or presence during a fire at Camp Doha, Kuwait, that involved detonation and burning of DU-containing munitions) was not as well understood as it is now. Consequently, veterans who reported their own DU exposure were working with the best available information and reported that they were exposed, whereas it is now understood that retention of DU-containing embedded shrapnel is the major source of increased DU exposure in military personnel. Furthermore, some people might have inhaled and cleared DU-containing particles or might have had embedded DU-containing shrapnel that was later removed; therefore, they would not have increased urinary uranium even though they had been exposed at least briefly. That exposure cannot be measured objectively, but there is a possibility that such exposures were linked with acute or later health effects.

GENERAL COHORT STUDIES (PREVALENCE STUDIES)

As discussed above, most studies rely on self-reported exposures for which validation is not possible. Some of them simply report on the prevalence of self-reported exposures, and others compare self-reported exposures with exposures of control groups (that is, nondeployed veterans or veterans deployed elsewhere, such as in Bosnia). For example, Kroenke et al. (1998), in their study of 18,495 veterans involved in the DOD Comprehensive Clinical Evaluation Program, indicate self-reported exposures to "fuel" in 89% of respondents, to tent heaters in 73%, to oil-fire smoke in 72%, and to DU in 16%. Similarly, Proctor (1998) describes self-reported exposures to a variety of agents (number who were exposed and number who were exposed and felt sick at the time of exposure) and includes measures on the Expanded Combat Exposure Scale regarding exposure to chemical- and biologic-warfare agents. Exposures to anti-nerve-gas pills, pesticides, Scud debris, chemical- and biologic-warfare agents, oil-fire smoke, vehicle exhaust, smoke from tent heaters, and smoke from burning human waste were all considered. Relatively high exposure prevalence was reported for all those agents except pesticides and chemical- and biologic-warfare agents. Cherry et al. (2001) describe the prevalence of self-reported exposures among the UK cohort (n = 7,971). Questionnaire results on 14 specific exposures are summarized after stratification by type of military service (Army, Navy, and Air Force). Overall, 30% reported more than six inoculations, 61% reported exposure to oil-fire smoke, 60% used nerve-agent prophylaxis for more than 14 days, and only 7% reported handling pesticides. Those studies clearly indicate that a high frequency of Gulf War veterans report a large number of exposures but provide little information regarding the validity or magnitude of the exposures.

Studies that compared self-reported exposure frequencies with exposure of control groups are more informative. For example, in their study of 527 active-duty Gulf War veterans and 970 nondeployed Seabees, Gray et al. (1999a) provide comparisons of multiple exposures between those deployed in the Gulf War and those not deployed. Many kinds of exposures were compared, and many were higher in deployed Gulf War veterans. For example, there were 2.3 times more diesel exposure, 3.8 times more tent-heater exposure, 2.5 times more DU exposure, and 117 times more oil-fire exposure in the deployed than in the nondeployed.

Unwin et al. (1999) conducted a cross-sectional postal survey of a random sample of veterans deployed in the Gulf War and of veterans deployed in Bosnia and veterans deployed in places other than the Persian Gulf during the Gulf War era (that is, military personnel who served during the period of the Gulf War but were not deployed in the gulf) to determine whether there was a relationship between ill health and self-reported exposures. Of 16 kinds of exposures assessed, all except exposure to dead animals and exhaust from heaters or generators were more frequent in the Gulf War-deployed cohort than in the control groups. More than 70% of the Gulf War-deployed cohort reported exposure to oil-fire smoke, to exhaust from heaters or generators, and to the sound of chemical alarms, and over 80% reported exposure to PB, NBC (nuclear, biologic, chemical) suits, and to diesel or petrochemical fumes.

STUDIES USING SIMULATION TO ASSESS THE POTENTIAL MAGNITUDE OF EXPOSURES

Simulation studies were used to assess the potential magnitude of exposures encountered under specific circumstances. In some cases, the simulations were later incorporated into fate and transport models.

Tent Heaters

Tent heaters are used to heat tents and other semienclosed environments. Tent heaters, whether vented to the outside or unvented, produce potentially dangerous chemicals as a result of combustion. Carbon monoxide (CO) is the most hazardous byproduct of combustion. Fuels, primarily kerosene-based aviation fuel, were used in the gulf for heaters, cooking stoves, and portable generators. The fumes and exhaust produced by those fuels, particularly when used in unventilated tents, might have exposed people to benzene, toluene, xylene, ethyl benzene, and combustion products, including CO, sulfur dioxide (SO_2), carbon dioxide (CO_2), nitrogen dioxide (NO_2), particulate matter (PM), lead, and other pollutants.

Cheng et al. (2001) performed simulation studies to measure emissions associated with tent heaters that used different fuel types. High concentrations were measured when tent doors were closed (air-exchange rates in this situation were 1.0-1.4 hr^{-1}), and concentrations were below the limits of detection when tent doors were open. Kerosene heaters were associated with higher concentrations of combustion products (Cheng 1999). Concentrations of PM_{10}-containing organic carbon, elemental carbon (soot), sulfate, and ammonium were as high as 850 $\mu g/m^3$. In comparison, under the National Ambient Air Quality Standards, the US Environmental Protection Agency has set a standard of 150 $\mu g/m^3$ for PM_{10}. That is a 24-hour standard that is not to be exceeded more than once a year (US Environmental Protection Agency 2006).

Many of the prevalence studies list self-reported exposures to tent-heater emissions, but there is no specification of heater or fuel types or of whether tent flaps were typically open or closed, so it is not possible to determine the magnitude of exposures. Given that the simulation studies demonstrate the potential for high exposures to combustion products (some of which are similar to those in the oil-fire smoke plume), exposures to tent-heater emissions could modify or potentially confound the assessment of associations between oil-well fire smoke exposure and respiratory outcomes.

Khamisiyah Demolition and Potential Exposure to Sarin and Cyclosarin

During a cease-fire period in March 1991, troops from the US 37th and 307th engineering battalions destroyed enemy munitions throughout the occupied areas of southern Iraq (PAC 1996a). A large storage complex at Khamisiyah, Iraq, which contained more than 100 bunkers, was destroyed. Two sites in the complex—one of the bunkers and another site called the pit—contained stacks of 122-mm rockets loaded with sarin and cyclosarin (Committee on Veterans Affairs 1998). According to estimates, 371 kg of sarin and cyclosarin combined was released (Winkenwerder 2002). US troops performing demolitions were unaware of the presence of nerve agents because their detectors, sensitive only to lethal or near-lethal concentrations of nerve agents (CDC 1999), did not sound any alarms before demolition. It was not until October 1991 that inspectors from the UN Special Commission (UNSCOM) confirmed the presence of a mixture of sarin and cyclosarin at Khamisiyah (Committee on Veterans Affairs 1998).

No air monitoring was conducted at the time of the Khamisiyah demolition. At the request of the Presidential Advisory Committee, Robert Walpole, of the Central Intelligence Agency (CIA), and Bernard Rostker, of DOD, used models to estimate ground-level concentrations of sarin and cyclosarin as a function of distance and direction from the detonation sites and then to estimate the extent of potential exposure of US military personnel to the nerve agents (PAC 1996a). The models produced a series of geographic maps of the Khamisiyah area that overlay known troop-unit locations with the projected path of the sarin–cyclosarin plume. Initially, however, because of the complexity of the modeling that needed to be done, CIA-DOD estimated that any noticeable effects of sarin and cyclosarin would possibly have been seen within 25 km of the demolition site. The CIA-DOD report estimated, on the basis of troop locations, that about 10,000 US troops had been within 25 km and thus might have been exposed to sarin or cyclosarin over a period of hours (CIA-DOD 1997). Given the uncertainties in that estimate, CIA-DOD doubled the distance and, again on the basis of unit locations, estimated that roughly 20,000 troops were within 50 km. In 1997, DOD mailed a survey to the 20,000 troops who might have been within 50 km of Khamisiyah; of the 7,400 respondents, more than 99% reported no acute effects that could be correlated with exposure to sarin or cyclosarin (CIA-DOD 1997). The survey was attached to a letter from the secretary of veterans affairs indicating that chemical weapons had been present at Khamisiyah at the time of the demolition. The letter also urged survey recipients to call the Gulf Incident Hotline with any additional information about the Khamisiyah incident or to report illnesses that they attributed to their service in the Gulf War.

The CIA-DOD (1997) models integrated four components:

- UNSCOM reporting and intelligence summaries of the amount, purity, and type of chemical-warfare agents stored at Khamisiyah.

- Results of experiments[3] performed later at Dugway Proving Ground to simulate the demolition at Khamisiyah and thus estimate the amount of sarin and cyclosarin released, the release rate, and the type of release (instantaneous, continuous, or flyout).

- A combination of dispersion models that incorporated meteorologic conditions at the time (including wind direction) to simulate the transport and diffusion of the plume so that agent concentrations downwind could be estimated.

- Unit location information to determine the position of troops in relation to the plume's path.

Potential exposure was categorized as a "first-noticeable-effects" level and a "general-population" level. At the first-noticeable-effects level, for which the lower limit was set at 1 mg-min/m^3, the estimated exposure would be high enough to cause watery eyes, runny nose, chest tightness, sweating, muscle twitching, or other early signs of exposure to organophosphorus compounds. The general-population level of exposure, for which the upper limit was set at 0.01296 mg-min/m^3, was the "dosage below which the general population, including children and older people, could be expected to remain 72 hours with no effects". Between those two was the "area of low-level exposure" (CIA-DOD 1997). The models indicated that the plume with concentrations at the first-noticeable-effects level would have dispersed to below 1 mg-min/m^3 within 3 days of the demolition. The plume with concentrations in the low-level range dispersed

---

[3]The experiments used a substitute chemical (triethyl phosphate) to simulate chemical-warfare agents and measured agent release concentrations after replicating the rockets in the pit, terrain, original warhead design, stacking of rockets, and other relevant information.

to be in the general-population level within 5 days of the demolition. Taking the potential first-noticeable-effects exposures and the potential low-level exposures into account and eliminating the counting of the same troops on multiple days, CIA-DOD estimated that nearly 99,000 troops might have been exposed to sarin or cyclosarin above the general-population level over the course of 4 days after the demolition of the storage pit at Khamisiyah. Those CIA-DOD findings were challenged in a US Senate report (Committee on Veterans Affairs 1998). The Senate report took issue with the method, especially the reconstruction of the pit site; with the nature of the demolition; and with the number of exposed troops.

At the request of the Senate Committee on Veterans Affairs, the Air Force Technical Applications Center (AFTAC) prepared another exposure model. The AFTAC report summary indicates that AFTAC used models different from those used by CIA-DOD to simulate atmospheric chemistry (Committee on Veterans Affairs 1998). The report indicated additional geographic areas of low-level exposure not modeled by CIA-DOD. The CIA-DOD model was reviewed by an expert panel in 1998; as stated in the Khamisiyah narrative, "this panel approved of the DOD/CIA modeling methodology but recommended a number of improvements, including revisions to the computer models used. The Special Assistant initiated improvements to the 1997 model process to obtain the highest quality of hazard area definition possible. Modeling improvements continued throughout 1998 and 1999 and culminated in redefined potential hazard areas in January 2000" (Winkenwerder 2002).

A second CIA-DOD model, a peer-reviewed revision of the first, was completed in 2000 (Rostker 2000), and a final report was released in 2002 (Winkenwerder 2002). The second CIA-DOD model differed from the first in that it incorporated updated unit-location and personnel data, revised the meteorologic models, reduced the estimates of nerve-agent release, combined the toxicity of sarin and cyclosarin (the first model used only sarin), and adjusted the general-population level to account for a briefer duration of troops' potential exposure. Troops were considered exposed to sarin at 0.0432 mg-min/m^3 and to cyclosarin at 0.0144 mg-min/m^3.

Neither of the models found any troops to have been exposed to concentrations above first-noticeable-effects levels, that is, concentrations that would have been high enough to induce a particular type of chemical alarm to sound and to produce visible signs of the acute cholinergic syndrome among troops. No medical reports by the US Army Medical Corps at the time of the release were consistent with signs and symptoms of acute exposure to sarin (PAC 1996a). That is in accordance with the result of the 1997 DOD survey completed by 7,400 troops within 50 km of Khamisiyah: no reports of cholinergic effects (CIA-DOD 1997).

Two other storage sites in central Iraq, Muhammadiyat and Al Muthanna, sustained damage from air attacks during the Gulf War. Munitions containing 2.9 metric tons of sarin–cyclosarin and 1.5 metric tons of mustard gas were damaged at Muhammadiyat, and munitions containing 16.8 metric tons of sarin–cyclosarin were damaged at Al Muthanna (PAC 1996a). Atmospheric modeling by CIA-DOD determined that the nearest US personnel—400 km away—were outside the range of contamination (PAC 1996a).

To estimate potential exposures to sarin or cyclosarin released from the demolition at Khamisiyah, a number of estimation and modeling procedures have been used. Table 2.1 and the chronology that follows describe the evolution of those approaches to estimate potential exposures. It must be noted, however, that in no cases were sarin and cyclosarin measured, although a Czechoslovak chemical-decontamination unit did detect sarin in areas of northern Saudi Arabia within the timeframe of the Khamisiyah demolition, which suggested that sarin was released into the air. Furthermore, the General Accounting Office (GAO) report (GAO 2004) suggests that multiple reports of sarin detection indicate that the spatial extent of sarin release was far greater than that predicted by the various models.

TABLE 2.1 Models of Sarin and Cyclosarin Release

Model	No. Troops Potentially Exposed	Features	Epidemiologic Studies Using Models to Assess Exposure to Sarin/Cyclosarin
50-km radius Rostker 1997	20,000	Circular radius around site; binary determination of exposure status	McCauley et al. 2002
1997–Model 1 CIA-DOD 1997	98,910	Dispersion model; exposure estimated for sarin; estimated four strata of exposure	Gray et al. 1999b
2000–Model 2 (2000 interim Report; 2002 final report) Rostker 2000 Winkenwerder 2002	101,752	Revision of Model 1, including improved meteorologic models, degradation parameters, increased location specificity, included Air Force personnel, increased toxicity of cyclosarin, modified thresholds for high and low exposure	Bullman et al. 2005; Davis et al. 2004; Smith et al. 2003; Smith et al. 2002b

Chronology (excerpted and modified from GAO 2004 and CIA-DOD 1997):

1996

- August: CIA asked Lawrence Livermore National Laboratory to perform atmospheric-dispersion calculations using a hypothetical release scenario.

- August: CIA, from the above modeling, stated that an area around Khamisiyah as large as 25 km downwind and 8 km wide could have been contaminated.

- September: DOD estimated that 10,000 troops had been within 25 km of Khamisiyah.

- October: DOD extended the distance to 50 km and estimated that 20,000 US troops had been within this zone.

- October: The deputy secretary of defense sent a memorandum to 20,000 veterans who had been identified as being within 50 km of Khamisiyah.

1997

- July: DOD and CIA jointly announced the results of Khamisiyah dispersion modeling (Model 1).

- July: DOD sent written notices to 98,910 veterans in the potential hazard area and about 10,000 notices to those who had received a survey and letter from the deputy secretary of defense but were not in the potential hazard area.

2000

- Revised model (Model 2) completed. Revision included updated meteorologic and dispersion models incorporating dilution, deposition, and degradation terms; revised emissions estimates from the CIA, and improved troop location information and specificity (Model 1 determined location at battalion level—500-900 soldiers; Model 2 determined location at the unit level—under 200 soldiers. That resulted in generally smaller geographic areas to be considered as exposure areas).

- DOD's estimate of the troop number possibly exposed increased by about 2,000. About 35,000 troops who had been notified of possible exposure were no longer in the possible hazard area, whereas about 37,000 newly identified troops probably were in the hazard area.

- Additional analyses resulted in change of potential number of exposed US troops to 101,752. The differences between the 1997 and 2000 models with respect to geographic coverage can be seen in the images below as excerpted from Rostker (2000). In general, the 2000 model estimates dispersion over a smaller area.

Day 1 – 1997 Model

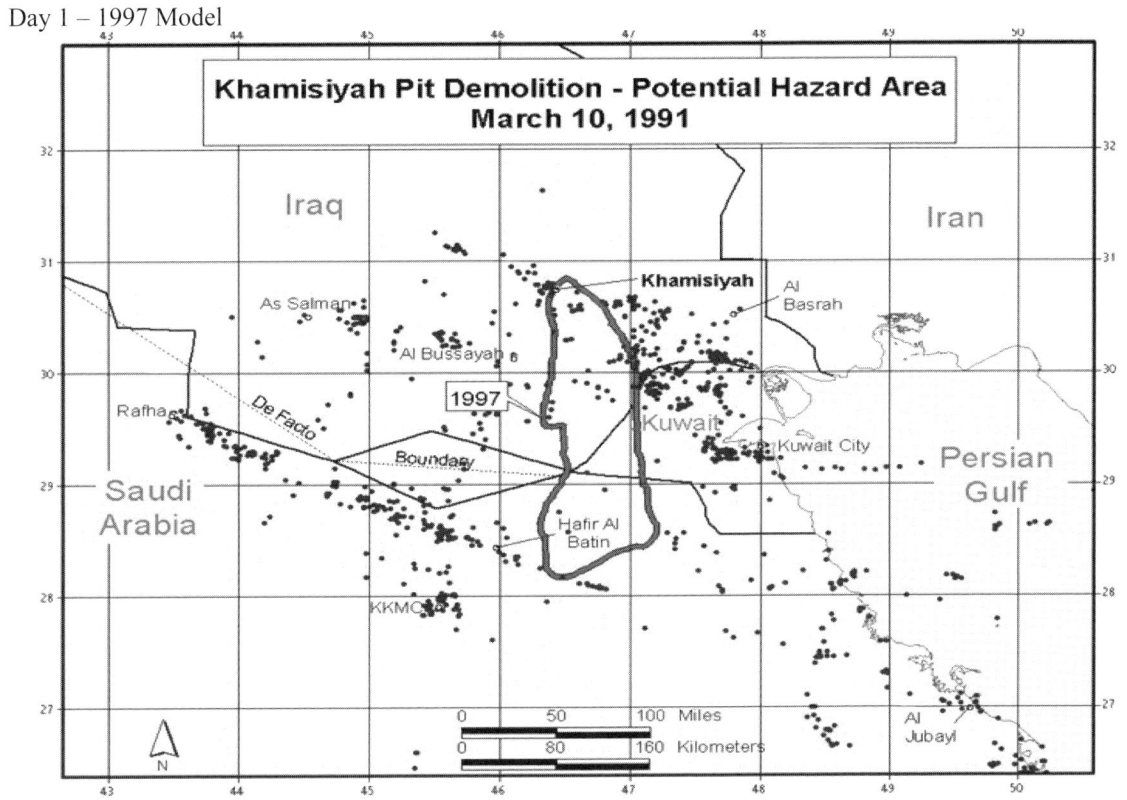

**Khamisiyah Pit Demolition - Potential Hazard Area
March 10, 1991**

SOURCE: Rostker 2000.

Day 1 – 2000 Model

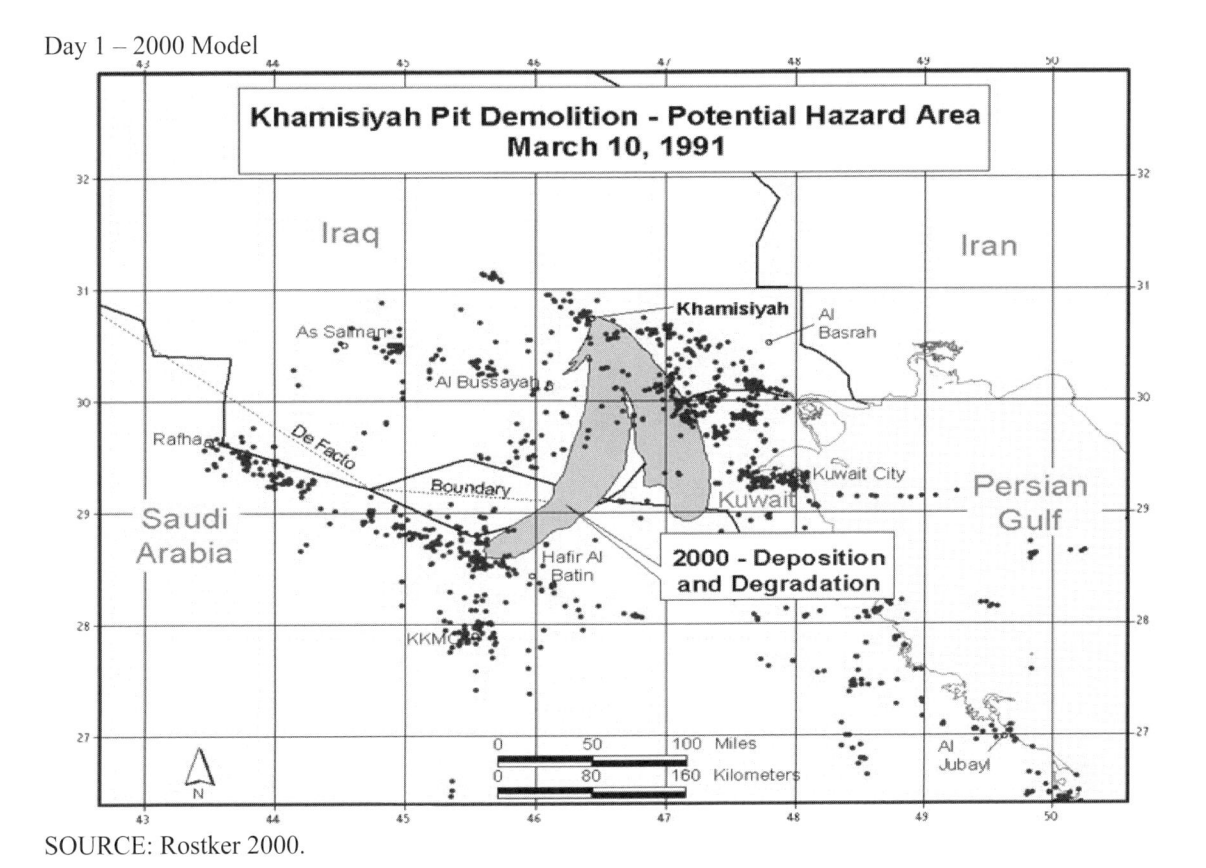

**Khamisiyah Pit Demolition - Potential Hazard Area
March 10, 1991**

SOURCE: Rostker 2000.

Day 2 – 1997 Model

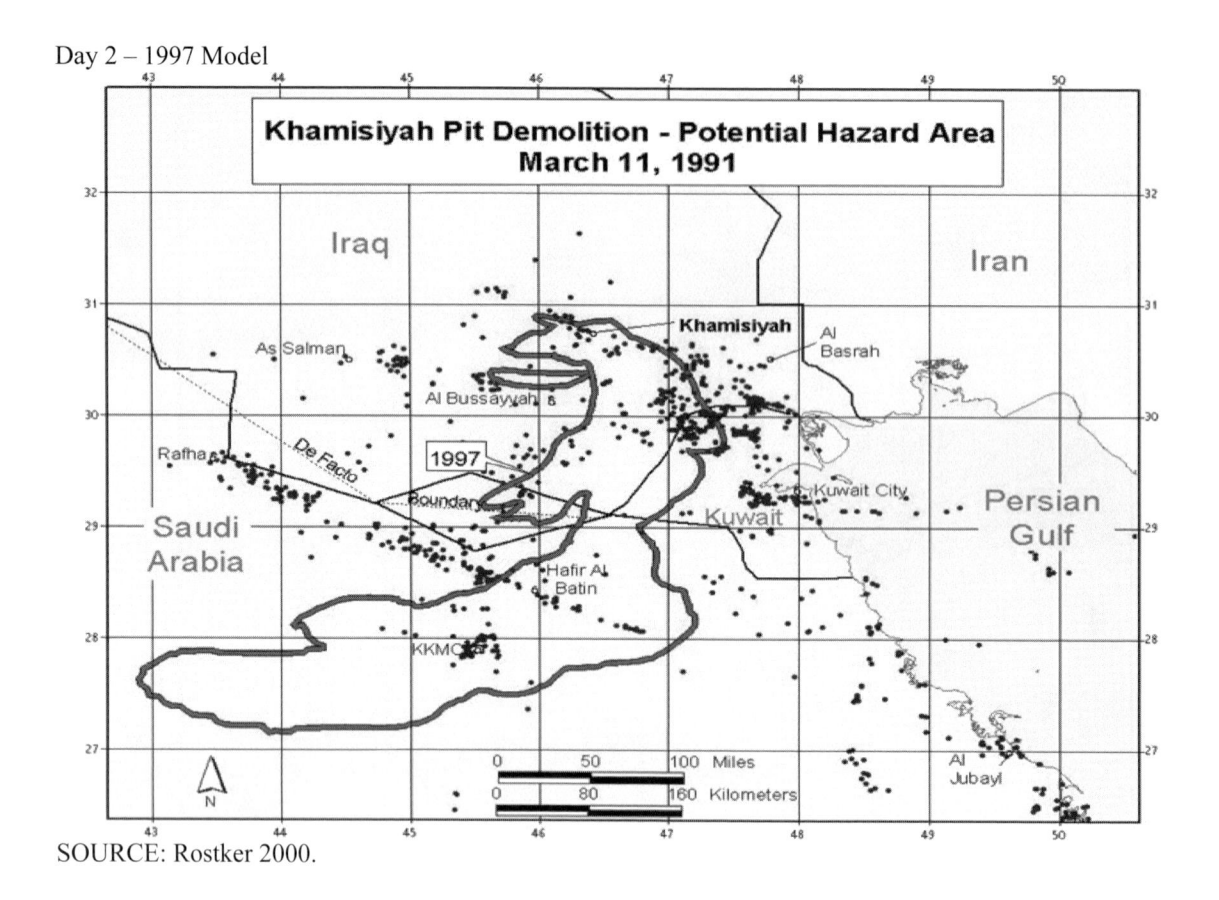

**Khamisiyah Pit Demolition - Potential Hazard Area
March 11, 1991**

SOURCE: Rostker 2000.

Day 2 – 2000 Model

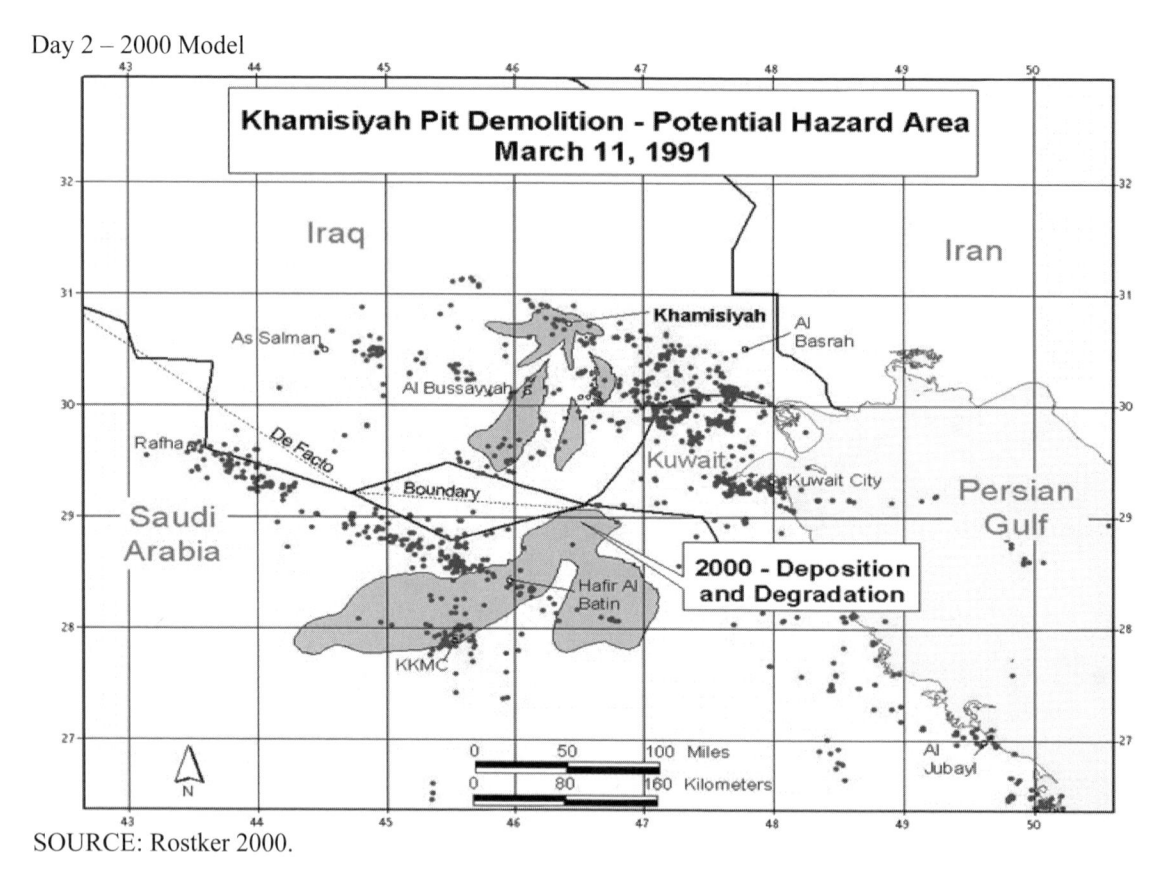

**Khamisiyah Pit Demolition - Potential Hazard Area
March 11, 1991**

2000 - Deposition and Degradation

SOURCE: Rostker 2000.

Day 3 – 1997 Model

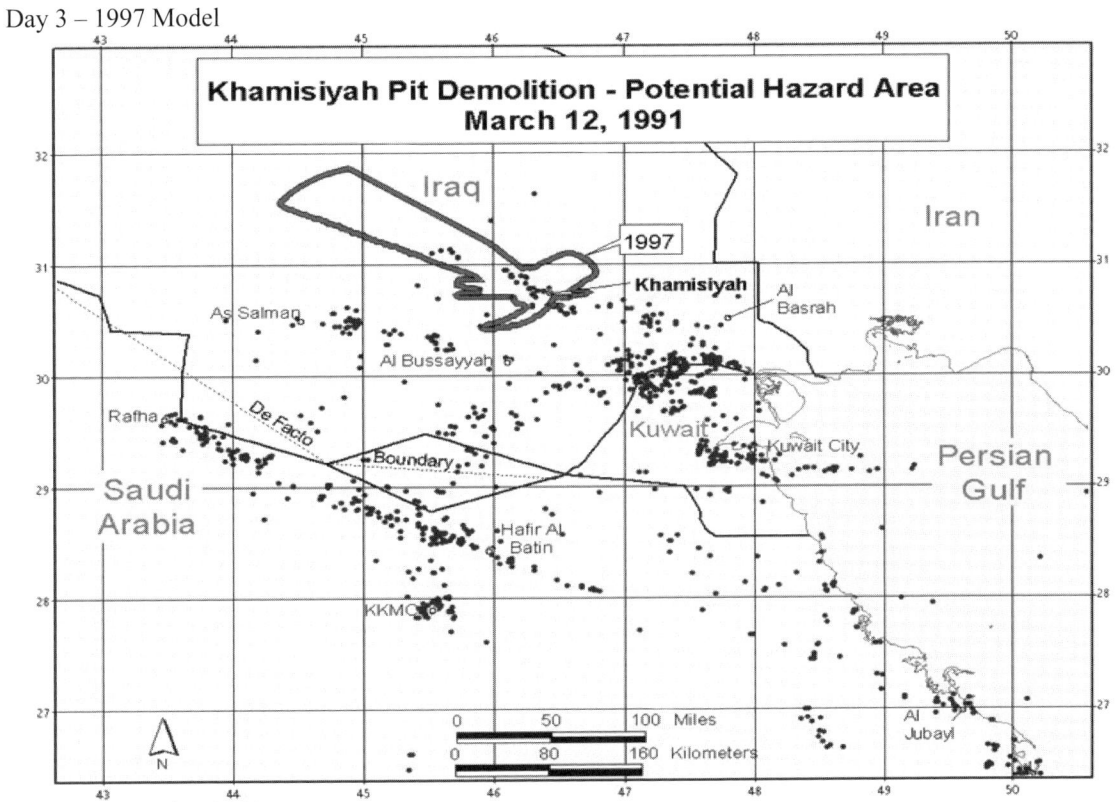

**Khamisiyah Pit Demolition - Potential Hazard Area
March 12, 1991**

SOURCE: Rostker 2000.

Day 3 – 2000 Model

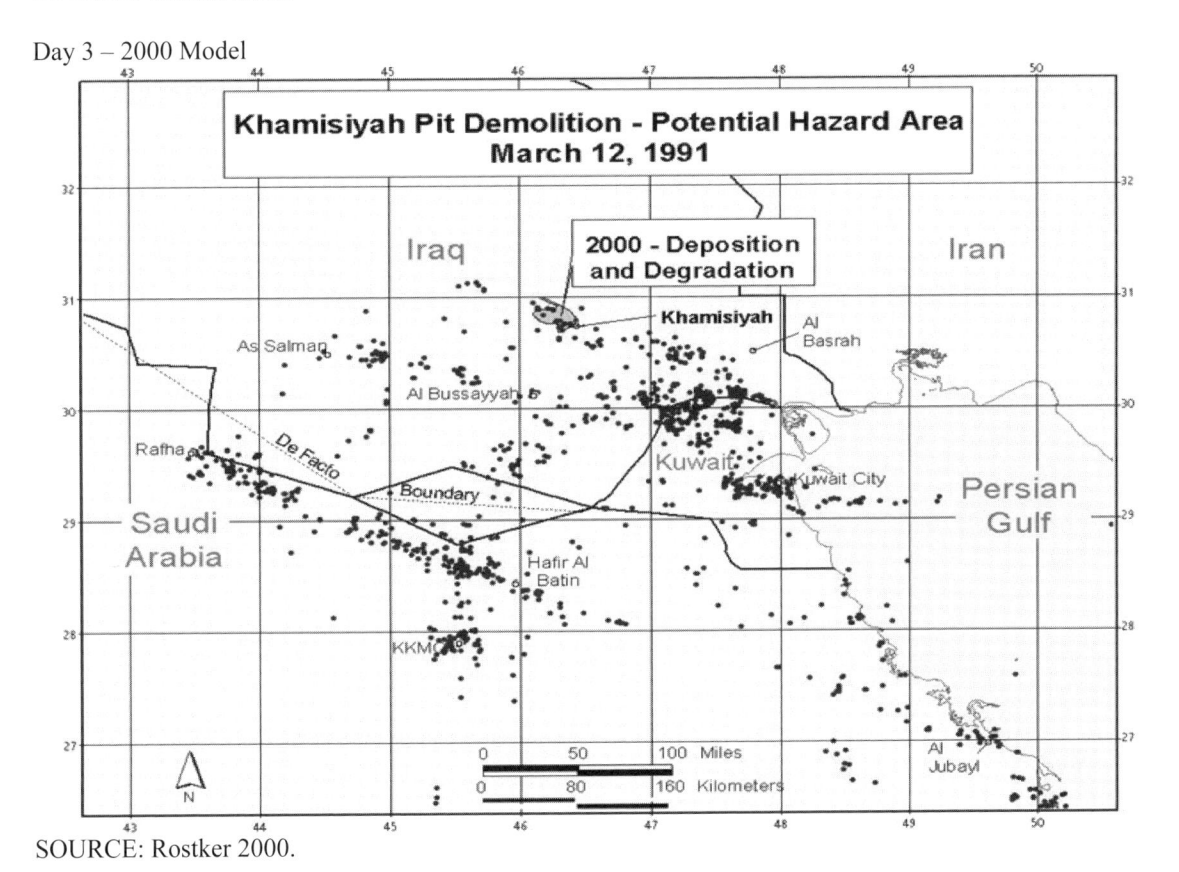

**Khamisiyah Pit Demolition - Potential Hazard Area
March 12, 1991**

SOURCE: Rostker 2000.

Day 4 – 1997 Model

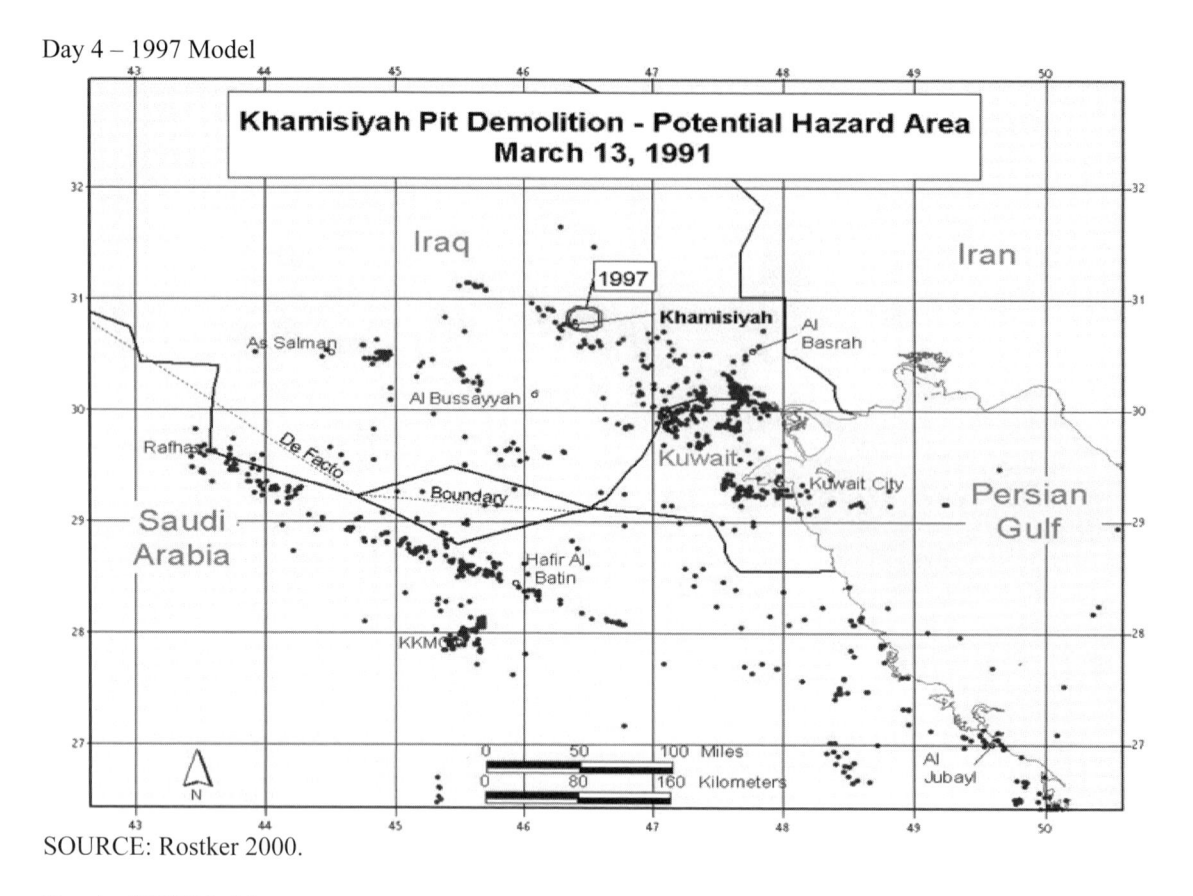

**Khamisiyah Pit Demolition - Potential Hazard Area
March 13, 1991**

SOURCE: Rostker 2000.

Day 4 – 2000 Model

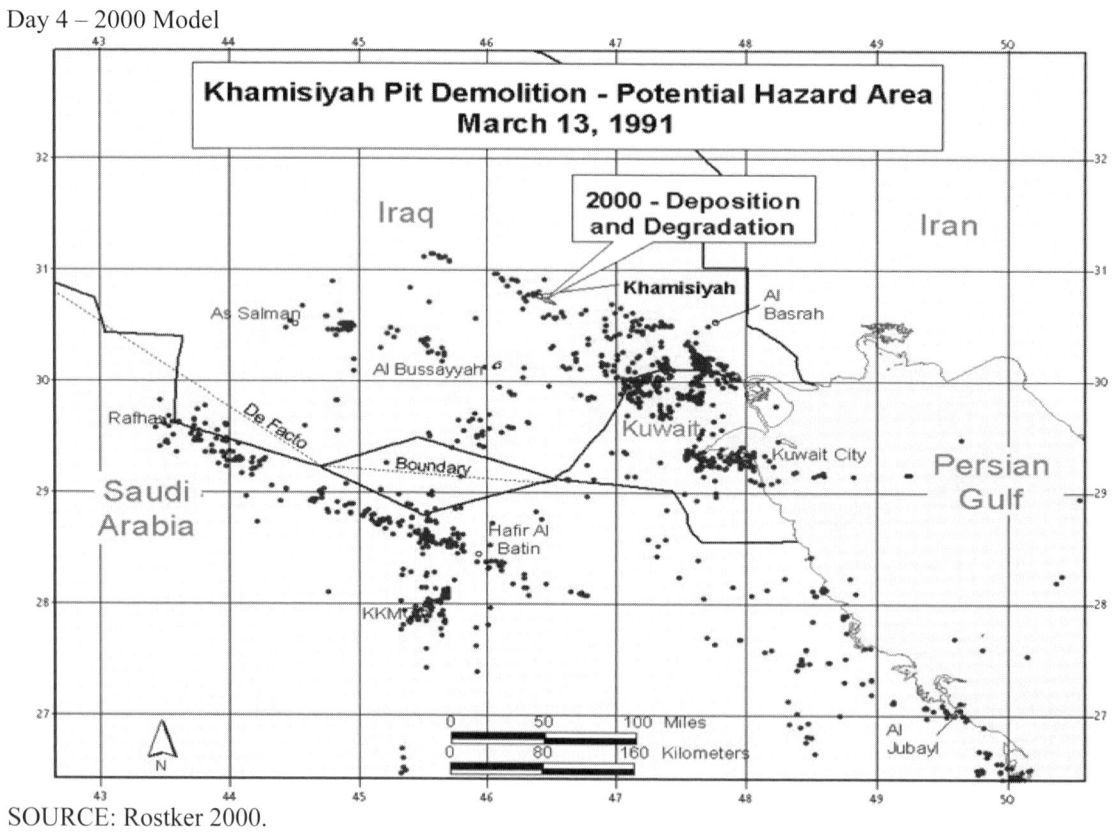

**Khamisiyah Pit Demolition - Potential Hazard Area
March 13, 1991**

SOURCE: Rostker 2000.

Epidemiologic Studies Using Fate and Transport Models to Assess Exposure to Sarin and Cyclosarin

On the basis of results of initial estimates of troops within 50 km who were potentially exposed, McCauley et al. (2002) compared those within 50 km of Khamisiyah with a nondeployed control group and a deployed control group that had not been determined to be within 50 km of the site. After completion of the study, results of the first dispersion model (Model 1) were released. On the basis of the revised model, roughly half those within 50 km were not under the plume.

Gray (1999b) used the original Khamisiyah dispersion model (1997, Model 1) to assess the potential relationship between exposure and postwar hospitalization. That model was linked to troop-movement data at the unit level (50-120 people, according to Gray et al.) in a geographic information system (GIS) framework. According to the authors, locations were not always recorded daily, and units were widely dispersed throughout the region. Additional interviews of 150 officers conducted by the DOD Office of the Special Assistant for Gulf War Illnesses (OSAGWI) were undertaken to optimize the troop-location information for the March 10-13 period of the Khamisiyah munitions destruction. It is not clear whether that resulted in any changes in location information used in the study. The dispersion model included estimates of emissions based on the numbers of rockets present at Khamisiyah and their sarin and cyclosarin content. Rocket-destruction modeling was developed by DOD and CIA to determine the amounts of the agents initially released and the rate of evaporative release. Several meteorologic models were combined with three dispersion models to generate five estimates of daily plume coverage related to the Khamisiyah demolition (detailed report described in DOD report (CIA-DOD 1997). The meteorologic models were validated with measurements of dispersion of smoke from the oil fires. That exercise resulted in a "notification plume" in which exposures were estimated to be at least a low level (0.0126 mg-min/m^3 for sarin) and, after consultation with an expert panel, an "epidemiologic plume" that combined results of the best meteorologic and dispersion models. Plumes were estimated for each day during March 10-13, 1991. Potentially exposed veterans (124,487) were classified into four groups: uncertain low exposure (n = 75,717); 0-0.01256 mg-min/m^3 (n = 18,952); 0.01257-0.09656 mg-min/m^3 (n = 23,061); and 0.09657-0.51436 mg-min/m^3 (n = 6,757). In addition, the study considered 224,804 nonexposed people. Overall, exposures of 349,291 veterans were estimated.

Bullman et al. (2005) used the revised model in their analysis of mortality (2000, Model 2). The revised model was also used by Smith et al. (2002b, 2003) and Davis et al. (2004). The revised model is based on recommendations from a DOD peer review of the initial model and additional information—specifically, more recent versions of the meteorologic models, updated information on the number of rockets damaged during the demolition (from 500 to 225), incorporation of gaseous deposition and nerve-agent decay in the modeling (which would lower exposure estimates and numbers exposed), and different thresholds for high and low exposure to sarin and cyclosarin. Most important, the revised model included more-specific information regarding the location of individual military units at the time of the Khamisiyah explosion and added information on Air Force personnel that was not included in the original estimates. In the earlier (1997) model, a person's location was determined largely by the location of the battalion (which comprised 500-900 people), whereas the revised model included location information on the company (which comprised 100-200 people). On the basis of a combination of the improved location data with revised 2000 modeling, estimates indicate that roughly 102,000 soldiers were within the low-exposure areas for a sarin dosage of 0.0432 mg-min/m^3 (0.00003 mg/m^3 times

1,440 minutes [24 hours]) and a cyclosarin dosage of 0.0144 mg-min/m^3, assuming that cyclosarin was more than 3 times as toxic as sarin—and no soldiers were in the high-exposure area. Of the 102,000, about 66,000 were considered to be exposed in the 1997 model. The increase in the number of potentially exposed people is based on the more accurate troop-location information.

Despite the continued improvement in the model, a recent GAO report (GAO 2004) remains critical of the models and argues that epidemiologic studies that use those models incorporate substantial exposure misclassification due to errors in estimation of troop locations combined with uncertainty regarding plume locations. The report identifies the following major problems with the modeling:

- Underestimation of plume heights and consequently underestimated areas of potential exposure.
- Source-emission estimates based on unrealistic simulation.
- In all models, large variability in plume location and size.

Specifically, the GAO report notes that the models used were not developed for long-range transport of chemical-warfare agents, included inaccurate and highly uncertain information regarding the numbers of sarin- or cyclosarin-containing rockets that were destroyed and the rate of release of the compounds from each rocket, and underestimated the height of plumes resulting from demolition. The GAO report also states that there is a large degree of variability between the models that were used regarding the size and path of a plume. The variability led to the use of a single composite model (also called the ensemble approach, in which results of different model combinations are averaged to approximate a variety of conditions—a relatively common approach in atmospheric science), but the results of any single model were excluded. That decision dramatically reduced the area of potential exposure. The GAO report notes that exposure of troops cannot be reliably estimated for epidemiologic studies.

Given the lack of any means to evaluate model performance, it is not possible to determine the accuracy of the final version of the model. One can conclude that any studies that used the earlier version of the model (Gray et al. 1999b; McCauley et al. 2002) or less specific exposure-estimation approaches are likely to be inaccurate according to the current understanding of the dispersion of agents from Khamisiyah. Even if the final model can be accepted as an accurate determination of exposure, an important problem for epidemiologic studies has been introduced by the numerous notifications to veterans regarding potential exposure to agents released from Khamisiyah. Given that the sequential notifications went to different people, under any of the more recent modeling scenarios there will be people who were previously notified as being exposed but who will be considered unexposed on the basis of model results. That might jeopardize the findings of studies that focused on Khamisiyah emissions, because knowledge of potential exposure might lead to biased reporting of health outcomes. Furthermore, the large differences between model estimates of who was and who was not exposed raises concern regarding the potential for exposure misclassification—even in the most recent model. Other characteristics of the model inputs (such as the amounts and purity of the sarin and cyclosarin that were estimated to be present before demolition) and dispersion parameters will affect the quantitative estimates of exposure and the extent of the area under the plume but are less likely to have major effects on a binary classification of exposed-unexposed.

STUDIES USING ENVIRONMENTAL FATE AND TRANSPORT MODELS FOR SPECIFIC EXPOSURES

At the end of the Gulf War, over 600 Kuwaiti oil wells were ignited by retreating Iraqi troops. Large plumes of smoke rose from the fires. Occasionally, the smoke remained near the ground and enveloped US military personnel. No systematic monitoring occurred from the initial deployment in 1990 to May 1991, when several independent teams from multiple US and international agencies (including the US Army Environmental Health Agency [AEHA] and the US Environmental Protection Agency [EPA]) went into Kuwait to monitor the ambient air contamination due to oil-well fire emissions (Spektor 1998). Smoke was sampled to improve understanding of the nature of the plumes generated by the burning oil wells. Most of the oil fires were still burning when measurements began.

Individual fires created distinct smoke plumes over short distances, but over longer distances the plumes merged into one "supercomposite" plume about 40 km wide south of Kuwait City. At the base of the plume, oil falling in droplet form or emitted from uncapped wells collected in pools on the desert; the pools sometimes were on fire as well (Hobbs and Radke 1992). The smoke plumes from individual fires varied in color and density. Black smoke plumes resulted from single well fires and had relatively high concentrations of carbon; they made up 60-65% of the fires. The densest black plumes were from the burning pools of oil. White smoke plumes, accounting for 25-30% of the fires, contained almost no carbon but had a higher concentration of inorganic salts; this is consistent with reports of the presence of brine solutions in the oil fields (Cofer et al. 1992; Spektor 1998).

The available monitoring data indicate that concentrations of nitrogen oxides, carbon monoxide, sulfur dioxide, hydrogen sulfide, other pollutant gases, and polycyclic aromatic hydrocarbons (PAHs) did not exceed those in the air of a typical US industrial city. PAH concentrations in the samples were low (PAC 1996b). High concentrations of particulate matter (PM) from sand and soot were often observed at multiple monitoring sites; an estimated 20,000 tons of soot, or fine-particle mass, was generated by the fires (Thomas et al. 2000) and made up about 23% of the PM in the Persian Gulf, often at concentrations twice those considered safe (Rostker 2000).

In addition to air monitoring, potential exposures of troops to smoke and combustion products from the oil-well fires were modeled (Draxler et al. 1994). Daily and seasonal normalized air concentrations due to emissions from the oil-well fires were computed with a modified Lagrangian transport, dispersion, and deposition model for February-October 1991. The highest normalized concentrations were near the coast between Kuwait and Qatar. Peak values moved farther west and inland with each season; that is, the smoke and combustion products moved from over the Gulf in spring to the west over the Saudi Peninsula by autumn.

Assessment of exposure to oil-fire smoke used self-reports of exposure and a detailed atmospheric model that was combined with troop-unit location information. For example, Kelsall (2004b) used self-reported exposures to oil-fire smoke and dust storms, including no exposure, low exposure (less than 5 hours/day or outside for less than 10 days), and high exposure (more than 5 hours/day or outside for 10 days or more). Most of the Australian veterans included in the study had either no (46.1%) or low (44.8%) exposure to oil-fire smoke. The timing of deployment relative to the beginning of the air war (January 17, 1991) and the ignition of oil wells was also used as a more objective estimate of exposure in this study.

The use of such modeling is supported by a number of measurements (ground-based and satellite) that provided detailed characterization of plume dispersion and dynamics. For example, Hobbs and Radke (1992) described the atmospheric properties of the Kuwaiti oil fires. Individual fires produced distinct, isolated plumes that merged beyond short distances and fanned outward horizontally. Smoke was never observed to rise above 6 km, and this prevented rapid transport over large distances. As a result of scavenging by clouds and precipitation, the residence time of a smoke particle in the atmosphere was relatively short (days). The composite plume from the north and south oil fields was 40 km wide south of Kuwait City and 0.5-2 km in altitude. Individual fires produced different plumes with different appearances (white and black smoke) due to different amounts of salt (more salt resulted in more whiteness). A composite plume consisted of about 30% salt, 15-20% soot (elemental carbon), 8% sulfate, and 30% organic carbon.

The oil-fire smoke model that was used in epidemiologic studies is described in detail in Smith et al. (2002a) and Lange et al. (2002). Briefly, the National Oceanic and Atmospheric Administration and the US Army Center for Health Promotion estimated 24-hour concentrations for 15 x 15-km grids (a total of 40,401 grids) in the Gulf War Theater. The Lagrangian HYSPLIT model was used with meteorologic data to estimate air concentrations of PM. The model estimates were compared with aircraft and surface measurements of carbon-soot smoke and sulfur dioxide.

The troop-location information was compiled from daily data provided by the US Armed Services Center for Unit Records Research as described in Gray et al. (1999b) and Smith et al. (2002a). Particle exposures were estimated for each troop unit on each day by using the closest HYSPLIT grid point. The unit-based exposures were then linked to DOD personnel files to identify individual service members in each unit and to assign exposures to each member. In their analysis, Smith et al. categorized exposure on the basis of estimated concentrations and durations: average daily exposure at 1-260 $\mu g/m^3$ for 1-25 days (33.7%), for 25-50 days (5.7%), and for over 50 days (0.9%); and average daily exposure at over 260 $\mu g/m^3$ for 1-25 days (16.8%), for 26-50 days (17.0%), and for over 50 days (9%). The reference category of no exposure accounted for 16.8% of the 405,142 study subjects (75% of the active-duty deployed Gulf War veterans).

It is important to note that all the above studies (Cowan et al. 2002; Gray et al. 1999b; Lange et al. 2002; Smith et al. 2002a) used the same exposure model and that the model was developed with ground-based measurement data and satellite data. The use of satellite data in particular improves confidence in the model's ability to describe the spatial extent of the oil-fire plumes accurately. Furthermore, the troop-unit (about 120 people per unit) location information was derived from handwritten records of GPS-based location information. For each unit, a single location was derived for each day; unit subgroups were identified (and assigned different exposures) if multiple locations for a unit were present for at least 3 consecutive days. Lange et al. (2002) also describe a quality-control check of the unit-location information. The results of the epidemiologic studies with this model are discussed in later chapters. There are limitations in any assessment of exposure that does not include individual measurements, but this model used state-of-the-art approaches and was based, wherever possible, on measurement data.

STUDIES USING BIOLOGIC MONITORING FOR SPECIFIC EXPOSURES

Depleted Uranium

Potential exposure to DU in the Gulf War resulted from the generation of uranium-dust particles and shrapnel when DU munitions penetrated a target. The penetration leads to the generation of fragments and particles that ignite easily. Exposure might occur through inhalation of particles or via embedded shrapnel. DU is almost entirely U-238 and has only 60% of the radioactivity of natural uranium (U) because it has been depleted of much of the more radioactive isotopes (U-235 and U-234). Uranium decay results in the release of alpha particles which travel short distances. Consequently, the principal radiation hazard associated with DU exposure is to tissues in immediate contact with internalized DU fragments or particles (McDiarmid et al. 2004a).

DU was used in several types of armor-piercing munitions during the Gulf War because of its extreme density and effectiveness in penetrating armor. In addition, DU was used in tank armor. Potential exposures were typically associated with friendly-fire incidents in which US tanks mistakenly fired DU munitions into other US combat vehicles (GAO 2000). Those incidents reportedly exposed 102 service members, who were in or on the vehicles at the time of impact, to shrapnel that contained embedded DU or to inhaled or ingested DU particles. An additional 60 people involved in rescue and evacuation and 191 who entered DU-contaminated vehicles after impact might also have been exposed. A fire at Camp Doha, Kuwait, involved detonation and burning of DU-containing munitions and led to additional potential exposures. More minor exposures might have been experienced by other people from a variety of circumstances such as passing through areas with DU smoke and handling spent DU munitions.

In response to concerns regarding the potential health effects of exposure to DU, VA initiated a urinary-uranium testing program, which was open to any concerned Gulf War veteran and, with special notification, provided to those who were suspected of receiving the highest DU exposures. McDiarmid et al. (2004b) reported the results of urinary uranium measurements for a total of 446 Gulf War veterans, including 169 veterans with no known friendly-fire involvement but self-reporting of other potential DU exposures. It is important to note that there is not complete overlap between those involved in the screening program and those noted above with potential exposure.

Of the 446 veterans, 22 had high urinary uranium (0.05 µg/g of creatinine). Of the 22, 15 had a second sample tested for confirmation, and only six had high uranium in the second sample. Of the initial 22 samples, 21 were subjected to isotopic analysis to determine the source of the high urinary uranium. Three subjects had confirmed DU exposure, and the remaining 18 had isotope ratios consistent with natural-uranium exposure (McDiarmid et al. 2004b). The results of the surveillance also indicate no increased urinary uranium in those participating in the program compared with a sample of US adults in the National Health and Nutrition Examination Surveys. In a comparison of exposure scenarios with urinary-uranium measurements, only the presence of embedded shrapnel was associated with high urinary uranium. Those reporting that they were in or on a vehicle hit by friendly fire (29 of the 446) or in a vehicle hit by enemy fire (31 of the 446) also had somewhat increased urinary uranium, although the report did not provide details of the statistical comparisons. In the study, 19 exposure categories or scenarios determined from questionnaire responses to 30 questions were analyzed in relation to urinary-uranium measurements. The 30 scenarios were determined by a panel of military experts and health physicists and ranked by intensity of exposure by the DOD OSAGWI (Office of the Special Assistant for Gulf War Illness).

Focused analyses of veterans who reported being involved in friendly-fire incidents have also been conducted. For example, a group of 39 veterans reporting exposure to DU during friendly-fire incidents were assessed by McDiarmid et al. (2004a). All urinary-uranium measurements, except one, were above 0.10 µg/g of creatinine and were from people with retained DU shrapnel. They also reported good agreement between measurements made in 2001 and those made in 1997 and 1999. An earlier study (McDiarmid et al. 2001b) of 50 veterans who had been exposed to friendly fire and had retained shrapnel found that all but one were excreting high concentrations of urinary uranium (0.018-39.1 µg/g of creatinine).

Hooper et al. (1999) reported on 33 US soldiers who had been in or on a vehicle struck by DU munitions, selected from a DOD list of 68 military personnel who were wounded in February 1991 in friendly-fire incidents. Of the 68, 48 were contacted, and 33 agreed to participate. Retained shrapnel was identified with x-ray pictures and 24-hour urine samples were collected. Of the 33, 23 had been told that they were wounded with shrapnel; most were unsure whether it was removed. X-ray analysis detected shrapnel in 15 of the 23 and in two others who were not aware of any shrapnel; that suggests that DU can penetrate to soft tissue without superficial wounds. Subjects with retained shrapnel had urinary uranium 150 times as high as those with no evidence of shrapnel (4.47 vs 0.03 µg/g of creatinine). Subjects with reported shrapnel wounds but no retained shrapnel on x-ray pictures had low urinary-uranium measurements that were comparable with those in subjects without suspected shrapnel. In those with confirmed shrapnel, urinary uranium remained high 4 years after exposure and was highly correlated with previous measurements (3 years after exposure).

It is important to note that even in those with documented persistent DU exposure, no signs of classic uranium-related adverse outcomes have been observed. Furthermore, it is argued that, on the basis of experience with uranium miners and that of veterans with persistent DU exposure, veterans with normal urinary-uranium measurements are unlikely to develop any uranium-related toxicity in the future regardless of initial DU exposure (McDiarmid et al. 2001a).

In summary, results of detailed surveillance indicate persistently increased urinary uranium in a very small number of subjects with embedded DU-containing shrapnel. The vast majority of veterans who participated in DU-exposure surveillance, presumably because of potential exposure, did not show evidence of long-term DU exposure. It is likely that some of those veterans were initially exposed to DU that has since been cleared or largely sequestered in bone.

Oil-Well Fire Smoke

In light of concern about oil-well fire smoke and exposure to other agents related to Gulf War service, Poirier et al. (1998) conducted biologic monitoring and DNA-adduct monitoring for PAH exposure of 61 US soldiers before deployment (while in Germany), during their deployment in Kuwait, and after their deployment (back in Germany). Environmental monitoring indicated low PAH concentrations in those soldiers' areas of service in Kuwait. Concentrations of adducts and markers of PAH exposure were lowest during the period of Kuwaiti deployment, and there was no indication of increased PAH exposure while the soldiers were deployed in Kuwait. Additional biologic monitoring data on the same people for metals and volatile organic compounds (VOCs) are described by US AEHA (1994). The biologic measurements did not in general indicate any Gulf War deployment-related differences in exposure. In most cases, results were within the normal US population reference range determined by the CDC National Center for Environmental Health. The one exception was higher predeployment and postdeployment than deployment concentrations of PAH-DNA adducts; that implies very low PAH exposure of soldiers deployed in the Gulf War. The data do not suggest any exposure at high PAH concentrations during deployment in Kuwait. Similarly, low exposure to metals and VOCs during deployment was found.

SUMMARY AND CONCLUSIONS

The studies that used self-reports indicate that there is a higher prevalence of reporting of multiple biologic and chemical exposures by veterans than by their nondeployed counterparts, especially exposures to oil-well fire smoke, diesel, insecticides, and tent-heater fumes. Those studies also show evidence of reporting bias regarding vaccinations, and studies of PB generally show a high prevalence of underconsumption and, for a smaller subset, overconsumption of PB relative to recommended dosages.

There is little agreement between subjective and objective measurements of exposure to DU and oil-well fire smoke. DU, however, is the best-characterized exposure because of extensive database of urinary uranium measurements. Those data indicate that a very small number of people clearly have prolonged exposure to DU. The oil-fire smoke studies, although not based on individual measurements as in the case of DU, incorporate objective exposure assessment inasmuch as models have been evaluated with actual measurements and with troop-location information based on measurements. There are substantial problems with the assessment of exposure to the Khamisiyah demolition products (sarin and cyclosarin). Inconsistencies between models and lack of any model evaluation make it difficult to ascertain the accuracy of exposure assessment. The epidemiologic studies that used the initial models (50 km and Model 1—Table 2.1) are likely to have greater inaccuracy. More recent studies (using Model 2) are likely to be improved but still retain substantial uncertainty, which results in a potential for exposure misclassification.

REFERENCES

Bullman TA, Mahan CM, Kang HK, Page WF. 2005. Mortality in US Army Gulf War Veterans Exposed to 1991 Khamisiyah Chemical Munitions Destruction. *American Journal of Public Health* 95(8):1382-1388.

CDC (Centers for Disease Control and Prevention). 1999. *Background Document on Gulf War-Related Research for the Health Impact of Chemical Exposures During the Gulf War: A Research Planning Conference.* Atlanta, GA: Centers for Disease Control and Prevention.

Cheng YS, Zhou Y, Chow J, Watson J, Frazier C. 2001. Chemical composition of aerosols from kerosene heaters burning jet fuels. *Aerosol Science and Technology* 35(6):949-957.

Cheng Y. 1999. *Characterization of Emissions from Heaters Burning Leaded Diesel Fuel in Unvented Tents.* Washington, DC: Department of Defense.

Cherry N, Creed F, Silman A, Dunn G, Baxter D, Smedley J, Taylor S, Macfarlane GJ. 2001. Health and exposures of United Kingdom Gulf war veterans. Part II: The relation of health to exposure. *Occupational and Environmental Medicine* 58(5):299-306.

CIA-DOD (Central Intelligence Agency-Department of Defense). 1997. *Modeling the Chemical Warfare Agent Release at the Khamisiyah Pit*. Washington, DC: CIA-DOD.

Cofer WR, Stevens RK, Winstead EL. 1992. Kuwaiti oil fires: Compositions of source smoke. *Journal of Geophysical Research* 97(D17):14521-14525.

Committee on Veterans' Affairs, US Senate. 1998. *Report of the Special Investigation Unit on Gulf War Illnesses*. Washington, DC: 105th Congress, 2nd session.

Cowan DN, Lange JL, Heller J, Kirkpatrick J, DeBakey S. 2002. A case-control study of asthma among U.S. Army Gulf War veterans and modeled exposure to oil well fire smoke. *Military Medicine* 167(9):777-782.

Davis LE, Eisen SA, Murphy FM, Alpern R, Parks BJ, Blanchard M, Reda DJ, King MK, Mithen FA, Kang HK. 2004. Clinical and laboratory assessment of distal peripheral nerves in Gulf War veterans and spouses. *Neurology* 63(6):1070-1077.

Draxler RR, McQueen JT, Stunder BJB. 1994. An evaluation of air pollutant exposures due to the 1991 Kuwait oil fires using a Lagrangian model. *Atmospheric Environment* 28(13):2197-2210.

GAO (Government Accountability Office). 2000. *Understanding of Health Effects From Depleted Uranium Evolving but Safety Training Needed*. Washington, DC: GAO.

GAO. 2004. *Gulf War Illnesses: DOD's Conclusions About US Troops' Exposure Cannot Be Adequately Supported*. Washington, DC: GAO.

Gray GC, Kaiser KS, Hawksworth AW, Hall FW, Barrett-Connor E. 1999a. Increased postwar symptoms and psychological morbidity among U.S. Navy Gulf War veterans. *American Journal of Tropical Medicine and Hygiene* 60(5):758-766.

Gray GC, Smith TC, Knoke JD, Heller JM. 1999b. The postwar hospitalization experience of Gulf War Veterans possibly exposed to chemical munitions destruction at Khamisiyah, Iraq. *American Journal of Epidemiology* 150(5):532-540.

Hobbs PV, Radke LF. 1992. Airborne studies of the smoke from the Kuwait oil fires. *Science* 256(5059):987-991.

Hooper FJ, Squibb KS, Siegel EL, McPhaul K, Keogh JP. 1999. Elevated urine uranium excretion by soldiers with retained uranium shrapnel. *Health Physics* 77(5):512-519.

Hotopf M, David A, Hull L, Ismail K, Unwin C, Wessely S. 2000. Role of vaccinations as risk factors for ill health in veterans of the Gulf war: Cross sectional study. *British Medical Journal* 320(7246):1363-1367.

Keeler JR, Hurst CG, Dunn MA. 1991. Pyridostigmine used as a nerve agent pretreatment under wartime conditions. *Journal of the American Medical Association* 266(5):693-695.

Kelsall HL, Sim MR, Forbes AB, Glass DC, McKenzie DP, Ikin JF, Abramson MJ, Blizzard L, Ittak P. 2004a. Symptoms and medical conditions in Australian veterans of the 1991 Gulf War: Relation to immunisations and other Gulf War exposures. *Occupational and Environmental Medicine* 61(12):1006-1013.

Kelsall HL, Sim MR, Forbes AB, McKenzie DP, Glass DC, Ikin JF, Ittak P, Abramson MJ. 2004b. Respiratory health status of Australian veterans of the 1991 Gulf War and the effects of exposure to oil fire smoke and dust storms. *Thorax* 59(10):897-903.

Kroenke K, Koslowe P, Roy M. 1998. Symptoms in 18,495 Persian Gulf War veterans. Latency of onset and lack of association with self-reported exposures. *Journal of Occupational and Environmental Medicine* 40(6):520-528.

Lange JL, Schwartz DA, Doebbeling BN, Heller JM, Thorne PS. 2002. Exposures to the Kuwait oil fires and their association with asthma and bronchitis among Gulf War veterans. *Environmental Health Perspectives* 110(11):1141-1146.

Madsen JM. 1998. *Clinical Considerations in the Use of Pyridostigmine Bromide As Pretreatment for Nerve-Agent Exposure*. Aberdeen Proving Ground, MD: Army Medical Research Institute of Chemical Defense.

Mahan CM, Kang HK, Dalager NA, Heller JM. 2004. Anthrax vaccination and self-reported symptoms, functional status, and medical conditions in the National Health Survey of Gulf War Era Veterans and Their Families. *Annals of Epidemiology* 14(2):81-88.

McCauley LA, Lasarev M, Sticker D, Rischitelli DG, Spencer PS. 2002. Illness experience of Gulf War veterans possibly exposed to chemical warfare agents. *American Journal of Preventive Medicine* 23(3):200-206.

McDiarmid MA, Engelhardt SM, Oliver M. 2001a. Urinary uranium concentrations in an enlarged Gulf War veteran cohort. *Health Physics* 80(3):270-273.

McDiarmid MA, Squibb K, Engelhardt S, Oliver M, Gucer P, Wilson PD, Kane R, Kabat M, Kaup B, Anderson L, Hoover D, Brown L, Jacobson-Kram D, Depleted Uranium Follow-Up Program. 2001b. Surveillance of depleted uranium exposed Gulf War veterans: Health effects observed in an enlarged "friendly fire" cohort. *Journal of Occupational and Environmental Medicine* 43(12):991-1000.

McDiarmid MA, Engelhardt S, Oliver M, Gucer P, Wilson PD, Kane R, Kabat M, Kaup B, Anderson L, Hoover D, Brown L, Handwerger B, Albertini RJ, Jacobson-Kram D, Thorne CD, Squibb KS. 2004a. Health effects of depleted uranium on exposed Gulf War veterans: A 10-year follow-up. *Journal of Toxicology and Environmental Health Part A* 67(4):277-296.

McDiarmid MA, Squibb K, Engelhardt SM. 2004b. Biologic monitoring for urinary uranium in gulf war I veterans. *Health Physics* 87(1):51-56.

Office of the Secretary of Defense. January 30, 1998. Letter to Honorable Arlen Specter, Chairman Committee on Veterans Affairs United States Senate.

PAC (Presidential Advisory Committee). 1996a. *Presidential Advisory Committee on Gulf War Veterans' Illnesses: Final Report*. Washington, DC: US Government Printing Office.

PAC. 1996b. *Presidential Advisory Committee on Gulf War Veterans' Illnesses: Final Report*. Washington, DC: US Government Printing Office.

Poirier MC, Weston A, Schoket B, Shamkhani H, Pan CF, McDiarmid MA, Scott BG, Deeter DP, Heller JM, Jacobson-Kram D, Rothman N. 1998. Biomonitoring of United States Army soldiers serving in Kuwait in 1991. *Cancer Epidemiology, Biomarkers and Prevention* 7(6):545-551.

Proctor SP, Heeren T, White RF, Wolfe J, Borgos MS, Davis JD, Pepper L, Clapp R, Sutker PB, Vasterling JJ, Ozonoff D. 1998. Health status of Persian Gulf War veterans: Self-reported symptoms, environmental exposures and the effect of stress. *International Journal of Epidemiology* 27(6):1000-1010.

Rostker B. 1997. *US Demolition Operations at the Khamisiyah Ammunition Storage Point.* Washington, DC: Department of Defense.

Rostker, B. 2000. *US Demolition Operations at Khamisiyah.* [Online]. Available: http://www.gulflink.osd.mil/khamisiyah_ii/ [accessed August 6, 2004].

Schumm WR, Reppert EJ, Jurich AP, Bollman SR, Webb FJ, Castelo CS, Stever JC, Kaufman M, Deng LY, Krehbiel M, Owens BL, Hall CA, Brown BF, Lash JF, Fink CJ, Crow JR, Bonjour GN. 2002. Pyridostigmine bromide and the long-term subjective health status of a sample of over 700 male Reserve Component Gulf War era veterans. *Psychological Reports* 90(3 Pt 1):707-721.

Smith TC, Heller JM, Hooper TI, Gackstetter GD, Gray GC. 2002a. Are Gulf War veterans experiencing illness due to exposure to smoke from Kuwaiti oil well fires? Examination of Department of Defense hospitalization data. *American Journal of Epidemiology* 155(10):908-917.

Smith TC, Smith B, Ryan MA, Gray GC, Hooper TI, Heller JM, Dalager NA, Kang HK, Gackstetter GD. 2002b. Ten years and 100,000 participants later: Occupational and other factors influencing participation in US Gulf War health registries. *Journal of Occupational and Environmental Medicine* 44(8):758-768.

Smith TC, Gray GC, Weir JC, Heller JM, Ryan MA. 2003. Gulf War veterans and Iraqi nerve agents at Khamisiyah: Postwar hospitalization data revisited. *American Journal of Epidemiology* 158(5):457-467.

Spektor DM. 1998. *A Review of the Scientific Literature As It Pertains to Gulf War Illnesses, Volume 6: Oil Well Fires.* Santa Monica, CA: RAND.

Thomas R, Vigerstad T, Meagher J, McMullin C. 2000. *Particulate Exposure During the Persian Gulf War.* Washington, DC: Department of Defense.

Unwin C, Blatchley N, Coker W, Ferry S, Hotopf M, Hull L, Ismail K, Palmer I, David A, Wessely S. 1999. Health of UK servicemen who served in Persian Gulf War. *Lancet* 353(9148):169-178.

United States Environmental Protection Agency. *National Ambient Air Quality Standards (NAAQS).* [Online]. Available: http://www.epa.gov/air/criteria.html [accessed July 11, 2006].

US AEHA (United States Army Environmental Hygiene Agency). 1994. *Final Report: Kuwait Oil Fire Health Assessment: 5 May-3 December 1991.* Report No. 39.26-L192-91. Washington, DC: US Army Environmental Hygiene Agency.

Winkenwerder W. 2002. *US Demolition Operations at Khamisiyah.* Washington, DC: Department of Defense.

Wolfe J, Proctor SP, Erickson DJ, Hu H. 2002. Risk factors for multisymptom illness in US Army veterans of the Gulf War. *Journal of Occupational and Environmental Medicine* 44(3):271-281.

3

CONSIDERATIONS IN IDENTIFYING AND EVALUATING THE LITERATURE

This chapter reviews the approach that the committee took to identify and evaluate the health studies of Gulf War veterans. It discusses the major types of epidemiologic studies considered by the committee, the factor analysis used by many of the studies that were evaluated, and finally the committee's inclusion and evaluation criteria.

The committee limited its review of the literature primarily to epidemiologic studies of Gulf War veterans to determine the prevalence of diseases and symptoms in that population. The studies typically examine veterans' health outcomes in comparison with outcomes in their nondeployed counterparts. Because this report is a review of disease or symptom prevalence, no attempt is made to associate diseases or symptoms with specific biologic or chemical agents potentially encountered in the gulf. In a general way, however, the committee did examine studies that assessed exposures in veterans (Chapter 2) and the influence of exposure information on the interpretation of veterans' health.

The committee members identified numerous cohort studies (Chapter 4) and case-control studies (Chapter 5) that they objectively reviewed without preconceived ideas about health outcomes. To assist them in their work, they developed criteria to determine which studies to include in their review.

TYPES OF EPIDEMIOLOGIC STUDIES

The committee focused on epidemiologic studies because epidemiology deals with the determinants, frequency, and distribution of disease in human populations. A focus on populations distinguishes epidemiology from medical disciplines that focus on the individual. Epidemiologic studies examine the relationship between exposures to agents of interest in a population and the development of health outcomes (in this review, deployment is the exposure). Such studies can be used to generate hypotheses for study or to test hypotheses posed by investigators. This section describes the major types of epidemiologic studies considered by the committee.

Cohort Studies

A cohort study is an epidemiologic study that follows a defined group, or cohort, over a period of time. It can test hypotheses about whether exposure to a specific agent is related to the development of disease and can examine multiple disease outcomes that might be associated with exposure to a given agent (or, for example, to deployment). A cohort study starts with people who are free of a disease (or other outcome) and classifies them according to whether they have been exposed to the agent of interest. It compares health outcomes in people who have been exposed to the agent in question with those who have not.

Cohort studies can be prospective or retrospective. In a prospective cohort study, investigators select a group of subjects and determine who has been exposed and who has not been exposed to a given predictor (independent) variable. They then follow the cohort to determine the rate or risk of the disease (or other health outcome) in the exposed and comparison groups.

A retrospective (or historical) cohort study differs from a prospective study in temporal direction; investigators look back to classify past exposures in the cohort and then track the cohort forward to ascertain the rate of disease.

Cohort studies can be used to estimate a risk difference or a relative risk, two statistics that measure association between the exposure groups. The risk difference, or attributable risk, is the rate of disease in exposed persons minus the rate in unexposed persons, representing the number of extra cases of disease attributable to the exposure. The relative risk is determined by dividing the rate of those who develop the disease in the exposed group (for example, the deployed group) by the rate of those developing the disease in the nonexposed group (for example, the nondeployed group). A relative risk greater than 1 suggests an association between exposure and disease onset; the higher the relative risk, the stronger the association.

Cohort studies have several advantages and disadvantages. Generally, the advantages outweigh the disadvantages if the study is well designed and executed. The advantages of cohort studies include the following:

- The investigator knows that the predictor variable preceded the outcome variable.

- Exposure can be defined and classified at the beginning of the study and subjects can be selected based on exposure definition.

- Information on potential confounding variables can be collected in a prospective cohort study so that they may be controlled in the analysis.

- Rare or unique exposures (such as Gulf War exposures) can be studied, and the investigators can study multiple health outcomes.

- Absolute rates or risk of disease incidence and prevalence can be calculated.[1]

Disadvantages of cohort studies include the following:

- They are often expensive because of the long periods of followup (especially if the disease has a delayed onset, for example, cancer), attrition of study subjects, and delay in obtaining results.

- They are inefficient for the study of rare diseases or diseases of long latency.

---

[1] Incidence is the occurrence of new cases of an illness or disease in a given population during a specified period. Incidence rate uses person-time in the denominator and cumulative incidence uses number of people at risk in the denominator. Prevalence is the number of cases of an illness or disease existing in a given population at a specific time.

- There is a possibility of the "healthy-worker effect"[2] (Monson 1990), which might introduce bias and can diminish the true disease-exposure relationship.

Case-Control Studies

In a case-control study, subjects (cases) are selected on the basis of having a disease; controls are selected on the basis of not having the disease. Cases and controls are asked about their exposures to specific agents. Cases and controls can be matched with regard to such characteristics as age, sex, and socioeconomic status to eliminate those characteristics as causes of observed differences, or those variables can be controlled in the analysis. The odds of exposure to the agent among the cases are then compared with the odds of exposure among controls. The comparison generates an odds ratio,[3] which is a statistic that depicts the odds of having a disease among those exposed to the agent of concern relative to the odds of having the disease among an unexposed comparison group. An odds ratio greater than 1 indicates that there is a potential association between exposure to the agent and the disease; the greater the odds ratio, the greater the association.

Case-control studies are useful for testing hypotheses about the relationships between exposure to specific agents and disease. They are especially useful and efficient for studying the etiology of rare diseases. Case-control studies have the advantages of ease, speed, and relatively low cost. They are also valuable for their ability to probe multiple exposures or risk factors. However, case-control studies are vulnerable to several types of bias, such as recall bias, which can dilute or enhance associations between disease and exposure. Other problems include identifying representative groups of cases, choosing suitable controls, and collecting comparable information about exposures on both cases and controls. Those problems might lead to unidentified confounding variables that differentially influence the selection of cases or control subjects or the detection of exposure. For the reasons discussed above, case-control studies are often the first approach to testing a hypothesis, especially one related to a rare outcome.

A nested case-control study draws cases and controls from a previously defined cohort. Thus, it is said to be "nested" inside a cohort study. Baseline data are collected at the time that the cohort is identified and insures a more uniform set of data on cases and controls. Within the cohort, individuals identified with disease serve as cases and a sample of those who are disease-free serve as controls. Using baseline data, exposure in cases and controls is compared, as in a regular case-control study. Nested case-control studies are efficient in terms of time and cost in reconstructing exposure histories on cases and on only a sample of controls rather than the entire cohort. Additionally, because the cases and controls come from the same previously established cohort, concerns about unmeasured confounders and selection bias are decreased.

Cross-Sectional Studies

The main differentiating feature of a cross-sectional study is that exposure and disease information is collected at the same point (period) of time. The selection of people for the

---

[2] The healthy-worker effect arises when a healthy employed population experiences lower mortality than the general population, which consists of a mix of healthy and unhealthy people. A population with increased external traumatic causes of death (such as Gulf War veterans), however, might be different from many occupational populations.
[3] An odds ratio is a good estimate of relative risk when the disease under study is rare.

study—unlike selection for cohort and case-control studies—is independent of both the exposure to the agent under study and disease characteristics. Cross-sectional studies seek to uncover potential associations between exposure to specific agents and development of disease. In a cross-sectional study, effect size is measured as relative risk, prevalence ratio, or prevalence odds ratio. It might compare disease or symptom rates between groups with and without exposure to the specific agent. Several health studies of Gulf War veterans are cross-sectional studies that compare a sample of veterans who were deployed to the Gulf War with a sample of veterans who served during the same period but were not deployed to the Gulf War.

Cross-sectional studies are easier and less expensive to perform than cohort studies and can identify the prevalence of diseases and exposures in a defined population. They are useful for generating hypotheses, but they are much less useful for determining cause-effect relationships, because disease and exposure data are collected simultaneously (Monson 1990). It might also be difficult to determine the temporal sequence of exposures and symptoms or disease.

General Remarks

Epidemiologic studies can establish statistical associations between exposure to specific agents or situations (for example, deployment to the Gulf War) and health effects, and associations are generally estimated by using relative risks or odds ratios. Epidemiologists seldom consider a single study sufficient to establish an association. It is desirable to replicate findings in other studies to draw conclusions about an association. Results of separate studies are sometimes conflicting. It is sometimes possible to attribute discordant study results to such characteristics as soundness of study design, quality of execution, and the influence of different forms of bias. Studies that result in a statistically precise measure of association suggest that the observed result is unlikely to have been due to chance. When the measure of association does not show a statistically precise effect, it is important to consider the size of the sample and whether the study had the power to detect an effect of a given size.

DEFINING A NEW SYNDROME

As the committee reviewed the literature on the health of Gulf War veterans, one fundamental question arose regarding whether the constellation of veterans' unexplained symptoms constitutes a syndrome. If so, is the symptom constellation best studied and treated as a new syndrome or as a variant form of a known syndrome (IOM 2000)? Identification of a new set of unexplained symptoms in a group of patients does not automatically mean that a new syndrome has been found. Rather, it constitutes the beginning of a process to demonstrate that the patients are affected by a clinical entity that is distinct from other established clinical entities.

The process of defining a new syndrome usually begins with establishment of a case definition that lists criteria for distinguishing the potentially new patient population from patients with known clinical diagnoses. Developing the first case definition is a vital milestone intended to spur research and surveillance. More like a hypothesis than a conclusion, definition is an early step in the process; it is often revised as more evidence comes to light. Case definitions usually are a mix of clinical, demographic, epidemiologic, and laboratory criteria.

A case definition leads to the creation of a more homogeneous patient population, another step in the eventual establishment of a new syndrome. A potential disadvantage of any case definition—if it is inaccurate—is the mislabeling or misclassification of a condition, which can thwart medical progress for years, if not decades (Aronowitz 2001). Classification of a new patient population also stimulates further understanding of prevalence, treatment, natural history, risk factors, and ultimately etiology and pathogenesis. As more knowledge unfolds about etiology and pathogenesis, an established syndrome can rise to the level of a disease. The renaming of a syndrome as a disease implies that the etiology or pathology has been identified.

STATISTICAL TECHNIQUES USED TO DEVELOP A CASE DEFINITION

Two statistical techniques have been used by investigators to identify symptom clusters that could potentially be used to develop case definitions suggestive of a new syndrome: factor analysis and cluster analysis. Many of the studies reviewed by the committee use those techniques. The aims of the techniques are different: factor analysis seeks to identify groups of individuals' most prominent symptoms, whereas cluster analysis seeks to identify people who have similar symptoms. Stated in another way, factor analysis analyzes patterns of symptoms, and cluster analysis categorizes people on the basis of their symptoms.

Factor analysis has been used far more frequently than cluster analysis in the major cohort studies. It has been used to identify groups of symptoms that might potentially point to a new syndrome. However, factor analysis by itself cannot definitively identify a new syndrome. That requires more research about the putative syndrome's clinical features, natural history, genetics, response to treatment, etiology, and pathogenesis (Robins and Guze 1970; Taub et al. 1995).

Factor analysis seeks to identify a small number of groups of highly related variables among a much larger number of measured variables. In the context of Gulf War research, the measured variables are the symptoms that veterans report in surveys. Factor analysis aggregates veterans' symptoms into smaller groups to discern more fundamental, yet immeasurable, variables, which are referred to as factors. The idea is that a factor and the group of symptoms that it represents are somehow related pathophysiologically and that the symptoms within a factor are different symptomatic manifestations of the same underlying disease process. In a research context, the factors could be used, for instance, as clinical criteria for a new syndrome.

In the Gulf War literature, the key issue is whether the factors identified by factor analysis are exclusive to deployed veterans vs a comparison population, usually nondeployed veterans. Finding factors peculiar to deployed veterans would imply that they might have a new syndrome with specific symptoms that could indicate biologic plausibility or a common pathophysiology, presumably triggered by an exposure that occurred in the Gulf War Theater. The names given to the new factors—such as neurologic factor or cognitive factor—are merely descriptive labels. Their purpose is to convey what the investigators believe the nature of a new syndrome might be, depending on which symptoms group or "load" onto the new factor, and, in the absence of more research, to establish the putative syndrome's features.

There are several key characteristics by which one can evaluate the methodology or findings of a particular factor analysis. Several are straightforward, such as sample size, sample population, type of symptom reporting (for example, interval, ordinal, or dichotomous) and what particular symptoms are subsumed under each factor. Several other characteristics, such as the method, rotation, factor-loading cutoff, number of factors isolated, and percentage of variance would require explanation.

Factor analysis examines how closely symptoms on a questionnaire are related in a study population. Studies typically elicit responses to long lists of potential symptoms (such as numbness and tingling in the extremities) and how severe the symptoms are. Conventional factor analysis correlates either the presence or severity of a symptom with the presence or severity of all other symptoms. Factor analysis not only identifies symptoms that correlate with each other but also identifies them as relatively uncorrelated with other variables. For instance, in acute gastroenteritis, severe nausea might correlate with severe vomiting and severe diarrhea but not with cough. Classically, factor analysis examines the severity of symptoms rather than just their presence or absence, with severity scored on a continuous (for example, 100-point) scale. Many studies of Gulf War veterans, however, have used either dichotomous scales ("present" vs "absent") or ordinal scales ("none", "mild", "moderate", or "severe") without using statistical techniques designed to deal with these types of variables. That can result in underestimation of the strength of a correlation (Muthen 1984).

In conducting a factor analysis, the first issue is which of the many symptoms on a questionnaire should be retained in the factor, that is, how strongly do symptoms need to be intercorrelated to be part of the factor? A commonly used method is "factor loading" or use of a "factor coefficient". That method quantifies how well an individual symptom correlates or does not correlate with a potential factor. There are different statistical techniques (such as principal factor, principal component, iterated principal components, and maximal likelihood factor analyses) to determine how well symptoms load onto a particular factor.

Once a preliminary list of factors is determined, the next step is to decide their relative importance and how well they can explain the universe of symptoms collected from the study population. Several investigators have taken an additional step and have examined potential factors to determine how meaningful or plausible they are clinically or pathophysiologically and to discard the implausible ones.

The investigator examines two statistics to determine how well the proposed factors describe the array of symptoms in the study population. The first is the "factor load", which refers to how strongly individual symptoms correlate with a factor. Factor loads are expressed on a scale of 0-1, and loads of greater than 0.4 are conventionally used as a cutoff point between factors that are strongly associated and factors that are not. The second statistic, the "eigenvalue", is a measure of how well each factor can explain or fit the observed relationships among all the symptoms in the study population (technically, the proportion of variance in the study). Eigenvalues have numeric values. An eigenvalue of more than 1.0 is conventionally taken to mean that a factor should be retained. Taken together, those two statistics can be used in an iterative fashion to establish the best fit of the data. A related technique, "rotation", allows for easier interpretation of factor loadings.

Once a relatively robust number of factors are identified, then how well the individual symptoms correlate with each other can be examined. Finally, the proportion of the variance that each factor explains is plotted by the individual factors arranged by decreasing eigenvalues. Investigators look for a sharp dropoff in the curve as an arbitrary cut point between important factors and minor or weak factors. The result is a small number of factors that can explain a large proportion of the variance observed in participants' answers to the symptom questionnaires.

A somewhat related technique, cluster analysis, has also been used in several of the Gulf War studies (see Chapter 5) to determine how groups of patients with certain symptoms might relate to one another. A k-means cluster analysis partitions study subjects into k clusters of individuals based on their reported symptoms. Individual participants are then reassigned in an iterative fashion until a best fit is reached and the uniqueness of the clusters is not improved by further reassignments. For instance in a group of persons who were exposed to Staphylococcal enterotoxin at a picnic, there would likely be two clusters—one cluster of persons with severe vomiting and nausea and one cluster with mild or even absent symptoms. While taking a different statistical approach to examining symptom complexes among individuals, studies using cluster analyses, like factor analyses, are ultimately dependent on representative samples and accuracy of self-reported data and can suffer from both selection (or participation) bias and recall bias to the extent that persons are more or less willing to participate and have greater or lesser recall of symptoms.

INCLUSION CRITERIA

The committee's evaluation included studies that would enable it to respond to its charge "to help inform the Department of Veterans Affairs of illnesses among Gulf War veterans that might not be fully appreciated". The committee included studies that would answer the question, What does the literature tell us about the health status of Gulf War veterans? To that end, the committee searched the literature and included descriptive epidemiologic studies of health outcomes in military personnel that served in the Gulf War Theater. The studies were not restricted to US personnel. A study also needed to demonstrate rigorous methods (for example, was published in a peer-reviewed journal, included details of methods, had a control or reference group, had the statistical power to detect effects, and included reasonable adjustments for confounders), include information regarding a persistent health outcome, and have a medical evaluation, conducted by a health professional, and use laboratory testing as appropriate. Those types of studies constituted the committee's primary literature. The committee did not evaluate studies of acute trauma, rehabilitation, or transient illness.

Studies reviewed by the committee that did not necessarily meet all the criteria of a primary study are considered secondary studies. Secondary studies are typically not as methodologically rigorous as primary studies and might present subclinical findings, that is, studies of altered functioning consistent with later development of a diagnosis but without clear predictive value.

Another step that the committee took in organizing its literature was to determine how all the studies were related to one another. Numerous Gulf War cohorts have been assembled, from several different countries; from those original cohorts many derivative studies have been conducted. The committee organized the literature into the major cohorts and derivative studies because they did not want to interpret the findings of the same cohorts as though they were results from unique groups (Chapter 4).

Finally, in assessing the descriptive studies, the committee was especially attentive to potential sources of bias, confounding, chance, and multiple comparisons, as discussed in the next section.

ADDITIONAL CONSIDERATIONS

In addition to determining the primary and secondary literature that would be used to draw conclusions, the committee considered other characteristics of the studies. These characteristics had to do with the methods used by researchers in designing and conducting studies and include bias and chance.

Bias

Bias refers to systematic, or nonrandom, error. Bias causes an observed value to deviate from the true value, and can weaken an association or generate a spurious association. Because all studies are susceptible to bias, a goal of the research design is to minimize bias or to adjust the observed value of an association by correcting for bias. There are different types of bias, such as selection bias which occurs when systematic error in obtaining participants results in a potential distortion of the true association between exposure and outcome.

Information bias results from the manner in which data are collected and can result in measurement errors, imprecise measurement, and misdiagnosis. Those types of errors might be uniform in an entire study population or might affect some parts of the population more than others. Information bias might result from misclassification of study subjects with respect to the outcome variable or from misclassification of exposure. Other common sources of information bias are the inability of study subjects to recall the circumstances of their exposure accurately (recall bias) and the likelihood that one group more frequently reports what it remembers than another group (reporting bias). Information bias is especially harmful in interpreting study results when it affects one comparison group more than another.

Confounding

Confounding occurs when a variable or characteristic otherwise known to be predictive of an outcome and associated with the exposure (and not on the causal pathway) can account for part or all of an apparent association. A confounding variable is an uncontrolled variable that influences the outcome of a study to an unknown extent, and makes precise evaluation of its effects impossible. Carefully applied statistical adjustments can often control for or reduce the influence of a confounder.

Chance

Chance is a type of error that can lead to an apparent association between an exposure to an agent and a health effect when no association is present. An apparent effect of an agent on a health outcome might be the result of random variation due to sampling in assembly of the study population rather than the result of exposure to the agent. Standard methods that use confidence intervals, for example, allow one to assess the role of chance variation due to sampling.

Multiple Comparisons

When an investigator initiates a large number of investigations simultaneously on the same dataset, multiplicity of comparisons poses a problem. When looking at so many different comparisons, the investigator is bound to find something of note by chance alone. For example, in many Gulf War veteran studies, the investigators are comparing multiple outcomes and multiple exposures. There are, however, ways to correct for multiple comparisons in studies. One way is to use a Bonferroni correction, a statistical adjustment for multiple comparisons. It effectively raises the standard of proof needed when an investigator looks at a wide array of hypotheses simultaneously.

Assignment of Causality

In addition to general considerations of research quality, the assessment of the studies reviewed raises a complex set of issues related to the assignment of causality (Pearl 2000). For purposes of this study, the committee scrutinized the degree to which studies were likely to provide strong causal evidence. To that end, the committee was guided by the Bradford Hill criteria (Hill 1965). In the spirit of those criteria (Phillips and Goodman 2004), the inferences expressed in this report are based on the totality of evidence reviewed and the committee's collective judgment.

LIMITATIONS OF GULF WAR VETERAN STUDIES

The studies to date have provided valuable information regarding the health of Gulf War veterans; however, many of the studies have limitations that hinder accurate assessment of the veterans' health status. Chapter 4 discusses the limitations. The issues under discussion include the possibility that study samples do not represent to the entire Gulf War population, low rates of participation in studies, self-reporting of symptoms and exposures, narrowness of studies in assessment of health status, insensitivity of instruments for detecting abnormalities in deployed veterans, and the use of period of investigation that is too brief to detect health outcomes that have long latency, such as cancer. In addition, many of the US studies are cross-sectional and this limits the opportunity to learn about symptom duration and chronicity, latency of onset, and prognosis. Finally, the problem of multiple comparisons that is common in many of the Gulf War studies results in confusion over whether the effect is real or occurring by chance. Those limitations make it difficult to interpret the results of the findings, particularly when several well-conducted studies produce inconsistent results.

SUMMARY

The committee reviewed and evaluated studies from the scientific and medical literature that were identified with searches of bibliographic databases and other methods. The committee adopted a policy of using only peer-reviewed published literature as the basis of its conclusions. Publications that were not peer-reviewed had no evidentiary value for the committee, that is, they were not used as evidence for arriving at the committee's conclusions about the prevalence of health effects. The process of peer review by fellow professionals promotes high standards of quality, although it does not guarantee the validity or generalizability of a study's findings.

Committee members read each article critically. In some instances, nonpeer-reviewed publications provided background information for the committee and raised issues that required further research. The committee, however, did not collect original data, nor did it perform any secondary data analysis. In its evaluation of the peer-reviewed literature, the committee considered several important issues, including quality and relevance; error, bias, and confounding; and the diverse nature of the evidence and the research.

REFERENCES

Aronowitz RA. 2001. When do symptoms become a disease? *Annals of Internal Medicine* 134(9 Pt 2):803-808.

Hill AB. 1965. The Environment and disease: Association or causation? *Proceedings of the Royal Society of Medicine* 58(10):295-300.

IOM (Institute of Medicine). 2000. *Gulf War and Health, Volume 1. Depleted Uranium, Sarin, Pyridostigmine Bromide, Vaccines.* Washington, DC: National Academy Press.

Monson R. 1990. *Occupational Epidemiology.* 2nd ed. Boca Raton, FL: CRC Press, Inc.

Muthen B. 1984. A general structural equation model with dichotomous, ordered, categorical, and continuous latent variable indicators. *Psychometrika* 49(1):115-132.

Pearl J. 2000. *Causality: Models, Reasoning and Inference.* Cambridge, UK: Cambridge University Press.

Phillips CV, Goodman KJ. 2004. The missed lessons of Sir Austin Bradford Hill. *Epidemiologic Perspectives and Innovations* 1(1):3.

Robins E, Guze SB. 1970. Establishment of diagnostic validity in psychiatric illness: Its application to schizophrenia. *American Journal of Psychiatry* 126(7):983-987.

Taub E, Cuevas JL, Cook EW 3rd, Crowell M, Whitehead WE. 1995. Irritable bowel syndrome defined by factor analysis. Gender and race comparisons. *Digestive Diseases and Sciences* 40(12):2647-2655.

4

MAJOR COHORT STUDIES

This chapter provides an overview of many of the major cohort studies of Gulf War veterans, discusses the general limitations of the studies, and summarizes the findings from each. Some of the cohorts were brought together in the first few years after the Gulf War; others were assembled more recently. Most of the studies compare sizable groups of deployed veterans with groups of nondeployed veterans or with veterans who were deployed to locations other than the Persian Gulf (for example, Bosnia).

The major cohort studies are important for understanding the health of Gulf War veterans; those studies' findings, on particular health outcomes, are evaluated thoroughly in Chapter 5 along with additional studies' findings on smaller samples of Gulf War veterans. The largest studies of Gulf War veterans have been conducted in countries that were members of the Gulf War coalition including: the United States, Canada, Denmark, Australia, and the United Kingdom.

Most major cohort studies address several fundamental questions about Gulf War veterans' health: What are the nature and prevalence of veterans' symptoms and diagnoses? Do symptoms that do not fit conventional medical diagnoses, and are therefore unexplained, warrant classification as a new syndrome? Are exposures to specific biologic, chemical, and radiologic agents during the Gulf War associated with veterans' symptoms and illnesses? Those questions are designed to guide the reader through a complex body of research.

Most major cohorts, once established, led to numerous studies that examined more detailed questions about Gulf War veterans' health; the committee refers to those studies as derivatives. Table 4.1, at the end of this chapter, provides information about each original cohort—for example, method of assembly, the eligible population, the specific study methods, the study population, and the percentage of subjects who were enrolled—and includes the derivative studies. The table lists a derivative study under the original cohort from which it drew its study population and provides additional information, including its purpose, design, enrollment of its subjects, sample size, response rates,[1] and other cohort characteristics. The table was vital in guiding the committee through its analysis and evaluation of the studies discussed in Chapter 5. The information helped the committee to identify the populations that have been studied and enabled the committee to understand which studies were independent of each other;

---

[1] Table 4.1 contains the figures given in each study publication, *except response rates*. For uniformity, the committee calculated response rates with this formula: response rate = number of study participants who responded divided by the number of people who were located (rather than the number of eligible people).

that is important because the committee did not want to factor in the health outcomes occurring in the same people repeatedly.

GENERAL LIMITATIONS OF GULF WAR COHORT STUDIES AND DERIVATIVE STUDIES

The 24 major cohort studies of Gulf War veterans and their derivative studies have contributed greatly to our understanding of veterans' health, but they are beset by limitations that are commonly encountered in epidemiologic studies, including lack of representativeness, selection bias, lack of control for potential confounding factors, self-reports of health outcomes, outcome misclassification, and self-reports of exposure. The committee members read each study carefully and noted the findings and limitations of each study.

The foremost limitation is lack of representativeness, which limits one's ability to generalize results to the entire population of interest; for example, about half the cohorts focus on groups of veterans that are selected for study according to where they served in the military (a military-unit-based study). Military-unit studies are not representative of all Gulf War veterans with respect to their duties and location during deployment, their military status during the war (active duty, reserves, or National Guard), their military status after the war (active duty, reserves, or discharged), their branch of service (Army, Navy, Air Force, or Marines), or ease of ascertainment (IOM 1999b). The most representative studies are population-based: the cohorts are selected on the basis of where their members reside. In population-based studies of Gulf War veterans, the cohort might be the entire deployed population, as in studies of Canadian and Australian veterans, or a random selection from the population of interest, as in several studies of US and British veterans. The committee, in evaluating major cohort studies, gave greater weight to Gulf War studies that were population-based.

A study's representativeness, even if it is population-based, can be compromised by low participation rates. Low participation rates can introduce selection bias, for example, when Gulf War veterans who are symptomatic choose to participate more frequently than those who are not symptomatic. Nondeployed veterans, who might be healthier, might be less inclined to participate. In some studies, researchers not only try to measure selection bias by comparing participants with nonparticipants from both deployed and nondeployed populations, but also make adjustments to overcome it, for example, by oversampling nondeployed populations as in the study by Eisen and colleagues (2005).

Selection bias might also occur through the so-called healthy-warrior effect. That form of bias has the potential to occur in most of the major cohorts that compare deployed veterans with nondeployed personnel. The healthy-warrior effect is a form of selection bias insofar as chronically ill or less fit members of the armed forces might be less likely to have been deployed than more fit members. That is, there might have been nonrandom assignment of those selected and not selected for deployment. Some of the best studies attempt to measure the potential for selection bias and adjust for it in the analysis.

A recurrent limitation is that most cohort studies rely on self-reporting of symptoms on questionnaires. Symptom self-reporting potentially introduces reporting bias, which occurs when the group being studied (such as deployed veterans) reports more frequently what it remembers than a comparison group (such as nondeployed veterans). Reporting bias, in this example, would lead to an overestimation of the prevalence of symptoms or diagnoses in the deployed population.

Symptom self-reporting might sometimes introduce another type of bias known as outcome misclassification, in which there are errors in how symptoms are classified into outcomes and analyzed. One Gulf War study sought to document outcome misclassification by comparing veterans' symptom reporting on questionnaires with clinical examination about 3 months later (McCauley et al. 1999b). The study found that the extent of misclassification depended on the type of symptom being reported; agreement between questionnaire and clinical examination ranged from 4-79%. The overall problem led the investigators to caution that questionnaire data, in the absence of clinical evaluation or adjustment, might lead to outcome misclassification. Another study also found poor reliability and validity of self-reported diagnoses when compared with medical records (Gray et al. 1999a). In contrast, a study by the Department of Veterans Affairs (VA) (Kang et al. 2000), which verified a random subset of self-reported conditions (n = 4,200) against medical records, found a strong correlation between the two (above 93%). Those data, however, were available only for the 45.2% who signed consent forms that allowed researchers to verify records.

The problem of symptom self-reporting is best addressed through medical evaluations, as was done by VA researchers (e.g., Eisen et al. 2005) and by several other investigators with the resources to conduct medical evaluations. Nevertheless, medical evaluations do not surmount the problem that some outcome measures being studied, such as chronic fatigue syndrome (CFS), are symptom-based syndromes that by definition lack a biologic "gold standard" with which symptoms can be validated. The lack of a diagnostic gold standard or other objective biologic markers poses a particular problem for veterans with fibromyalgia, CFS, and multiple chemical sensitivity (MCS) (IOM 1999a).

Another limitation of most major cohort studies is self-reporting of exposures. Self-reporting of exposures, like self-reporting of symptoms, introduces the possibility of recall bias, the tendency for participants who are symptomatic to overestimate (or underestimate) their exposures compared with those who are not symptomatic. Indeed, a major study from the UK found that Gulf War veterans with more symptoms were likely to report more exposures than those not deployed to the gulf (Unwin et al. 1999). Other complicating factors are exposures often cannot be validated by objective means, often occurred years earlier, and might have been perceived rather than actual. For example, because of the sensitivity of the chemical-warfare monitors, many false alarms might have been perceived by veterans as actual exposures. Enhanced recordkeeping and monitoring of the environment during and after the Gulf War would have averted that problem. Indeed, many expert panels have recommended efforts to improve recordkeeping and environmental monitoring in future deployments (e.g., IOM 1999b; NRC 2000a; NRC 2000b; NRC 2000c).

Other limitations of the body of evidence are that studies might be too narrow in their assessment of health status, the measurement instruments might have been too insensitive to detect abnormalities that affect deployed veterans, and the period of investigation has been too brief to detect health outcomes that have a long latency or require many years to progress to the point where disability, hospitalization, or death occurs. Virtually all US studies are cross-sectional, and this limits the opportunity to learn about symptom duration and chronicity, latency of onset (especially for health conditions with a long latency, such as cancer), and prognosis.

ORGANIZATION OF THIS CHAPTER

This chapter organizes numerous major cohort studies by a key feature of study design—how the cohort was assembled. Roughly half the chapter covers cohorts that are population-based, and the rest includes cohorts that are military-unit based.

For each major cohort, we use a uniform format. The cohort methods and major findings regarding symptoms and diagnoses are described first. Then we turn to how symptoms, if unexplained, cluster together (under the heading "Symptom Clustering"). The next section reviews findings of the medical evaluation, if one was conducted. The final section describes what symptom-exposure relationships were found.

This chapter does not cover studies whose sample population is drawn from any of the Gulf War registries, because they lack comparison groups. Registries have been set up in the United States, by the Department of Defense (DOD) and VA, and in the UK by the Ministry of Defence for UK Gulf War. Registries are self-selected case series of veterans who presented for care, so they cannot be and were not intended to be representative of the symptoms and diagnoses of the entire group of Gulf War veterans. Nor were registries designed with control groups or with diagnostic standardization across the multiple sites at which examinations took place (Joseph 1997; Roy et al. 1998). Finally, registries relied on standard diagnostic classifications and were not designed to probe for novel diagnoses[2] or to search for biologic correlates. Thus, because of their methodologic limitations, registry studies cannot stand alone as a basis of conclusions or of the conduct of research. But they do provide a glimpse into veterans' symptoms and the difficulties of fitting those symptoms into standard diagnoses. Registry programs have been a valuable source of information for generating hypotheses that have been tested in rigorous epidemiologic studies with control groups to estimate the health status of Gulf War veterans.

POPULATION-BASED STUDIES

The Iowa Study

The "Iowa study", a major population-based study of US Gulf War veterans, was a cross-sectional survey of a representative sample of 4,886 military personnel who listed Iowa as their home of record at the time of enlistment (Iowa Persian Gulf Study Group 1997). The study examined the health of military personnel in all branches of service who were still serving or had left service. The sample was randomly selected from and representative of 28,968 military personnel. Of the study subjects who were contacted, 3,695 (90.7%) completed a telephone interview. Study subjects were divided into four groups: Gulf War-deployed regular military, Gulf War-deployed National Guard or Reserve, non-Gulf War-deployed regular military, and non-Gulf War-deployed National Guard or Reserve. Trained examiners using standardized questions, instruments, and scales interviewed the subjects.[3] When compared with the groups not

---

[2]Registries rely on the ICD-9-CM (Joseph 1997; Murphy et al. 1999).
[3]Sources of questions included the National Health Interview Survey, the Behavioral Risk Factor Surveillance Survey, the National Medical Expenditures Survey, the Primary Care Evaluation of Mental Disorders, the Brief Symptom Inventory, the CAGE questionnaire (for alcoholism), the PTSD (Posttraumatic Stress Disorder)

deployed to the Persian Gulf, the two groups of Gulf War military personnel reported roughly twice the prevalence of symptoms suggestive of fibromyalgia, cognitive dysfunction, depression, alcohol abuse, asthma, posttraumatic stress disorder (PTSD), sexual discomfort, and chronic fatigue.[4] In a separate analysis, the prevalence of MCS symptoms was about twice the prevalence in the comparison population (Black et al. 2000). Furthermore, in the main cohort study, which used a standardized instrument for assessing functioning (the Medical Outcome Study's 36-item questionnaire known as the Short Form-36, or SF-36), Gulf War veterans displayed significantly lower scores on all eight subscales for physical and mental health. The subscales profile aspects of quality of life. The subscales for bodily pain, general health, and vitality showed the greatest absolute differences between deployed and nondeployed veterans. In short, this large, well-controlled study demonstrated that some sets of symptoms were more frequent and quality of life poorer among Gulf War veterans than among nondeployed military controls.

Symptom Clustering

The Iowa study was the first major population-based study to group sets of symptoms into categories suggestive of known syndromes or disorders, such as fibromyalgia or depression. Its finding of considerably higher prevalence of symptom groups suggestive of fibromyalgia, depression, and cognitive dysfunction among Gulf War veterans motivated other researchers to examine, through factor analysis, the potential for a new syndrome that would group and classify veterans' symptoms. Several years later, the Iowa investigators performed a factor analysis on their cohort (Doebbeling et al. 2000). They identified three symptom factors in deployed veterans—somatic distress, psychologic distress, and panic—but the factors were not exclusive to deployed veterans. Thus, the study did not support the existence of a new syndrome (see Chapter 3 for a discussion of factor analysis).

Exposure-Symptom Relationships

The Iowa study assessed exposure-symptom relationships by asking veterans to report on their exposures in the Gulf War. Researchers found that many of the self-reported exposures were significantly associated with health conditions. For example, symptoms of cognitive dysfunction were found to have been associated with self-reports of exposure to solvents or petrochemicals, smoke or combustion products, lead from fuels, pesticides, ionizing or nonionizing radiation, chemical-warfare agents, use of pyridostigmine bromide (PB), infectious agents, and physical trauma. A similar set of exposures were associated with symptoms of depression or fibromyalgia. The study concluded that no exposure to any single agent was related to the conditions that the authors found to be more prevalent in Gulf War veterans (Iowa Persian Gulf Study Group 1997).

Women's Health

The Gulf War was among the first wars to see a sizable fraction of women in the military. About 7% of military personnel serving in the Persian Gulf were women (Joseph 1997). The Iowa study was one of the few population-based US studies that investigated the health of

---

Checklist—Military, the Centers for Disease Control and Prevention Chronic Fatigue Syndrome Questionnaire, the Chalder Fatigue Scale, the American Thoracic Society questionnaire, and the Sickness Impact Profile.

[4]The conditions listed were not diagnosed, because no clinical examinations were performed. Rather, before conducting their telephone survey, researchers grouped sets of symptoms from their symptom checklists into a priori categories of diseases or disorders. After a veteran identified himself or herself as having the requisite set of symptoms, researchers analyzing responses considered the veteran as having symptoms "suggestive" of or consistent with a particular disorder but not as having a formal diagnosis of the disorder.

women separately (Carney et al. 2003). Women were less likely to participate in combat than men, but 71% of women had at least one combat exposure. Women also reported similar rates of exposure to environmental agents, such as diesel fuel and smoke from oil-well fires. Their patterns of health-care use varied from that of men: they had significantly more outpatient, as well as inpatient, health care 5 years after the war. They were also more likely than men to receive VA compensation (17% vs 7%), although their level of disability was similar.

Department of Veterans Affairs Study

A major population-based study of US veterans was mandated by Public Law 103-446. The study is a retrospective cohort design conducted by VA. Its purpose is to estimate the prevalence of symptoms and other health outcomes (including reproductive outcomes in spouses and birth defects in children) in Gulf War veterans vs non-Gulf War veterans. This population-based survey had three phases. In the first, a questionnaire was mailed to 30,000 veterans. The second phase validated self-reported data with medical-record review and analyzed characteristics of those who did not respond to the mailed survey. The third phase was a comprehensive medical examination and laboratory testing of a random sample of 2,000 veterans drawn from the Gulf War population and a comparison group.

The study was designed to be representative of the nearly 700,000 US veterans sent to the Persian Gulf and 800,680 non-Gulf War veterans of the same era. Questionnaires were mailed to a stratified random sample of 15,000 Gulf War and 15,000 non-Gulf War veterans identified by DOD and representing various units and branches of the military. The questionnaires contained a list of 48 symptoms and questions about chronic medical conditions, functional limitations, and other items from the National Health Interview Survey. A questionnaire about exposures was also included. The overall response rate was about 70%.

Survey Findings (Phases I and II)

The investigation found significantly higher symptom prevalence of all 48 symptoms among Gulf War veterans (Kang et al. 2000). Four of the most frequently reported symptoms were runny nose, headache, unrefreshing sleep, and anxiety (Table 4.2). Numerous chronic medical conditions—such as sinusitis, gastritis, and dermatitis—were reported more frequently among Gulf War veterans; many were reported twice as often. Ten symptoms and 12 medical conditions were remarkably similar in prevalence to those in a UK cohort (Unwin et al. 1999). Gulf War veterans reported significantly higher rates of functional impairment (27.8% vs 14.2%), limitations of employment (17.2% vs 11.6%), and health-care use as assessed by clinic visits (50.8% vs 40.5%) or hospitalizations (7.8% vs 6.4%) compared with nondeployed veterans. In a randomly selected subset of veterans, medical-record reviews verified more than 90% of self-reported reasons for clinic visits or hospitalizations. A separate analysis of the VA cohort found that 10% of them, compared with 4% of controls, met symptom-based criteria for PTSD, and 4.9% (vs 1.2%) met symptom-based criteria for CFS (Kang et al. 2003).

Symptom Clustering

The VA study searched for potentially new syndromes through factor analysis. A separate article by Kang and colleagues (2002) found that 47 symptoms reported by veterans yielded six factors, only one of which contained a cluster of neurologic symptoms that did not load on any factors in the non-Gulf War deployed veterans. The symptoms in the cluster were loss of balance or dizziness, speech difficulty, blurred vision, and tremors or shaking. A group of

277 deployed veterans (2.4%) vs 43 nondeployed veterans (0.45%) met a case definition subsuming all four symptoms. The authors interpreted their findings as suggesting a possible unique neurologic syndrome related to Gulf War deployment that requires objective supporting clinical evidence.

TABLE 4.2 Results of VA Study

Most Common Self-Reported Symptoms[a]	Prevalence in Gulf War Veterans (%)	Prevalence in Non-Gulf War Veterans (%)
Runny nose	56	43
Headache	54	37
Unrefreshing sleep	47	24
Anxiety	45	28
Joint pain	45	27
Back pain	44	30
Fatigue	38	15
Ringing in ears	37	23
Heartburn	37	25
Difficulty in sleeping	37	21
Depression	36	22
Difficulty in concentrating	35	13
5 Most Common Self-Reported Chronic Medical Conditions[a]	**Prevalence in Gulf War Veterans (%)**	**Prevalence in Non-Gulf War Veterans (%)**
Sinusitis	38.6	28.1
Gastritis	25.2	11.7
Dermatitis	25.1	12.0
Arthritis	22.5	16.7
Frequent diarrhea	21.2	5.9

[a] Subjects were asked whether symptoms were recurring or persistent during previous 12 months. Differences in prevalence are all statistically significant ($p < 0.05$).
SOURCE: Kang et al. 2000.

Exposure-Symptom Relationships

A nested case-control analysis (see Chapter 3) was performed on those who met the case definition for the possible neurologic syndrome to determine which of 23 self-reported exposures were more common among cases than among controls (not deployed to the Gulf War) (Kang et al. 2002). Exposures to a variety of chemical agents were reported to be higher among cases than controls; the exposures noted were to chemical-agent-resistant compound paint, depleted uranium, nerve gas, food contaminated with oil or smoke, and bathing in or drinking water contaminated with oil or smoke. Dose-response relationships were not studied because of the nature of the dataset regarding self-reported exposure.

Another cohort study (Kang et al. 2000) did not assess exposure-symptom relationships. It reported on exposures only by compiling the percentages of veterans who reported each of 23 environmental exposures and nine vaccine or prophylactic exposures (such as to PB). The five most common environmental exposures reported by more than 60% of survey participants were to diesel, kerosene, or other petrochemical fumes; to local food other than that provided by the armed forces; to chemical protective gear; to smoke from oil-well fires; and to burning trash or feces.

Medical Evaluation Findings (Phase III)

Three studies have reported on physical examinations of a subsample of the cohort that assayed for general medical status (Eisen et al. 2005), distal symmetric polyneuropathy (Davis et al. 2004), and pulmonary function (Karlinsky et al. 2004). The examinations were conducted in 2001, about 10 years after the Gulf War.

Eisen and colleagues (2005) examined 12 primary health outcome-measures and physical functioning on SF-36. Outcome measures were chosen by the authors to cover the most common symptoms reported by veterans, such as musculoskeletal pain, fatigue, rashes, and neuropathy (as noted in Kang et al 2000).

The study evaluated 1,061 Gulf War and 1,128 non-Gulf War veterans who had been randomly selected from 11,441 Gulf War-deployed and 9,476 non-Gulf War-deployed veterans who previously had participated in a 1995 questionnaire survey (Kang et al. 2000). Researchers were blind to deployment status. Despite three waves of recruitment into the study, the participation rate in the 2005 study was low: only 60.9% of Gulf War veterans and 46.2% of non-Gulf War veterans participated. To determine nonparticipation bias, the study authors obtained previously collected findings from participants and nonparticipants from the DOD Manpower Data Center and gathered sociodemographic and self-reported health findings from the 1995 VA study (Kang et al. 2000).

Four of 12 conditions were more prevalent among GW veterans: fibromyalgia (2.0% vs 1.2%; odds ratio [OR] 2.32, 95% confidence interval [CI] 1.02-5.27), CFS (1.6% vs 0.1%; OR 40.6, 95% CI 10.2-161.15), dermatologic conditions (34.6% vs 26.8 %; OR 1.38, 95% CI 1.06-1.80), and dyspepsia (9.1% vs 6.0%; OR 1.87, 95% CI 1.16-2.99). Fibromyalgia was diagnosed according to the 1990 criteria developed by the American College of Rheumatology (Wolfe et al. 1990). CFS was diagnosed according to the case definition developed by the International Chronic Fatigue Syndrome Study Group (Fukuda et al. 1994). The rate of CFS in the nondeployed veterans was similar to that of the US population. For dermatologic diagnoses, the study created two categories, one of which had a higher OR (see discussion in Chapter 5). A dyspepsia diagnosis required a history or symptoms of frequent heartburn and recurrent abdominal pain, and the use of antacids or other medications.

Gulf War veterans reported worse physical health on the SF-36 (49.3 vs 50.8; $p < 0.001$), but the magnitude of the difference, although statistically significant, was not clinically significant. The analyses adjusted for age, sex, race, years of education, cigarette smoking history, duty type (active vs reserves or National Guard), service branch (Army or Marines vs Navy or Air Force), and rank (enlisted vs officer). The limitations of the study were its performance 10 years after the 1991 Gulf War, which precludes diagnoses that have already resolved, and low participation rates (60.9% Gulf War and 46.2% non-Gulf War), which introduce the possibility of participation bias.

In the study by Davis et al. (2004), the presence of distal symmetric polyneuropathy was evaluated with a history, physical examination, and standardized electrophysiologic assessment of motor and sensory nerves in 1,061 deployed veterans and 1,128 nondeployed veterans. Spouses of deployed and nondeployed veterans were also used as controls. A population of 244

Khamisiyah-exposed deployed veterans was also tested. Blood studies were performed to rule out metabolic causes of neuropathy. The diagnosis of peripheral neuropathy was defined as a distal sensory or motor neuropathy identified on the basis of the neurologic examination, nerve conduction study, or both. No difference in adjusted population prevalence of distal symmetric polyneuropathy between deployed and nondeployed veterans was found with electrophysiology (3.7% vs 6.3%; $p = 0.07$), neurologic examination (3.1% vs 2.6%; $p = 0.60$), or the two methods combined (6.3% vs 7.3%; $p = 0.47$). The prevalence of distal symmetric polyneuropathy in the spouses of deployed and nondeployed veterans did not differ (2.7% vs 3.2 %; $p = 0.64$). Veterans exposed to the Khamisiyah ammunition-depot explosion did not differ significantly from nonexposed deployed veterans in prevalence of polyneuropathy.

Karlinsky and colleagues (2004) reported results of pulmonary-function tests (PFTs) on the same VA population as Eisen and colleagues. PFT results were classified into five categories: normal pulmonary function, nonreversible airway obstruction, reversible airway obstruction, restrictive lung physiology, and small-airway obstruction. The pattern of PFT results was similar in deployed and nondeployed veterans, with no statistically significant differences. The pattern of PFT results was also reported to be similar in those exposed and not exposed (according to DOD exposure estimates developed in 2002) to nerve agents from destruction of munitions at the storage site at Khamisiyah in 1991. Prevalences of self-reported pulmonary symptoms were higher in deployed veterans; however, self-reported diagnoses, use of asthma medications, and self-reported physician visits and hospitalizations for pulmonary conditions were similar in deployed and nondeployed. Although no adjustments were made for covariates, demographic variables were similar in the two groups, and a history of tobacco-smoking was more common in deployed than in nondeployed (51.1% vs 44.4%; $p = 0.03$).

Oregon and Washington Veteran Studies

Veterans from Oregon or Washington were studied in a series of analyses by investigators of the Portland Environmental Hazards Research Center (McCauley et al. 1999b). A questionnaire was sent to a random sample (n = 2,343) of 8,603 Gulf War veterans who listed Oregon or Washington as their home state of record at the time of deployment, according to data provided by the DOD Manpower Data Center. The response rate was 48.4%. The study found high rates (21-60%) of self-reported symptoms, including cognitive-psychologic symptoms, unexplained fatigue, musculoskeletal pain, gastrointestinal complaints, and rashes. However, in the next phase of the study, the clinical-examination component, the first 225 participants displayed differences between the symptoms they reported on questionnaires and the symptoms they reported at clinical examination. The greatest differences were in rash or lesions (4% agreement between questionnaire and clinical examination), gastrointestinal complaints (20% agreement), and musculoskeletal pain (35% agreement). The authors interpreted those findings as suggesting the likelihood of outcome misclassification when self-administered questionnaires were relied on.

Symptom Clustering

Investigators studied clusters of unexplained symptoms by creating a new case definition for unexplained illness (Storzbach et al. 2000). Using questionnaire data, potential cases were identified as those reporting at least one of the following symptoms: musculoskeletal pain; cognitive-psychological changes, gastrointestinal complaints; skin or mucous membrane lesions; or unexplained fatigue. Veterans whose symptom clusters remained unexplained at clinical examination (after exclusion of established diagnoses) were defined as constituting cases. Controls were those who at the time of clinical examination had no history of case-defining symptoms during or after their service in the Gulf War. In an analysis of the 241 cases vs 113 controls, investigators found, at medical evaluation, small but statistically significant deficits in cases on some neurobehavioral tests of memory, attention, and response speed. Cases also were significantly more likely to report increased distress and psychiatric symptoms (Storzbach et al. 2000). Finally, more than half the veterans with unexplained musculoskeletal pain met symptom-based criteria for fibromyalgia, and a large proportion met symptom-based criteria for CFS (Bourdette et al. 2001). Bourdette and colleagues also undertook a factor analysis, which yielded three symptom-based factors: cognitive-psychologic, mixed somatic, and musculoskeletal. These case-control studies and others from this cohort are reviewed further in Chapter 5.

Exposure-Symptom Relationships

Another nested case-control analysis of the population-based cohort examined exposures that might account for cases of unexplained illness (Spencer et al. 2001). The sample consisted of 241 veterans with unexplained illness and 113 healthy controls. In multivariate analysis, exposures most highly associated with unexplained illness were combat conditions, heat stress, and having sought medical attention during the Gulf War. Exposure to PB, insecticides and repellents, and stress was not statistically significantly associated with unexplained illness when multiple simultaneous exposures were controlled for. Those findings led investigators to conclude that unexplained illnesses were not associated with cholinesterase-inhibiting neurotoxic chemicals. One strength of this study was its elimination of numerous self-reported exposures (such as anthrax and botulinum toxoid vaccines) with questionable validity as determined by lack of test-retest reliability or time-dependent information (for example, chemical weapon exposure reported by precombat veterans or postcombat veterans who could not have been so exposed) (McCauley et al. 1999a).

Kansas Veteran Study

The state of Kansas established the Kansas Persian Gulf War Veterans Health Initiative to determine the patterns of veterans' health problems in the state. Using lists of eligible veterans from DOD, Steele and colleagues (2000) conducted a population-based survey of veterans who listed Kansas as their home state of record. A stratified random sample of 3,138 was selected, from which 2,396 were located with instate contact information. The survey, mailed out in 1998, asked about 16 specific medical or psychiatric conditions, 37 symptoms, service branch, locations during the Gulf War (including whether the veterans were notified about the Khamisiyah demolitions), and vaccinations. Kansas Gulf War veterans, in comparison with Kansas nondeployed veterans, reported greater prevalence of 10 physician-diagnosed conditions: skin conditions, stomach or intestinal conditions, depression, arthritis, migraine headaches, CFS, bronchitis, PTSD, asthma, and thyroid condition. Using their own definition of Gulf War illness, which was similar to that used by the Centers for Disease Control and Prevention (CDC) (Fukuda et al. 1998), the investigators found that its prevalence was most associated with the period and location in the gulf in which veterans served. It was least prevalent in the period before the war, for example. Overall, the multisymptom illness was found in 34% of deployed, 12% of nondeployed who had received vaccines, and 4% of nondeployed who did not receive vaccines. The study concluded that excess morbidity is tied to characteristics of Gulf War service and that vaccine exposure might contribute to onset of multisymptom illness.

Canadian Veteran Study

The findings of a 1997 survey (Goss Gilroy Inc. 1998)[5] mailed to the entire cohort of Canadian Gulf War veterans were similar to those of the Iowa study. Respondents from Canada who had been deployed to the Gulf War (n = 3,113) were compared with respondents deployed elsewhere (n = 3,439) during the same period. Of the Gulf War veterans responding, 2,924 were male, 189 female. Deployed forces had higher rates of self-reported chronic conditions and symptoms of a variety of clinical outcomes than controls. Those outcomes and symptoms include chronic fatigue, cognitive dysfunction, MCS, major depression, PTSD, chronic dysphoria, anxiety, and respiratory diseases. The greatest differences between deployed and nondeployed forces were in the first three. The symptom grouping with the highest overall prevalence was cognitive dysfunction, which occurred in 34-40% of Gulf War veterans and 10-15% of control veterans. Gulf War veterans also reported significantly more visits to health-care practitioners, greater dissatisfaction with their health status, and greater health-related reductions in recent activity.

Symptom Clustering

The Canadian study did not search for potentially new syndromes.

Exposure-Symptom Relationships

In Canadian Gulf War veterans, the greatest number of symptom groupings was associated with self-reported exposures to psychologic stressors and physical trauma. Several symptom groupings also were associated with exposure to chemical-warfare agents, absence of routine immunizations, sources of infectious diseases, and ionizing or nonionizing radiation. Nevertheless, a subset of Canadian veterans who, because they were based at sea, could not have been exposed to many of the agents reported symptoms as frequently as did land-based veterans.

United Kingdom Veteran Studies

The UK sent over 53,000 personnel to the Gulf War. From the pool of veterans, two teams of researchers each studied a separate, nonoverlapping, stratified random sample of Gulf War veterans. The first team was from the University of London (Guy's, King's, and St. Thomas's Medical Schools), the second team from the University of Manchester. A third team of researchers from the London School of Hygiene and Tropical Medicine surveyed the entire cohort of 53,000 veterans for a more narrowly focused study of birth defects and other reproductive outcomes.

University of London Veteran Studies

Unwin and collaborators (1999) at the University of London investigated the health of servicemen from the UK in a population-based study. The study used a random sample of the entire UK contingent deployed to the Gulf War[6] and two comparison groups. One of the comparison groups was deployed to the conflict in Bosnia (n = 2,620); this made the study the only one to use a comparison population with combat experience during the time of the Gulf War.

---

[5]In January 1997, Goss Gilroy Inc. was contracted by the Canadian Department of National Defence to carry out an epidemiologic survey of Canadians who served in the Gulf War to establish the overall health status of Gulf War personnel.

[6]UK military personnel in the Gulf War were somewhat different from US personnel in demographics, combat experience, and exposures to particular agents (UK Ministry of Defence, 2000).

The second comparison group (n = 2,614) was deployed to other noncombat locations outside the UK in the same period. As opposed to what was done in some studies, this nondeployed control group was recruited from among the subset of nondeployed service members who were fit for combat duty and thus avoided selection bias related to the healthy-warrior effect. Through a mailed questionnaire, the investigators asked about symptoms (50 items), medical disorders (39 items), exposure history (29 items), functional capacity, and other topics. The findings on the Gulf War cohort and comparison cohorts were compared through calculation of ORs. The study controlled for potential confounding factors (including sociodemographic and lifestyle factors) by logistic regression analysis. Only male veterans' results were analyzed, because female veterans' roles and symptoms were distinct enough to warrant separate consideration.

The Gulf War-deployed veterans (n = 2,961) reported higher prevalences of symptoms and diminished functioning than did either comparison group. Gulf War veterans were 2-3 times more likely than comparison subjects to have met symptom-based criteria for chronic fatigue, posttraumatic stress reaction, and "chronic multisymptom illness", the label for the first case definition[7] developed by CDC researchers to probe for the existence of a potential new syndrome among Gulf War veterans (Fukuda et al. 1998). It should be noted, that the Bosnia cohort, which also had been deployed to a combat setting, reported fewer symptoms than the Gulf War cohort suggests that combat deployment itself does not account for higher symptom reporting.

A separate analysis of this UK Gulf War cohort found that the prevalence of self-reported symptoms of MCS[8] was 1.3%, statistically significantly greater than in the comparison groups. The prevalence of CFS, 2.1%, was not statistically significantly greater than in the nondeployed Gulf War-era cohort but was greater than in the Bosnian cohort (Reid et al. 2001). Results of this and other studies with respect to, for example, PTSD and other psychiatric disorders, are discussed and evaluated in Chapter 5.

A followup study using a postal survey was sent 11 years after the war to a stratified random sample of participants from the first study. The followup study found modestly lower prevalence of fatigue symptoms and psychologic distress but slightly higher prevalence of physical symptoms on the SF-36 in the Gulf War cohort than in the earlier study. Gulf War veterans were still more symptomatic than nondeployed controls (Hotopf et al. 2003a).

Symptom Clustering

In a companion study using the UK dataset, Ismail and colleagues (1999) set out to determine whether the symptoms that occurred with heightened prevalence in UK Gulf War veterans constituted a new syndrome. By applying factor analysis, they identifed three fundamental factors, which they classified as related to mood, the respiratory system, and the peripheral nervous system. The pattern of symptom reporting by Gulf War veterans differed little from the patterns by Bosnia and Gulf War-era comparison groups, although the Gulf War cohort had a higher frequency of symptom reporting and greater severity of symptoms. Furthermore, the study did not identify in this cohort the six factors characterized by Haley and colleagues (1997b) in their factor-analysis study described in the next section. The UK authors interpreted their results as evidence against the existence of a unique Gulf War syndrome. Nevertheless, in a

---

[7] A case is defined as having one or more chronic symptoms in at least two of these three categories: fatigue, mood-cognition (for example, depression or difficulty in remembering or concentrating), and musculoskeletal (joint pain, joint stiffness, or muscle pain). This case definition was developed as a research tool to organize veterans' unexplained symptoms into a potentially new syndrome.

[8] Based on criteria of Simon and colleagues (1993).

later study of veterans' beliefs, the authors found that 17.3% of UK Gulf War veterans believed that they had a condition known as Gulf War syndrome (Chalder et al. 2001).

Exposure-Symptom Relationships

In the UK Gulf War cohort, most self-reported exposures were associated with all the health outcomes; that was also true in the two comparison cohorts (Unwin et al. 1999). The authors interpreted that finding as evidence that the exposures were not uniquely associated with Gulf War illnesses. Veterans with symptoms, regardless of deployment status, were more likely to report a wide variety of exposures than those without symptoms. Within the Gulf War cohort, two vaccine-related exposures—vaccination against biologic-warfare agents and multiple vaccinations—were associated with the case definition of "chronic multisymptom illness" developed by CDC researchers (Fukuda et al. 1998). A later analysis of the data on a subcohort of UK veterans found that receiving multiple vaccinations during deployment was weakly associated with five of the six health outcomes examined, including "chronic multisymptom illness" as defined by CDC (Hotopf et al. 2003a). Another separate analysis of a subgroup of veterans meeting case criteria for MCS symptoms found that they were significantly more likely to report several types of pesticide exposures. Veterans meeting case criteria for CFS were not more likely to report pesticide exposure but were more likely to report combat-related injury (Reid et al. 2001).

University of Manchester Veteran Study

The University of Manchester study used a random sample of UK veterans 7 years after the Gulf War (Cherry et al. 2001a; Cherry et al. 2001b). The cohort was deliberately separate from that studied by Unwin and colleagues (1999). Two groups of veterans deployed to the Gulf War (n = 8,210, a main cohort and a validation cohort) were compared with veterans who were not deployed but whose health would not have prevented deployment (n = 3,981). Veterans were sent a questionnaire about the extent to which they were burdened by 95 symptoms in the previous month. By asking them to mark their answers on a visual analogue scale, investigators sought to determine the degree of symptom severity. Investigators also sought to determine areas of peripheral neuropathy by asking veterans to shade body areas on two mannequins in which they were experiencing pain or numbness and tingling. On almost all 95 symptoms, deployed veterans reported greater symptom severity. The overall mean symptom severity scores of the two Gulf War cohorts were similar and significantly greater than that of the non-Gulf War cohort. For 14 symptoms—including memory, concentration, and mood problems—the severity scores of deployed veterans were at least twice those of the nondeployed veterans. Numbness and tingling were reported by about 13% of deployed and about 7% of nondeployed. Widespread pain was also reported more frequently (12.2% vs 6.5%).

Symptom Clustering

Through factor analysis, the investigators identified seven factors which accounted for 48% of the variance. Deployed veterans' scores were significantly different on five factors: psychologic, peripheral, respiratory, gastrointestinal, and concentration. No difference was found in the neurologic factor; and appetite, the final factor, was significantly lower than in the non-Gulf War cohort. None of the factors was exclusive to Gulf War veterans, so the investigators concluded that their findings did not support a new syndrome (Cherry et al. 2001a).

Exposure-Symptom Relationships

The two UK Gulf War cohorts completed a second questionnaire with details of the dates when they were sent to each location and the exposures they had experienced. The exposure questionnaire contained 14 exposures. The main analysis involved a multiple regression of each of the seven factors identified through factor analysis on all exposures and other potential confounders. Many of the reported exposures correlated with one another. In the multivariate regression analysis, the number of days on which veterans handled pesticides was related to the overall severity score and to the peripheral and neurologic factors. The number of days on which they applied insecticide to their skin was related to severity and to the peripheral, respiratory, and appetite factors. The number of inoculations was associated with skin and musculoskeletal symptoms. There was a marked dose-response gradient for the association between insect repellents and the peripheral and respiratory factors. A dose-response gradient for the association of handling pesticides with the peripheral factor was present but less robust. The handling of pesticides and side effects of handling nerve-agent prophylaxis were associated with peripheral neuropathy (OR 1.26; $p < 0.001$), and the use of insect repellent was associated with widespread pain (OR 1.15; $p < 0.001$) (Cherry et al. 2001b).

London School of Hygiene and Tropical Medicine Veteran Study

The third British study was a very large mail survey conducted by researchers from the London School of Hygiene and Tropical Medicine (Maconochie et al. 2003; Simmons et al. 2004). It was designed largely to assess reproductive outcomes among Gulf War veterans, but it contained open-ended questions about their general health. The exposed cohort consisted of all UK Gulf War veterans, and the unexposed cohort consisted of a random sample of nondeployed UK military personnel from the same period. Although the numbers of surveys returned in the study were large (25,084 by Gulf War veterans and 19,003 by non-Gulf War veterans), the participation rates were low (47.3% and 37.5% of male and female Gulf War veterans, respectively, and 57.3% and 45.6% of male and female nondeployed veterans). The survey included a broad variety of items on reproductive and child health, exposure histories, current health, and health of sexual partners, and was supplemented by examination of medical records for pregnancies, live births, and outcomes. Maconochie et al. (2003) reported that 42-46% of participants had conceived or had attempted to conceive a child by 2001. In a subanalysis restricted to male respondents (24,379 Gulf War veterans and 18,439 nondeployed veterans), Simmons et al. (2004) reported that 61% of Gulf War veterans reported at least one new medical symptom or disease since 1990 compared with 37% of nondeployed veterans. The symptoms most strongly associated with Gulf War deployment were mood swings, memory loss or lack of concentration, night sweats, general fatigue and sexual dysfunction. Overall, 6% of the Gulf War veterans believed that they had "Gulf War syndrome"; that belief was associated with the highest reporting of new symptoms or diseases.

Danish Peacekeeper Studies

Military personnel from Denmark were involved primarily in peacekeeping or humanitarian roles after the end of the Gulf War. They were studied in a series of population-based cohort studies (Ishoy et al. 1999b; Suadicani et al. 1999). A total of 821 Danes were eligible by virtue of having been deployed at any time in the period between August 1990 to December 1997. The Gulf War veterans were matched by age, sex, and profession to 400 members of the Danish armed forces who were not deployed to the Gulf War. Symptom and exposure questionnaires and health and laboratory examinations were used. Findings of health examinations were not used in the study's analysis of exposure-symptom relationships.

Of 22 neuropsychologic symptoms, 17 were significantly more prevalent among Gulf War veterans than among controls. Many of the symptoms were correlated with one another. Headache and fatigue-related symptoms were present in about 20% of deployed vs up to 10% of nondeployed. Gastrointestinal symptoms and diseases and symptoms related to the skin or allergy were more frequent in deployed veterans, but gastrointestinal symptoms, which were suggestive of irritable bowel syndrome (Ishoy et al. 1999b), were no more prevalent in Gulf War veterans than in Danish troops that had been previously deployed overseas. The pattern of symptoms, except musculoskeletal symptoms (which were not more prevalent), was similar to the patterns seen in the UK, VA, and Canadian cohorts. The investigators also examined male participants for sexual dysfunction. Decreased libido or nonorganic erectile dysfunction was reported by 12% of Gulf War veterans and 3.7% of nondeployed troops. An extensive examination of serum sex hormones failed to detect clinically significant differences. Predictors of male sexual dysfunction were feeling threatened and bathing in or drinking contaminated water (Ishoy et al. 2001b). The investigators concluded that the overlap of symptoms between veterans deployed during and after the war indicated the existence of common risk factors independent of exposure to war itself.

Symptom Clustering

The authors did not use factor analysis, but they did use a multiple logistic regression analysis with adjustments for age and sex to find the most relevant neuropsychologic symptoms (Suadicani et al. 1999). Only five of the 17 symptoms remained significant after adjustment for the interrelationship of variables. About 21% of Gulf War veterans vs 6.2% of controls reported a clustering of three to five of the relevant symptoms (p < 0.001). Relevant symptoms included concentration or memory problems, repeated headache, balance disturbances or dizziness, abnormal fatigue not caused by physical activity, and problems in sleeping all night. The symptoms excluded from further analysis included numbness or tingling in hands and feet, suddenly diminished muscular power, and tingling or shivering of arms, legs, or other parts of the body.

Exposure-Symptom Relationships

One of the analyses investigated whether 22 neuropsychologic symptoms were associated with 18 self-reported environmental exposures[9] (Suadicani et al. 1999). Most exposures were significantly associated with three to five relevant neuropsychologic symptoms in a univariate analysis. One psychologic exposure ("having watched colleagues or friends threatened or shot at") and environmental exposures, especially "bathing in or drinking contaminated water (fumes, oil, chemicals)", remained significant after adjustment in a multiple logistic model that adjusted for associations of exposures with one another. A separate multivariate analysis of gastrointestinal symptoms found them to be associated with two exposures: burning of waste or manure and exposure to insecticide against cockroaches (Ishoy et al. 1999a).

Australian Veteran Studies

Investigators from Monash University conducted a cohort study of Australian service personnel who had or had not been deployed to the gulf as part of the multinational force. The exposed cohort comprised 1,456 participants, the nonexposed cohort 1,588. Participation rates

---

[9]Exposures did not include PB or vaccinations against chemical- or biologic-warfare agents, because Danish veterans had a peacekeeping role and thus were not at risk for chemical or biologic warfare.

were 80.5% and 56.8%, respectively (Kelsall et al. 2004a). In the Australian contingent sent to the Gulf War, members of the Navy were heavily overrepresented (86.5%). Very few experienced direct combat. Despite their lack of combat exposure, deployment was a stressful event: deployed veterans experienced higher rates of fear and threat of entrapment, attack (including nerve-agent warfare), and death or injury (Ikin et al. 2004). Participants completed a mailed questionnaire, which consisted of a physical and mental health screening questionnaire (SF-12), a test for nonpsychotic psychologic illness (GHQ-12), a PTSD checklist (PCL-S), and a questionnaire about military service and exposures. Thirty-one percent of Gulf War veterans had developed DSM-IV diagnoses since the Gulf War compared with 21% of non-Gulf War veterans. Significant excesses were seen in PTSD, depression, and substance-use disorders. A more recent study of Australian Navy Gulf War veterans noted that those veterans reported many stressful experiences, including fear of death and perceived threat of attack, more frequently in relation to the Gulf War than other military services (Ikin et al. 2005). The study population was the entire cohort of 1,579 veterans deployed to the 1991 Gulf War, but in the final analysis, results were restricted to 1,232 male participants.

Kelsall et al. (2004a) stated that participants in the exposed cohort reported a higher prevalence of all symptoms and reported more severe symptoms. McKenzie et al. (2004) reported that Gulf War veterans had poorer psychologic health and that the number of stressful exposures correlated with poorer scores on three standard instruments used to measure functioning and psychologic health.

Symptom Clustering

Forbes et al. (2004) used factor analysis to attempt to group symptom complexes for this cohort. Three factors emerged as more prominent in Gulf War veterans—psychophysiologic distress, somatic distress, and arthroneuromuscular distress and the symptoms in those complexes were more severe in Gulf War than in non-Gulf War veterans. This well-designed study confirms the extent and greater severity of symptoms in Gulf War veterans, even in a predominantly naval population with few direct military attacks, no deaths, and few casualties. The results suggest a deployment effect in the absence of actual combat.

Exposure-Symptom Relationships

Greater symptom severity was associated with 10 or more immunizations, use of PB, pesticides, insect repellents, presence in a chemical-weapons area, and reporting of stressful military service (Kelsall et al. 2004a).

MILITARY-UNIT-BASED STUDIES

Ft. Devens and New Orleans Cohort Studies

The symptom experience of two deployed cohorts of Gulf War veterans was studied by Boston-based researchers. One of the cohorts, an Army cohort based in Ft. Devens, Massachusetts, was surveyed longitudinally at three times (1991, 1993-1994, and 1997), and underwent psychiatric interviews and other clinical evaluations at the second time (e.g., White et al. 2001). A second deployed cohort from New Orleans was also studied at the second time, as was a non-Gulf War-deployed unit sent to Germany. The Germany-deployed unit was an air ambulance company of National Guard from Maine that had been deployed to Germany for handling wounded personnel evacuated from the gulf.

In comparison with veterans deployed to Germany during the Gulf War era, stratified random samples of both Gulf War cohorts (Ft. Devens and New Orleans) had increased prevalence of 51 of 52 items on a health-symptom checklist (Proctor et al. 1998). The greatest differences in prevalence of reported symptoms were of dermatologic symptoms (such as rash, eczema, and skin allergies), neuropsychologic symptoms (such as difficulty in concentrating and difficulty in learning new material), and gastrointestinal symptoms (such as stomach cramps and excessive gas). With a separate checklist, researchers found a higher prevalence of PTSD, according to the Clinician-Administered PTSD Scale (CAPS) (5% Ft. Devens, 7% New Orleans, and 0% Germany). The study's nearly 300 subjects represented a stratified random sample of 2,949 troops from Ft. Devens and 928 from New Orleans; both groups consisted of active-duty, reserve, and National Guard troops. The comparison group was Germany-deployed veterans from an air ambulance company (n = 48). The cohorts were also the focus of several studies of stress-related disorders, such as PTSD, depression, and substance abuse (see Chapter 5).

Symptom Clustering

Symptom clustering among the Ft. Devens cohort was studied in 1997 with CDC's case definition of multisymptom illness (Wolfe et al. 2002). The case definition was applied to findings from the 52-item health checklist. About 60% of respondents met the CDC case definition. That group was roughly evenly divided between "mild to moderate" and "severe" cases. On the basis of logistic regression, many of the exposures were associated with meeting the case definition, including anti-nerve-gas pills, anthrax vaccination, tent heaters, exposure to oil-fire smoke, and chemical odors.

Exposure-Symptom Relationships

In 1994-1996, Proctor (1998) surveyed the deployed cohorts (Ft. Devens and New Orleans) on about eight exposures, and asked respondents to rate each on a scale of 0-2, (0 = no exposure; 1 = exposed; 2 = exposed and felt sick at the time). Using standardized regression, they found the strongest associations between several exposures—debris from SCUDS, chemical and biologic warfare agents—and musculoskeletal, neurologic, neuropsychologic and psychologic symptoms.

Seabee Reserve Battalion Studies

Haley and collaborators (1997b) studied members of one battalion of naval reservists called to active duty for the Gulf War. The battalion was a mobile construction battalion for other branches of the military. More than half the battalion had left the military by the time of the study. Participants were recruited from those for whom investigators had addresses and from veterans' meetings. Of those participating, 70% reported having had a serious health problem since returning from the Gulf War. A telephone survey of a random sample of nonparticipants found that, while they were demographically similar to participants, fewer (43%) reported having serious health problems since the war. Eleven percent of participants and only 3% of nonparticipants were unemployed. Participation rate was low (41.1% of 606 males in the battalion; 58.0% of those located), and there was no comparison cohort of nondeployed veterans. All those features strongly suggest selection bias, which could lead to overestimation of health effects among participants.

Symptom Clustering

The study was the first to cluster symptoms into new syndromes by applying factor analysis. Through standardized symptom questionnaires and two-stage factor analysis, the investigators defined what they considered to be either six syndromes or six variants of a single syndrome, which they labeled impaired cognition, confusion-ataxia, arthromyoneuropathy, phobia-apraxia, fever-adenopathy, and weakness-incontinence. One-fourth of the veterans in this uncontrolled study (n = 63) were classified as having one of the six syndromes. The first three syndromes had the strongest factor clustering of symptoms.

In a followup study of the same cohort, Haley and colleagues (1997a) used a case-control design to examine neurologic function. They chose as cases the 23 veterans who had scored highest on the three syndromes with the strongest factor clustering. Controls consisted of two small groups of healthy veterans, of which one (n = 10) was deployed to the Gulf War and the other (n = 10) was not. The results of extensive neurologic and neurobehavioral testing demonstrated that cases had significantly greater evidence of neurologic dysfunction compared with controls. Investigators concluded that the three syndromes, derived from factor analysis of symptoms, might signify variant forms of expression of a generalized injury to the nervous system.[10] In a later study, cases with one of the three syndromes were more likely than healthy controls to exhibit vestibular dysfunction (Roland et al. 2000). Related research on the same subset of veterans found evidence of basal ganglia and brainstem neuronal loss through magnetic resonance spectroscopy (Haley et al. 2000b). Those studies are discussed further in Chapter 5.

Exposure-Symptom Relationships

The three syndromes identified by Haley and colleagues (1997a) were the focus of another case-control study that examined their relationship to self-reported exposures to neurotoxicants. The study tested the hypothesis that exposure to organophosphates and related chemicals that inhibit cholinesterase are responsible for the three nervous system-based syndromes (Haley and Kurt 1997). Each of the syndromes was associated with a distinct set of risk factors. The "impaired-cognition syndrome" was found, through multiple logistic regression, to be associated with jobs in security and the wearing of flea-and-tick collars. The "confusion-ataxia syndrome" was associated with self-reports of having been involved in a chemical-weapons attack and of having advanced adverse effects of PB. Finally, "arthromyoneuropathy" was associated with higher scores on the scale of advanced adverse effects of PB and with an index created by the investigators to enable veterans to self-report the amount and frequency of their use of government-issued insect repellent. The authors concluded that some Gulf War veterans had delayed chronic nervous system syndromes as a result of exposure to combinations of neurotoxic chemicals (Haley and Kurt 1997).

Another study by Haley and collaborators (1999) examined whether genetic susceptibility could play a role in placing some veterans at risk for neurologic damage by organophosphate chemicals. They hypothesized that neurologic symptoms in ill veterans might be explained by their having genetic polymorphisms (variations) in metabolizing enzymes. One set of polymorphisms could impair their ability to quickly detoxify organophosphorus compounds,

---

[10] Neuropsychologic or neurologic impairments have been the focus of several smaller studies as well. Some found subtle changes in nerve-conduction velocity and cold sensation (Jamal et al. 1996) and in some tests of finger dexterity and executive functioning (Axelrod and Milner 1997); others found no significant differences in measures of nerve conduction and neuromuscular functioning (Amato et al. 1997) or neuropsychologic performance (Goldstein et al. 1996).

such as sarin, soman, and some pesticides. The investigators studied 45 veterans: 25 with chronic neurologic symptoms identified through their earlier factor-analysis study and 20 healthy controls from the same battalion. They measured blood butyrylcholinesterase and two types, or allozymes, of paraoxonase/arylesterase-1 (PON1). The genotypes encoding the allozymes were also studied. The investigators found that veterans who were ill had blood butyrylcholinesterase concentrations similar to those of control subjects; however, ill veterans had lower type Q paraoxonase/arylesterase, the allozyme that hydrolyzes sarin rapidly. They also were more likely to have the type R genotype, which encodes the allozyme that has low hydrolyzing activity for sarin. The authors interpreted their findings as suggesting that reduced ability to detoxify organophosphorus chemicals might have contributed to the onset of neurologic symptoms in some Gulf War veterans. Contrary evidence was provided by Hotopf and colleagues (2003b), who did not find differences in PON1 activity among symptomatic vs healthy Gulf War veterans in a more representative, population-based sample.

Larger Seabee Cohort Studies

The first in a series of studies by Gray and collaborators (1999a) surveyed active-duty Seabees who remained on active duty for at least 3 years after the Gulf War. The Seabees were from 14 Seabee commands at two locations (Port Hueneme, California, and Gulfport, Mississippi). Those who were deployed to the Gulf War were in mobile construction battalions serving in the same tasks and at the same sites as did the reserve Seabee battalion studied by Haley and collaborators. During the Gulf War, Seabees built airports, supply points, and roads. Unlike Haley et al., Gray et al. excluded Gulf War veterans who were no longer active at the time of the study.

Gray and colleagues enrolled 1,497 study subjects, 527 of whom were Gulf War veterans and 970 nondeployed veterans. Study subjects filled out symptom and exposure questionnaires and answered additional questions screening for PTSD, CFS, and various psychologic symptom domains; they had serum tested for acute-phase reactants and had handgrip strength tested.

The deployed veterans reported greater prevalence of 35 of 41 symptoms. In a subset of veterans, symptom reporting was not reliable when retested several months later, and it lacked validity on the basis of checks with medical records. Gulf War veterans were more likely to report symptoms of PTSD (OR 1.8, 95% CI 1.3-2.5). They also had a small but significant decrease in handgrip strength.

Symptom Clustering

Knoke and colleagues (2000) reported a factor analysis of the active-duty Seabee study in response to a factor analysis conducted by Haley et al. (1997b). Knoke and colleagues found that three factors were more common among Seabees who had been deployed—somatization, depression, and obsessive-compulsive symptoms—and that they affected 20% of Gulf War veterans. Their findings were similar to those of Doebbling ct al. (2000), Fukuda (1998), and Ismail et al. (2002) and consistent with findings in a civilian population with CFS (Nisenbaum et al. 1998). They concluded, unlike Haley after the study of Seabee reservists from one reserve battalion, that there was no evidence of a unique spectrum of neurologic injury.

Gray et al. (2002) re-examined the question of symptoms and exposures by expanding their deployed and nondeployed cohorts to include all Seabees who had been on active duty during the time of the Gulf War regardless of whether they remained on active duty, were in the reserve, or had separated from service. There were 11,868 participants and a 67.4% participation rate. Participants were divided into three exposure groups: 3,831 who had been deployed to the Gulf War, 4,933 who had been deployed elsewhere, and 3,104 who had not been deployed. Those who had been deployed to the gulf reported more poor health, missed work, cognitive failure, hospitalizations, digestive diseases, and depression. They also reported a greater frequency of leishmaniasis, CFS, PTSD, MCS, and irritable bowel syndrome. Overall, 22.1% of Gulf War veterans met a working definition of Gulf War syndrome. Being defined as a case was associated with participation in a federal Gulf War veteran registry, being female, a reservist, or a member of two Seabee battalions, or having a nontraditional Seabee occupation. There were also weak associations (OR < 2.0) with 12 specific exposures.

Exposure-Symptom Relationships

Deployed veterans reported more frequent exposure to 26 of the 30 possible agents. The agents reported were somewhat different from those reported in a Seabee reserve battalion (Haley and Kurt 1997; Haley et al. 1997b). For example, the reserve battalion reported wearing pet flea collars and being exposed to chemical weapons, but the active-duty Seabees in this study did not. The authors chose to dichotomize the 26 exposures artificially; they carried out further analyses only on the subset of the 11 agents which the OR for exposure between Gulf War veterans and nondeployed veterans was greater than 3, and if more than 5% of Gulf War veterans reported the exposure. The study found many exposure-symptom associations with the 11 agents. The authors apparently carried out multivariate analysis, but they did not report its results. They stated that they "could not isolate or implicate specific war exposures" using their multivariate analysis. Other study limitations were recall bias, moderate-to-low response rate (53%), exclusion of veterans no longer in active service, and lack of representativeness of the entire Gulf War population.

Pennsylvania Air National Guard Study

A large study by Fukuda and colleagues (1998) used factor analysis and other methods to assess the health status of Gulf War veterans in response to requests from DOD, VA, and the state of Pennsylvania to assess the prevalence and causes of an unexplained illness in members of one currently active Air National Guard unit. By studying that unit and three comparison Air Force populations,[11] the investigators aimed to organize symptoms into a case definition and to carry out clinical evaluations on participants from the index Air National Guard unit (using a nested case-control design). All the units that were studied had a combination of deployed and nondeployed veterans. For purposes of assessing symptom prevalence, the investigators combined the four units and compared questionnaire responses of deployed and nondeployed. Of 3,723 participants surveyed, those deployed to the Gulf War experienced higher prevalence of chronic symptoms (33 of 35 symptoms of more than 6-month duration were reported to be more prevalent) than nondeployed veterans. For purposes of developing a case definition, the investigators focused, at first, solely on the Pennsylvania index unit. The authors used two broad methods to derive a case definition: (1) a clinical approach in which symptoms had to be reported for 6 months or longer, had to occur in 25% or more of Gulf War veterans, and had to occur at least 2.5 times more frequently in Gulf War veterans than in non-Gulf War veterans; and (2) factor analysis. The two approaches yielded similar case definitions.

---

[11] Air National Guard in Pennsylvania, US Air Force Reserve in Florida, and US Air Force active duty in Florida. Those comparison units were demographically similar to the index unit but had different primary missions.

Symptom Clustering

The authors defined a case of chronic multisymptom illness as having one or more chronic symptoms from at least two of three categories: fatigue, mood-cognition symptoms (for example, depression or difficulty in remembering or concentrating), and musculoskeletal symptoms (joint pain, joint stiffness, or muscle pain). Severe cases were defined as those in which each case-defining symptom had been reported as severe. According to that definition, 39% of Gulf War-deployed veterans and 14% of nondeployed veterans had mild-to-moderate cases, whereas 6% and 0.7%, respectively, had severe cases. On the basis of a total of 158 clinical examinations in one unit, there were no abnormal physical or laboratory findings that differentiated those who met the case definition and those who did not. Cases, however, reported significantly lower functioning and well-being.

A sizable fraction (14%) of nondeployed veterans also met the mild-to-moderate case definition. The investigators concluded that their case definition could not specifically characterize Gulf War veterans with unexplained illnesses. The study, however, had several limitations, the most important of which was its coverage of only current Air Force personnel several years after the Gulf War (including Air National Guard, Air Force Reserve, and active duty), which limits its generalizability to other branches of service and to those who left the service possibly because of illness.

Medical Evaluation

To assess risk factors, the authors performed clinical evaluations on a subset of participants (n = 158), all of whom volunteered for the evaluation and came from the index unit of the Pennsylvania Air Force National Guard. Of the members of this unit, 45% had been deployed to the Gulf War. Overall, there were few abnormal findings from blood, stool, and urine testing among those who met the case definition for chronic multisymptom illness. There were no differences between cases and noncases in the proportion that seroreacted to botulinum toxin, anthrax-protective antigen, leishmanial antigens, and other antigens. This was among the few studies to have assessed exposures (mostly to infectious diseases) via laboratory testing as opposed to self-reports, but the sample undergoing clinical evaluation was relatively small and restricted to Air Force National Guard members.

Exposure-Symptom Relationships

A nested case-control study of the same cohort (n = 1,002) sought to identify self-reported exposures associated with cases of chronic multisymptom illness (Nisenbaum et al. 2000). It found that meeting the case definition of severe and mild-to-moderate illness was associated with use of PB, use of insect repellent, and belief in a threat from biologic or chemical weapons. Having an injury requiring medical attention was also associated with having a severe case of chronic multisymptom illness.

OTHER COHORT STUDIES

The following studies are grouped because they are more narrowly focused than the major cohort studies and often lack analysis of symptom clustering or exposure-symptom relationships.

Hawaii and Pennsylvania Active Duty and Reserve Study

One of the first epidemiologic studies of US Gulf War veterans was a congressionally mandated study of more than 4,000 active-duty and reserve personnel from bases in Pennsylvania and Hawaii. The main purpose of the study was a focus on psychologic health, but the first publication also dealt with physical health (Stretch et al. 1995). It found that veterans deployed to the Gulf War reported higher prevalence of 21 of 23 symptoms on a symptom checklist than nondeployed veterans. Overall, deployed veterans were about 2-4 times more likely than nondeployed veterans to report each symptom. In a later publication (Stretch et al. 1996a), the authors reported that deployed veterans commonly reported significant levels of stress, including operating in desert climates, long duty days, extended periods in chemical-protective clothing, lack of sleep, crowding, lack of private time, physical workload, and boredom. Another publication examined PTSD as measured by the Impact of Event Scale and the Brief Symptom Inventory (Stretch et al. 1996b). The prevalence of PTSD symptoms in the deployed veterans was 8.0-9.2% vs 1.3-2.1% in the nondeployed. The low overall response rate, 30.6%, limits the generalizability of the three studies.

New Orleans Reservist Studies

A series of studies by Sutker and colleagues analyzed psychologic outcomes in a cohort of New Orleans reservists (n = 1,520). The cohort consisted of Louisiana National Guard and reservists from the Army, Air Force, and Navy. The overall response rate was 83.7%. Of the 1,272 who responded, 876 had been deployed and 396 had not been deployed. Deployed veterans, assessed by survey at an average of 9 months (time 1) after the war, had higher scores on depression and anxiety symptom scales[12] than nondeployed veterans (Brailey et al. 1998). Twenty-five months after the war (time 2), the researchers studied only the deployed group (n = 349), which represented 88.1% of the original group of deployed veterans studied at time 1. Responses at time 2 revealed higher rates of PTSD, depression, and hostility than at time 1. The increasing prevalence of PTSD was linked to symptom clusters of hyperarousal and numbing (Thompson et al. 2004). Hyperarousal and numbing were also associated with development of depression, anxiety, hostility, and physical symptoms. Higher wartime stress exposure vs low-stress exposure was related to PTSD and somatic problems (Brailey et al. 1998; Sutker et al. 1993). The personality and coping factors found to increase the likelihood of PTSD were low personality hardiness, high avoidance coping, and low perceived family cohesion (Sutker et al. 1995b).

Air Force Women Study

Female Air Force veterans were studied by Pierce (1997), who examined a stratified sample of 525 women (active-duty, National Guard, and reserve) drawn from all 88,415 women who served in the Air Force during the Gulf War era. Women deployed to the Gulf War reported rash, cough, depression, unintentional weight loss, insomnia, and memory problems more frequently than women deployed elsewhere. The pattern of symptom reporting was similar to that reported by men and women in other Gulf War studies (Carney et al. 2003; Unwin et al.

---

[12] The authors used the Beck Depression Inventory, the Brief Symptom Inventory for Anxiety and Depression, the PTSD checklist, and the Mississippi Scale for PTSD.

2002). In addition, women deployed to the Gulf War were more likely than controls to report sex-specific problems, such as breast cysts and lumps, and abnormal cervical cytology.

Connecticut National Guard

Southwick and colleagues studied two deployed units of the Connecticut National Guard (n = 240) (Morgan et al. 1999; Southwick et al. 1993; Southwick et al. 1995). The focus of the study was on trauma-related symptoms and the course of PTSD. The cohort was studied prospectively at 1 month, 6 months, 2 years, and 6 years. The study was unusual in its frequency of followup, but there was no nondeployed comparison group. Although 240 were eligible to participate, only 119 filled out the first questionnaire and 84 filled out a second questionnaire at the 6-month mark. From 1 month to 6 months, the prevalence of PTSD increased, as did the severity rating of some symptom clusters (for example, hyperarousal). The prevalence varied, depending on the symptom scale and cutoffs being used, but on the average rose from about 3% to 6.5%. The degree of combat exposure was associated with the degree of PTSD symptoms. All veterans who, at 1 month or 6 months, met the criteria for PTSD according to the Mississippi scale, also met the criteria at 2 years (Southwick et al. 1995).

TABLE 4.1 Major Cohort Studies (Shaded) and Derivative Studies

Reference	Eligible population	Type of study or methods	Date(s) of enrollment	Subgroup (n = eligible subjects)	Contacted or Located (% of eligible)	Responded or Enrolled (Response Rate)	Comments
Iowa Persian Gulf Study Group 1997	Iowa listed as home of record on initial military record, and service in regular military or activated National Guard/Reserve sometime from 8/2/1990–7/31/1991, identified by Defense Manpower Data Center, Monterey CA, DOD (n = 28,968)	Population based interview study Stratum random sample with proportional allocation—64 strata (GW, type of military, age, sex, race, rank, branch), Pilot study not eligible	9/1995–5/1996	Total 4,886	4,072 (83.3%)	3,695 (90.7%; 75.6% of eligible)	Limited assessment of selection bias

Derivatives from Iowa Persian Gulf Study Group 1997

Reference	Purpose	Study design	Population			Comments
			(where appropriate)			
			Eligible	Located	Enrolled (Response Rate)	
Black et al. 1999	Quality of life and health-services utilization among those with MCS	Cross-sectional survey	Population described in Iowa Persian Gulf Study Group 1997: n = 3,695			
Black et al. 2000	Prevalence of MCS syndrome	Cross-sectional survey	Population described in Iowa Persian Gulf Study Group 1997: n = 3,695			
Doebbeling et al. 2000	Definition of Persian Gulf War Syndrome	Factor analysis	Population described in Iowa Persian Gulf Study Group 1997: n = 3,695			
Zwerling et al. 2000	Prevalence of self-reported postwar injuries	Cross-sectional survey	Population described in Iowa Persian Gulf Study Group 1997: n = 3,695			
Barrett et al. 2002	Association between PTSD and self-reported physical health status	Cross-sectional survey	Population described in Iowa Persian Gulf Study Group 1997 excluding 13 with missing information Total: n = 3,682 GWVs: n = 1,889 NDVs: n = 1,793			
Black et al. 2004b	Prevalence and risk factors for anxiety	Cross-sectional survey	Population described in Iowa Persian Gulf Study Group 1997: n = 3,695			
Lange et al. 2002	Exposures to Kuwait oil fires and asthma and bronchitis	Cross-sectional survey	Population described in Iowa Persian Gulf Study Group 1997 excluding 336 with unknown exposure information: n = 1,560			

Reference	Purpose	Study design	Population Eligible	Located	Enrolled (Response Rate) (where appropriate)	Comments
Black et al. 2004a	Depression in deployed and nondeployed veterans Cases: report of depression, cognitive dysfunction, or chronic widespread pain Controls: without any of these conditions	Case-comparison study	Total population: 602 veterans interviewed in phase II from original study Iowa Persian Gulf Study Group 1997; cases defined as reporting depression, cognitive dysfunction or chronic widespread pain in phase I; controls defined as free of all 3 conditions			Not a case-control study; comparison of deployed and nondeployed depressed veterans

Reference	Eligible population	Type of study or methods	Date(s) of enrollment	Subgroup (n = eligible subjects)	Contacted or Located (% of eligible)	Responded or Enrolled (Response Rate)	Comments
Kang et al. 2000	Any person who served in the US military on active duty, in reserves, or in National Guard, irrespective of whether they were still in the service or separated (n = 693,826 deployed; 800,680 not deployed (~50% of all troops in military 9/1990–5/1991 but not in the gulf)	Cross-sectional study, population-based sample; 15,000 troops deployed, 15,000 troops not deployed; "National Health Survey of GW Era Vets and their families"		Total: 30,000 GWV: 15,000 NGV: 15,000	30,000 15,225 14,775	20,917 (70%) 11,441 (75%) 9,476 (64%)	Compared survey participants with VA health registry participants (n = 15,891) Did not list number of missing addresses From Kang et al. 2002 learned that 225 NGVs were actually GWVs Some misclassification in original sample of 15,000 each Assessed selection bias

Derivatives from Kang et al. 2000 – U.S

Reference	Purpose	Study design	Population (where appropriate)			Comments
			Eligible	Located	Enrolled (Response Rate)	
Kang et al. 2002	Factor analysis to define GWS	Factor analysis	Numbers used in factor analysis: GWVs: n = 10,423 NGVs: n = 8,960 Cases and noncase GWVs analyzed for self-reported exposures: n = 11,441; cases = 277			
Kang et al. 2003	Prevalence of PTSD and CFS	Cross-sectional cohort study	GWVs: n = 11,441 NGVs: n = 9,476			
Karlinsky et al. 2004	Prevalence of respiratory symptoms and pulmonary-function abnormalities	Cross-sectional medical evaluation survey (including pulmonary-function tests)	Recruited from 5,885 on list of matched GWVs and NDVs GWVs: n = 1,036 NDVs: n = 1,103			No details on participation rates
Davis et al. 2004	Prevalence of symptoms suggesting distal symmetric polyneuropathy in GWVs and spouses	Cross-sectional clinical and lab assessment "VA Medical Evaluation" Eligible population selected randomly from Kang population and likely to live close to examination center	GWVs: n = 1,996 Spouses: n = 745 NGVs: n = 2,883 Spouses: n = 846		GWVs: 1,061 (53.2%) Spouses: 484 (65%) NGVs: 1,128 (39.1%) Spouses: 533 (63%)	Total number of spouses eligible not given, number located not given, response rates from total eligible
Eisen et al. 2005	Prevalence of fibromyalgia, CFS, dermatologic conditions, dyspepsia, SF-36, hypertension, obstructive lung disease, arthralgias, peripheral neuropathy 10 years after deployment	Cross-sectional medical evaluation survey "VA Medical Evaluation" Eligible population selected randomly from Kang population	GWVs: n = 1,996 NGVs: n = 2,883	1,741 (87.2%) 2,444 (84.8%)	GWVs: 1,061 (60.9%; 53.2% of eligible) NGVs: 1,128 (46.2%; 39.1% of eligible)	Even though veterans with CFS were more likely to participate, bias was nondifferential Authors adjusted for problem of generalizability by correcting for overrepresentation

Reference	Eligible population	Type of study/methods	Date(s) of enrollment	Subgroup (n = eligible subjects)	Contacted or Located (% of eligible)	Responded or Enrolled (Response Rate)	Comments
Unwin et al. 1999	UK military who served in gulf region 9/1/1990-6/30/1991, excluding special forces (n = 53,462); comparison populations: personnel served in Bosnia 4/1/1992-2/6/1997 (n = 39,217); Era cohort, in military but not gulf 1/1/1991 (n = 250,000)	Cross-sectional survey, stratified random sample (service, sex, age, service status, rank, fitness, as appropriate for population), oversampled women	8-9/1997 through 11/11/1998	Total GWV Bosnia NGV	12,592 (12,744; 152 unknown addresses) 4,246 4,250 4,248	8,195 (65.1%) (calculated from numbers listed in text of study) 2,961 (69.7%) 2,620 (61.6%) 2,614 (61.5%)	Authors note that 800 Bosnia veterans later moved to GWV group; extensive assessment of selection bias

Derivatives from Unwin et al. 1999: UK Gulf War veterans – University of London

		Population *(where appropriate)*			
Reference	Purpose	Study design	Eligible	Located	Enrolled (Response Rate)	Comments
---	---	---	---	---	---	---
Ismail et al. 1999	Is there a GWS?	Factor analysis	Total cohort: n = 3,214			Says men only, from Unwin et al. looks as though women are included
Chalder et al. 2001	Prevalence of belief of GWS and comparison of health with other veterans	Cross-sectional cohort study	GWVs only: n = 2,961			
Reid et al. 2001	Prevalence of MCS and CFS, association with exposures and psychologic morbidity	Cross-sectional cohort study	GWVs: n = 3,531 Bosnia: n = 2,050 Era: n = 2,614			Report that they attempted 5,046 GWV's—more than reported in Unwin et al.
Reid et al. 2002	Prevalence of specific chemical sensitivities	Cross-sectional cohort study	GWVs: n = 3,531 Bosnia: n = 2,050 Era: n = 2,614			Case = report ≥1 trigger

Reference	Purpose	Study design	Population			Comments
				(where appropriate)		
			Eligible	Located	Enrolled (Response Rate)	
Ismail et al. 2002	Prevalence of psychiatric disorders in veterans with and without unexplained physical disability	(Two-phase) cohort study, medical assessment 1/1999-9/2000	Disabled GWVs: n = 406	166	111 (67%)	Report in text that 740 cohort members were eligible, but only 607 were contacted; does not explain why
			Nondisabled GWVs: n = 3,047	158	98 (62%)	
			Disabled Bosnia: n = 138	98	54 (55%)	
			Disabled Era: n = 278	184	79 (43%)	
Hotopf et al. 2003a	Describe changes in health	Cohort study	Eligible: all responders except 503 who refused future contact and 449 who did not complete relevant parts of questionnaire	GWVs: 1,472 Bosnia: 909 Era: 924	1089 (74.0%) 638 (70.2%) 643 (69.6%)	Report 8,196 reported to first survey (Unwin et al.: n = 8,195)
			Selected all women, all male veterans with fatigue score > 8, for GWVs 50% sample of those with scores 4-8, all 4-8s in Bosnia and Era, 1 in 8 sample of veterans with scores < 4			
Nisenbaum et al. 2004	Patterns of symptom reporting	Factor analysis	GWVs: n = 3,454 Bosnia: n = 1,979 Era: n = 2,577 US 1991 GWVs from four Air Force units: n = 1,163			US veterans from Fukuda et al. 1998

Reference	Purpose	Study design	Population (where appropriate)			Comments
			Eligible	Located	Enrolled (Response Rate)	
Everitt et al. 2002	Patterns of symptom reporting	Cluster analysis	500 veterans randomly selected from each group			Reports that there had been 3,529 responses in GWV group (as opposed to 3,531)
Hotopf et al. 2000	Ill health and vaccinations	Cross-sectional study	Servicemen who served in gulf: n = 3,284 Included those with vaccination records: n = 923			
David et al. 2002	Poor memory and concentration	Case-control study, medical evaluation – case definition Cutoff: 1st decile of distribution of SF36-PF subscale in Era cohort Recruited 12-18 months after phase 1	GWV: n = 406 Bosnia: n = 138 Era: n = 278 GWV well: n = 3,047	738 randomly selected and contacted	GWV ill: 111 (66.9%) Era ill: 78 (42.5%) Bosnia ill: 54 (56.8%) GWV well: 98 (62.4%)	Excluded three randomly selected eligibles with current serious physical illness

Reference	Eligible population	Type of study or methods	Date(s) of enrollment	Subgroup (n = eligible subjects)	Contacted or Located (% of eligible)	Responded or Enrolled (Response Rate)	Comments
Cherry et al. 2001a; Cherry et al. 2001b	UK military who served in gulf region 9/1990-6/1991, excluding special forces identified by MOD Comparison population—not deployed, in military 1/1/1991 No overlap with Unwin et al. population	Cross-sectional survey, Stratified by sex, age, service, rank; GWV stratum matched with randomly selected sample for NGVs, three stratified random samples; main and validation cohort selected; mail and personal visits depending on service group	1st site visits 12/1997, followup until 9/1999	Total	14,254	12,191 (85.5%); 11,914 usable questionnaires (83.6%)	Macfarlane et al. study of mortality on same group Analysis on 7,971 GWVs who reported having been in the gulf, useful questionnaires and contacted outside MOD medical assistance Assessed selection bias
				Gulf-Main	4,755	4,076 (85.7%)	
				Gulf-Validation	4,750	4,134 (87.0%)	
				Non-gulf	4,749	3,981 (83.8%)	

Reference	Eligible population	Type of study or methods	Date(s) of enrollment	Subgroup (n = eligible subjects)	Contacted or Located (% of eligible)	Responded or Enrolled (Response Rate)	Comments
Maconochie et al. 2003	UK military who served in gulf 8/1990-6/1991 excluding special services identified by MOD (n = 52,811) Comparison population – not deployed, appropriately fit, in military 1/1/1991 (n = 52,924)	Retrospective cohort study of reproductive outcomes Comparison group stratum matched on service, sex, age, serving status, rank	8/1998-3/2001	Total	105,735	44,087 (41.7%, 48.5% adjusted)	Adjusted response rate—accounts for undelivered Assessed selection bias
				GWV men	51,581	24,379 (47.3%, 53% adjusted)	
				GWV women	1,230	705 (57.3%, 72% adjusted)	
				NGV men	51,688	18,439 (35.7%, 42% adjusted)	
				NGV women	1,236	564 (45.6%, 60% adjusted)	

Derivatives from Maconochie et al. 2003: UK Gulf War veterans – Reproductive study

Reference	Purpose	Study design	Population Eligible	Located	Enrolled (Response Rate)	Comments
			(where appropriate)			
Doyle et al. 2004	Miscarriage, stillbirth, congenital malformation in offspring	Retrospective cohort study	As reported in Maconochie et al.			
Simmons et al. 2004	Incidence of self-reported adult ill health	Retrospective cohort study, comparison of deployed and nondeployed and among those who believe they have GWS	Men only: n = 42,818			

Reference	Eligible population	Type of study or methods	Date(s) of enrollment	Subgroup (n = eligible subjects)	Contacted or Located (% of eligible)	Responded or Enrolled (Response Rate)	Comments
Kelsall et al. 2004a	All Australian veterans served in gulf 8/2/1990-9/4/1991 Comparison group—randomly selected from Australian Defense Force personnel in	Postal questionnaire and comprehensive health assessment Comparison group frequency matched by	8/2000-4/2002	Total 4,795	4,604 recruitable (96.0%)	3,044 (66.1% of recruitable; 63.5% of eligible)	Assessed selection bias using telephone-survey-only results
				GWV	1,808	1,456 (80.5% of	

Reference	Eligible population	Type of study or methods	Date(s) of enrollment	Subgroup (n = eligible subjects)	Contacted or Located (% of eligible)	Responded or Enrolled (Response Rate)	Comments
	operational units at time but not deployed (n = 26,411)	service type, sex, age, rank		1,871	recruitable (96.6%)	recruitable; 77.8% of eligible); 1,414 with both instruments (78.2% of recruitable; 75.6% of eligible)	
				NGV 2,924	NGV 2,796 recruitable (95.6%)	1,588 (56.8% of recruitable; 54.3% of eligible); 1,411 with both instruments (50.5% of recruitable; 48.3% of eligible)	

Derivatives from Kelsall et al. 2004a: Australian cohort

			Population			
				(where appropriate)		
Reference	Purpose	Study design	Eligible	Located	Enrolled (Response Rate)	Comments
Forbes et al. 2004	Self-reported symptoms	Factor analysis	GWVs with complete data on 62 symptoms: n = 1,322 NGVs with complete data: n = 1,459			
McKenzie et al. 2004	Psychologic health, SF-12	Cross-sectional study	Male GWVs: n = 1,424; 1,374 with complete data Male NGVs: n = 1,548; 1,513 with complete data			
Ikin et al. 2004	Psychologic disorders and association with exposure to GW-related psychologic stressors	Cross-sectional study	Male GWVs: n = 1,424; 1,381 with psychologic health interview) Male NGVs: n = 1,548; 1,377 with psychologic health interview)			
Kelsall et al. 2004b	Respiratory health status, exposure to oil-fire smoke and dust storms	Cross-sectional study	Full cohort as described in Kelsall et al. 2004a			

Reference	Eligible population	Type of study or methods	Date(s) of enrollment	Subgroup (n = eligible subjects)	Contacted or Located (% of eligible)	Responded or Enrolled (Response Rate)	Comments
McCauley et al. 1999b	Veterans deployed to gulf 8/1/1990-7/31/1991 who listed Oregon or Washington as home state of record at deployment and currently reside in either state (n = 8,603), identified from ODSS database, DOD	Population-based case-control study Phase I: mail survey of randomly selected population, reservists oversampled, all women selected Phase II: case-control study	Mail survey: 11/1995-1/1998, this report is through 6/1997	Total 2,343 (by 6/1997 only mailed: 1651)	1,396 (84.6% of eligible mailed by 6/1997)	675 (48.4%; 40.9% of total eligible); 454 eligible for clinical study (32.5%; 27.5% of total eligible)	Report only on first 225 clinical examinations of potential cases Limited assessment of selection bias
				GWVs with unexplained illnesses (cases)	Potential cases of unexplained illnesses: 297	158 (53.2%)	
				Healthy GWVs (controls)	Potential controls: 130	67 (51.5%)	

Derivatives from McCauley et al. 1999b

		Population (where appropriate)			
Reference	Purpose	Study design	Eligible	Located	Enrolled (Response Rate)	Comments
---	---	---	---	---	---	---
Spencer et al. 1998	Differential exposures and persistent unexplained illness	Case-control study		1,084 returned survey by 6/1/1998, 567 (52.3%) agreed to participate Cases: n = 244 Controls: n = 113		Case and control numbers same as finals listed in Bourdette et al. 2001
Anger et al. 1999	Neurobehavioral deficits	Case-control study		Tested by 12/1996 Cases: n = 66 Controls: n = 35		Appears to be from before survey was completed
Binder et al. 1999	Subjective cognitive complaints, affective distress, objective cognitive performance among cases of unexplained illness	Cross-sectional study		Cases with psychologic complaints: n = 100		

87

Reference	Purpose	Study design	Population Eligible	Located	Enrolled (Response Rate)	Comments
Storzbach et al. 2000	Psychologic differences between veterans with and without GW unexplained symptoms	Case-control study	Cases: n = 241 Controls: n = 113			Number of cases reported as both 244 and 241
Storzbach et al. 2001	Neurobehavioral deficits, defined by ODTP	Case-control study	Cases: n = 239 with ODTP data available Controls: n = 112 with ODTP data available			
Binder et al. 2001	CFS and cognitive deficits on computerized cognitive testing	Case-control study	Cases: those who met revised definition of CFS and had AFQT data available: n = 32 Controls: with AFQT data available: n = 62			
Ford et al. 2001	PTSS and unexplained illness	Case-control study	Cases: n = 237 Controls: n = 112			
McCauley et al. 2002a	Chronic fatigue	Case-control study	Final numbers from Bourdette et al. 2001 Cases: n = 103 who met fatigue case definition (ICF n = 59, CFS-94 n = 44) Controls: n = 113			Analyzed subgroup of cases from main study, specifically looking at chronic fatigue

Reference	Eligible population	Type of study or methods	Date(s) of enrollment	Subgroup (n = eligible subjects)	Contacted or Located (% of eligible)	Responded or Enrolled (Response Rate)	Comments
Bourdette et al. 2001	As above	As above, including symptom factor analysis Case definition: ≥ 1 of (a) cognitive or psychologic changes including memory loss, confusion, inability to concentrate, mood swings, and/or somnolence; (b) gastrointestinal distress; (c) fatigue; (d) muscle and joint pain; (e) skin or mucous-	Mail survey: 11/1995-1/1998	Total 2,343 sample; 2,022 eligible GWVs with unexplained illnesses (cases)	1,760 (75.1% of total sample; 87.0% of total eligible) Potential: 336	1,119 (63.6%; 55.3% of total eligible) (799 eligible for clinical study, 443 examined) 244 met definition	Subject numbers differ for derivative studies: Binder et al. 2001: 801 eligible for clinical study Storzbach et al. 2000: 517 cases and 213 controls contacted by telephone

88

Reference	Eligible population	Type of study or methods	Date(s) of enrollment	Subgroup (n = eligible subjects)	Contacted or Located (% of eligible)	Responded or Enrolled (Response Rate)	Comments
		membrane lesions Symptoms began during or after deployment, persisted ≥1 month, occurred during 3 months before to recruitment		Healthy GWV (controls)	Potential: 107	113 met definition	Misclassified some cases and controls Limited assessment of selection bias—clinical sample not representative of nonresponders

Reference	Eligible population	Type of study or methods	Date(s) of enrollment	Subgroup (n = eligible subjects)	Contacted or Located (% of eligible)	Responded or Enrolled (Response Rate)	Comments
McCauley et al. 2002b	Active or reserve Army or National Guard deployed in gulf or on active duty but not deployed 1/1/1991-3/31/1991 and residents of Oregon, Washington, California, Georgia, or North Carolina Khamisiyah-exposed GWVs (n = 5,328) Non-Khamisiyah-exposed GWVs (n = 143,910) NDVs (n = 814,331)	Cohort study, telephone survey	1999	Total 3,219	2,918 (90.6%)	Final sample: 1,779 (61.0%; 55.3% of eligible)	Khamisiyah exposure defined as in 50-km radius of Khamisiyah 3/4/1991-3/13/1991 Enrolled=interviewed and met exposure definition No formal assessment of selection bias (previously reported on selection bias-(McCauley et al. 1999b)
				Khamisiyah-exposed GWVs 923	846 contacted	653 (77.2% of contacted; 70.7% of eligible)	
				Non-Khamisiyah-exposed GWVs 927	841 contacted	610 (72.5% of contacted; 65.8% of eligible)	
				NDVs 1,369	1,231 contacted	516 (41.9% of contacted; 37.7% of eligible)	

Reference	Eligible population	Type of study or methods	Date(s) of enrollment	Subgroup (n = eligible subjects)	Contacted or Located (% of eligible)	Responded or Enrolled (Response Rate)	Comments
Steele 2000	Kansas residents on active duty 8/1990-7/1991, separated or retired from military, or currently serving in reserve component (total n = 16,566; GWVs: n = 6,235; NGVs: n = 10,331)	Telephone interview, cross-sectional survey	2/1998-8/1998	Total 3,138	2,396 (76.3%) located, 2,211 (70.5%) invited	2,030 (91.8% of invited; 64.7% of total)	No assessment of selection bias
				GWVs	1,665 invited	1,548 (93% of invited)	
				NGVs	546 invited	482 (88% of invited)	

Reference	Eligible population	Type of study or methods	Date(s) of enrollment	Subgroup (n = eligible subjects)	Contacted or Located (% of eligible)	Responded or Enrolled (Response Rate)	Comments
Proctor et al. 1998	Devens cohort: US Army active, reserve and National Guard veterans followed since return to United States; initial survey spring 1991 (n = 2,949, or 60% of 4,915 per Wolfe et al. 1999b); Second survey in winter 1992/spring 1993 (n = 2,313); those who completed HSC eligible for this study (n = 2021) New Orleans cohort: active, reserve, and National Guard, US Army, Navy, Marine, and Air Force troops deployed to gulf (n = 928), initial survey within 9 months of return in 1991; those who completed	Cross-sectional study from larger cohorts followed longitudinally Stratified random sample from GWV cohorts designed to give equal representation of higher and lower symptom reporters; oversampled women Self-reported symptoms	Devens: spring 1994-fall 1996 New Orleans: summer 1994-fall 1995 Germany: spring 1995	Total 645	444 (68.8%)	343 (77.3%; 53.2% of eligible) participated 300 (67.6%; 46.5% of eligible) completed questionnaires 332 (74.8%; 51.5% of eligible) environmental interview 254 (57.2%; 39.4% of eligible) in-person neuropsychologic testing and psychologic diagnostic interview	Germany group not longitudinal, only studied at time 2. 300 who completed Devens questionnaire were analyzed in this study Assessed selection bias

Reference	Type of study or methods	Eligible population	Date(s) of enrollment	Subgroup (n = eligible subjects)	Contacted or Located (% of eligible)	Responded or Enrolled (Response Rate)	Comments
		HSC eligible for study (n = 818)		Devens 353	259 (73.3%)	220 (84.9% of located and contacted; 62.3% of eligible) 186 (71.8%; 52.7% of eligible) completed questionnaire, 213 (82.2%; 60.3% of eligible) environmental interview, 148 (57.1%; 41.9% of eligible) psychologic test	
	Germany-deployed cohort: Maine National Guard air ambulance unit sent to Germany during GW (12/1990-8/1991), personnel whose intended mission was handling and transport of wounded from GW			New Orleans 194	126 (64.9%)	73 (58% of located and contacted; 37.6% of eligible) 66 (52.4%; 34.0% of eligible) completed questionnaire, 71 (56.3%; 36.6% of eligible) environmental interview, 58 (46.0%; 29.9% of eligible) psychologic test	

Reference	Eligible population	Type of study or methods	Date(s) of enrollment	Subgroup (n = eligible subjects)	Contacted or Located (% of eligible)	Responded or Enrolled (Response Rate)	Comments
				Germany 98	59 (60.2%)	50 (84.7% of located and contacted; 51.0% of eligible) 48 (81.4%; 49.0% of eligible) completed questionnaire, environmental interview and psychologic test	

Derivatives from Proctor et al. 1998

Reference	Purpose	Study design	Population			Comments
			(where appropriate)			
			Eligible	Located	Enrolled (Response Rate)	
Wolfe et al. 1999a	Relationship of psychiatric status with health problems	Cross-sectional study	Complete data from diagnostic interviews and questionnaires Devens cohort: n = 148 New Orleans cohort: n = 56 Germany cohort: n = 48			
White et al. 2001	Neurobehavioral effects	Cross-sectional study	Devens = 142 New Orleans cohort: n = 51 Germany cohort: n = 47			
Proctor et al. 2001	Chemical sensitivity and chronic fatigue	Cross-sectional study	Devens cohort: n = 180 Germany cohort: n = 46			
Lindem et al. 2003b	Neuropsychologic performance, traumatic stress symptomatology and exposure to chemical and biologic-warfare agents	Cross-sectional study	Participants who completed all evaluations (including full neuropsychological and PTSD evaluations) Devens cohort: n = 141 New Orleans cohort: n = 37 Germany cohort: n = 47			

Reference	Purpose	Study design	Population (where appropriate)			Comments
			Eligible	Located	Enrolled (Response Rate)	
Lindem et al. 2003a	Neuropsychologic performance compared with neuropsychologic symptom reporting	Cross-sectional study	Devens cohort: n = 142 New Orleans cohort: n = 51 Germany cohort: n = 47			Does not describe how numbers were obtained
Studies using Devens time 1 and time 2 survey responders only						
Wolfe et al. 1999b	Course and predictors of PTSD	Longitudinal cohort study	Time 1- 1991: Devens survey responders: n = 2,949 (60.0% of 4,915) Time 2- 1993-1994: Devens survey: n = 2,313 (78.4% of 2,949)			Nonresponders at time 2 were more likely to be younger, member of minority group, and deployed from active duty, but there were no differences in PTSD rates, indicating a lack of selection bias at time 2
Wagner et al. 2000	PTSD impact on physical health	Cross-sectional study	Devens survey responders with complete data on PTSD (main dependent variable) on time 2 survey: n = 2,301			
Wolfe et al. 2002	Multisymptom illness	Cross-sectional study	Devens survey responders: n = 2,949 (contact information on 2,903 available (98.4%)) Responders to new questionnaire (3/1997-3/1998): n = 1,290 (full data n = 945 (73.3%))			

Reference	Eligible population	Type of study or methods	Date(s) of enrollment	Subgroup (n = eligible subjects)	Contacted or Located (% of eligible)	Responded or Enrolled (Response Rate)	Comments
Ishoy et al. 1999b	Danish Gulf War Study All Danish Gulf War veterans stationed in gulf 8/2/1990-12/31/1997, peacekeeping UN force, officers, noncommissioned officers, enlisted privates Comparison group—members of Danish Armed Forces employed according to contract who could have been but had not been deployed in gulf	Cross-sectional study with medical examinations Comparison group matched on sex, age, profession, randomly selected at end of examinations Enrolled by questionnaire, health examination	Enrollment comparison group in 1996, health examinations 2/1997-1/1998	GWVs: 821 NGVs: 400		686 (83.6%) 231 (57.8%)	No explanation for why 400 NGVs were selected as comparison group Limited assessment of selection bias (spot tests showed that most frequent reason for not participating was lack of time)

Derivatives from Ishoy et al. 1999b: Danish Peacekeeping Veterans

			Population (where appropriate)			
Reference	Purpose	Study design	Eligible	Located	Enrolled (Response Rate)	Comments
Ishoy et al. 1999a	Gastrointestinal symptoms attributable to physical, chemical, or biologic exposures	Cross-sectional study	GWVs: n = 686 NGVs: n = 257			26 more NGVs than reported in Ishoy et al. 1999b, sample size reported in abstract only, appears that only data on GWVs were analyzed
Suadicani et al. 1999	Determinants of long-term neuropsychologic symptoms	Cross-sectional study	GWVs: n = 686 (667 in tables) NGVs: n = 257			See above
Ishoy et al. 2001a	Male reproductive-health characteristics	Cross-sectional study, serum levels of reproductive hormones	Male GWVs: n = 661 Male NGVs: n = 215			
Ishoy et al. 2001b	Male sexual problems	Cross-sectional study	Male GWVs: n = 661 Male NGVs: n = 215			

Reference	Eligible population	Type of study or methods	Date(s) of enrollment	Subgroup (n = eligible subjects)	Contacted or Located (% of eligible)	Responded or Enrolled (Response Rate)	Comments
Haley et al. 1997b	Active and retired members of 24th Reserve Naval Mobile Construction Battalion, called to active duty in GW, living in Alabama, Georgia, Tennessee, South Carolina, North Carolina in 11/1994 (n = 606)	Cross-sectional survey, factor analysis	1994, 1995	606	429 located with address and telephone number (70.8%); approximately 350 contacted	249 (58.0% of located; 41.1% of eligible)	No assessment of selection bias

Derivatives from Haley et al. 1997b

Reference	Purpose	Study design	Population			Comments
			(where appropriate)			
			Eligible	Located	Enrolled (Response Rate)	
Haley and Kurt 1997	Self-reported exposure to neurotoxic chemical combinations and association with factor-analysis-defined syndrome	Cross-sectional study	GWVs: n = 249			
Haley et al. 2001	Structural equation modeling to test construct validity of case definition of GWS	Validation study	Original sample: n = 249 (as described in Haley et al. 1997b) Validation sample: GWVs living in north Texas, registered with Gulf War clinic of Dallas VA Medical Center and responders to advertisements in area (n = 335)			

Reference	Purpose	Study design	Population (where appropriate)			Comments
			Eligible	Located	Enrolled (Response Rate)	
Haley et al. 1997a	Neurologic function in cases and controls	Case-control study, controls of similar age, sex, education to syndrome 2	Cases: three strong syndromes, veterans with highest factor scores on factor 1 (impaired cognition), on factor 2 (confusion-ataxia), on factor 3 (arthro-myo-neuropathy): n = 23 (5 with syndrome 1, 13 with syndrome 2, 5 with syndrome 3) Deployed controls: GWVs who reported no serious health problems: n = 10 selected from 70 eligible Nondeployed controls: members of the battalion not deployed: n = 10 selected from 150 eligible			Not true nested case-control studies; selection of cases was, appropriately, from original cohort, but selection of controls was not; 10 of 20 controls were from 150 newly discovered members of battalion who had not been deployed; those 10 were not from original cohort and there is no indication that they were tested for "caseness"
Hom et al. 1997	Neuropsychologic correlates of GWS	Case-control study, controls of similar age, sex, education as cases	Cases: highest factor scores from six syndrome factors: n = 26 Controls: n = 20 (10 deployed who reported no serious health problems, 10 nondeployed)			Assume same population as prior studies, plus 3 new cases; no information on how selected
Haley et al. 1999	Association of low PON1 type Q arylesterase activity with neurologic symptom complexes	Case-control study	n = 45 of 46 in case-control study with blood samples available (missing 1 case)			Assume same population as prior studies, plus 3 new cases; no information on how selected
Roland et al. 2000	Vestibular dysfunction in GWS	Case-control study	Cases: n = 23 Controls: n = 20 As described in Haley et al. 1997a			
Haley et al. 2000b	Effect of basal ganglia injury on central dopamine activity	Case-control study	Cases and controls from Haley et al. 1997a Cases with "confusion ataxia": n = 12 Deployed controls: n = 8 Nondeployed controls: n = 7			

Reference	Purpose	Study design	Population (where appropriate)		Comments
			Eligible	Located Enrolled (Response Rate)	
Haley et al. 2000a	Brain abnormalities in GWS	Case-control study, subjects with proton (hydrogen 1) magnetic resonance spectroscopy	As described in Haley et al. 1997a Cases: n = 22 (from the 23, 1 excluded because of multiple myeloma) Deployed controls: n = 9 (from the 10, 1 declined) Nondeployed controls: n = 9 (from the 10, 1 declined)		
Haley et al. 2004	Measurements of abnormalities of autonomic nervous system	Case-control study	As described in Haley et al. 1997a Cases: n = 22 Deployed controls: n = 9 Nondeployed controls: n = 9 (17 controls and 21 cases in final analysis because of Holter availability and illness)		
Kurt 1998	GW illness and exposures to anticholinesterases	Review	Total: n = 249 23 cases, 10 deployed controls, 10 nondeployed controls		

Reference	Eligible population	Type of study or methods	Date(s) of enrollment	Subgroup (n = eligible subjects)	Contacted or Located (% of eligible)	Responded or Enrolled (Response Rate)	Comments
Fukuda et al. 1998	Index population: ANG unit in Lebanon, Pennsylvania Comparison populations: 3 Air Force populations: Unit A: ANG unit from Pennsylvania, similar demographics to index but different primary mission Unit B: US Air Force Reserve and Unit C: active-duty Air Force, similar missions to index, from Florida Any member on base was eligible	Cross-sectional survey, in-person interviews, factor analysis Clinical evaluation of GWVs from index unit (4–5/1995)	1–3/1995	Total Index Unit A Unit B Unit C	6,151 1,083 1,520 1,141 2,407	3,723 (60.5%) 667 (61.6%) 538 (35.4%) 838 (73.4%) 1,680 (69.8%)	Numbers deployed to gulf: index, 47%; A, 22%; B, 32%; C, 28%; Total, 31% Not clear who was on active duty during war Started out as cluster investigation in Lebanon "Located" is everyone on base when survey conducted No assessment of selection bias

Reference	Eligible population	Type of study or methods	Date(s) of enrollment	Subgroup (n = eligible subjects)	Contacted or Located (% of eligible)	Responded or Enrolled (Response Rate)	Comments
Gray et al. 2002	All regular and Reserve Navy personnel who had served on active duty in Seabee commands for ≥30 days 8/1/1990-7/31/1991 (n = 18,945)	Cross-sectional survey	5/1997-7/1/1999	Total 18,945	17,599 (92.9%)	12,049 (68.6%; 63.6% of total eligible); 11,868 actually completed (67.4%; 62.6% of total eligible)	No information on response rates by subgroup Assessed selection bias (telephone survey)
				GW Seabees		3,831	
				Deployed elsewhere		4,933	
				Nondeployed		3,104	

Reference	Eligible population	Type of study or methods	Date(s) of enrollment	Subgroup (n = eligible subjects)	Contacted or Located (% of eligible)	Responded or Enrolled (Response Rate)	Comments
Gray et al. 1999a	Active-duty Seabees in Navy after war and serving in one of two large Seabee centers (Port Hueneme, California and Gulfport, Mississippi)	Cross-sectional survey	Clinical evaluations late 1994, early 1995	Total		1,497 (53%)	Assessed selection bias
				GWVs		527	
				NDVs		970	

Derivatives from Gray et al. 1999a: Seabee study

Reference	Purpose	Study design	Population			Comments
			Eligible	Located	Enrolled (Response Rate)	
			(where appropriate)			
Knoke et al. 2000	Self-reported symptoms to identify syndrome	Factor analysis	Same population as Gray et al. 1999a GWVs: n = 528 NDVs: n = 968			Sample size not exactly same as Gray et al. 1999a

97

Reference	Eligible population	Type of study or methods	Date(s) of enrollment	Subgroup (n = eligible subjects)	Contacted or Located (% of eligible)	Responded or Enrolled (Response Rate)	Comments
Gray et al. 1996	Active-duty Army, Navy (including Marine Corps), or Air Force deployed in GW ≥1 day 8/8/1990-7/31/1991 (n = 579,931) Comparison group of randomly selected active-duty military personnel on rosters as of 9/30/1990 but not in PG before 7/31/1991 (n = 700,000)	Retrospective cohort study, hospitalization experience obtained from computerized hospitalization records of DOD	4 years of study before war: 10/1/1988-7/31/1990; 1991: 8/1/1991-12/31/1991; 1992: 1-12/1992; 1993: 1-9/30/1993	Numbers for each period = those on whom data complete and on active duty on 1st day of period Before war 1,279,931 1991 1,165,411 1992 1,075,430 1993 839,389			Selection bias not an issue (hospitalization records)

Derivatives from Gray et al. 1996: Hospitalization study

Reference	Purpose	Study design	Population				Comments
			(where appropriate)				
			Eligible	Located	Enrolled (Response Rate)		
Knoke et al. 1998	Testicular cancer	Retrospective cohort study, followup through 4/1/1996 (reported by 4/1/1997)	Males on active duty GWVs: n = 517,223 NDVs: n = 1,291,323				Expanded study period from Gray et al. 1996
Gray et al. 1999b	Hospitalization of GWVs possibly exposed to Khamisiyah	Retrospective cohort study, followup 3/1991-9/1995	Army veterans regular and reserve, deployed in GW 3/10-3/13/1991: n = 349,291 Not exposed: n = 224,804 Uncertain low dose: n = 75,717 Specific estimated subclinical exposure—three levels: n = 48,770				Expanded study period from Gray et al. 1996

		Population *(where appropriate)*				
Reference	Purpose	Study design	Eligible	Located	Enrolled (Response Rate)	Comments
Smith et al. 2003	Khamisiyah, postwar hospitalizations revisited	Retrospective cohort study, followup 3/10/1991-12/31/2000	Regular active-duty and reserve Army and Air Force personnel in theater in March 1991: n = 431,762 Active duty: n = 333,382; Reserve and National Guard: n = 84,690 Not exposed: n = 318,458 Possibly exposed: n = 99,614	Demographic and exposure data available on 418,072 (96.8%)		Expanded study period from Gray et al. 1996; hospitalization info limited to active-duty, so Reserves and National Guard only in study until 6/10/1991 (estimated return to civilian operations)

Reference	Eligible population	Type of study or methods	Date(s) of enrollment	Subgroup (n = eligible subjects)	Contacted or Located (% of eligible)	Responded or Enrolled (Response Rate)	Comments
Knoke and Gray 1998	All active-duty Army, Air Force, Navy, Marine Corps, Coast Guard deployed to GW ≥1 day 8/8/1990-7/31/1991 or not deployed but on active duty for at least part of GW period and remained on active duty at end of period	Retrospective cohort study, admissions (after GW deployment period and before 4/1/1996) to US military hospitals worldwide reported to DOD computerized database	Reported by 10/1/1996	Deployed 552,111 Nondeployed 1,479,751		1st hospitalizations: 6,672 1st hospitalizations: 18,823	Selection bias not an issue (hospitalization records)

Reference	Eligible population	Type of study or methods	Date(s) of enrollment	Subgroup (n = eligible subjects)	Contacted or Located (% of eligible)	Responded or Enrolled (Response Rate)	Comments
Dlugosz et al. 1999	All regular, active-duty Army, Navy, Marine Corps, Air Force personnel who served ≥1 month active duty 8/2/1990-7/31/1991 (n = 1,984,996)	Retrospective cohort study, hospitalizations for mental disorders, reported to DOD (n = 30,539), deployed =1+ days 8/2/1990-7/31/1991; followup 6/1/1991-9/30/1993	6/1/1991-9/30/1993	30,539			Selection bias not an issue (hospitalization records)

Reference	Eligible population	Type of study or methods	Date(s) of enrollment	Subgroup (n = eligible subjects)	Contacted or Located (% of eligible)	Responded or Enrolled (Response Rate)	Comments
Sutker et al. 1995b	Troops assigned to Marine, Air Force, Navy, Army Reserve, and National Guard units deployed to combat who underwent psychologic debriefing within 1 year of return (n = 1,423), restricted to those who saw war-zone assignment (n = 808)	Discriminant-function analysis	Within 1 year of return from GW	GWVs with appropriate assessment instruments: 775 PTSD cases No distress		97 484	No explanation of how study subjects were selected No assessment of selection bias

Derivatives from Sutker et al. 1995b and other small studies by same research group

			Population (where appropriate)		
Reference	Purpose	Study design	Eligible	Located	Enrolled (Response Rate)	Comments
Benotsch et al. 2000	Longitudinal relationship between resources and emotional distress	Longitudinal study	GWVs: n = 348			
Brailey et al. 1998	Characterizations of war-zone stressors; documentation of acute and psychologic sequelae to war-zone participation; identification of risk factors predictive of psychologic symptoms	Longitudinal study	GWVs: n = 349			
Sutker et al. 1995a	Ethnicity and sex comparisons in assessment of psychologic distress	Cross-sectional study	Total: 912 of 1,423 described in Benotsch et al. 2000 (excluded on basis of completion of instruments and measure of intelligence) Wartime deployed: n = 653 Stateside duty: n = 259			
Thompson et al. 2004	Early symptom predictors of chronic distress in GWVs	Longitudinal study	GWVs: n = 348 described in Benotsch et al. 2000			

Reference	Purpose	Study design	Population		Comments
			\multicolumn{2}{c	}{*(where appropriate)*}	
			Eligible Located	Enrolled (Response Rate)	
Sutker et al. 1993	War-zone trauma and stress-related symptoms	Cross-sectional study		Participants drawn from five National Guard and Army Reserve units deployed in GW and assessed by invitation of Louisiana Army and ANG and US Army Reserve as part of evaluation and debriefing program: n = 306; 215 with complete information Total responses for analysis: n = 275 (118 of 215 reported up to three of most stressful conditions or events)	
Sutker et al. 1994a	Psychologic symptoms and psychiatric diagnoses in GW troops serving graves registration duty	Clinical report		Army reservists served on war-zone graves registration duty: n = 24 (of 35-member company)	
Sutker et al. 1994b	Pyschopathology in war-zone-deployed and -nondeployed GWVs assigned graves registration duty	Clinical report		Army reservists assigned to three quartermaster companies: n = 207 eligible; n = 124 completed assessment and debriefing exercises; n = 63 selected randomly from 207; n = 60 in final sample (3 excluded because of inconsistent responses); of 60, 40 deployed, 20 stateside	
Sutker et al. 2002	Exposure to war trauma, war-related PTSD and psychologic impact of Hurricane Andrew	Cross-sectional study		Recruited from Marine, Air Force, Army Reserve, National Guard, and Coast Guard units in six southeastern Louisiana parishes; activated in GW (66% saw GW duty); resided in southern Louisiana during 1992 Hurricane Andrew (n = 312)	
Vasterling et al. 1997	Relationship of intellectual resources to PTSD	Cross-sectional study		Volunteers recruited from enrollment lists of local military units that had mobilized for GW (n = 95); 41 met study eligibility criteria and completed full assessment For this study, 18 PTSD diagnosed and 23 psychopathology-free veterans selected	
Vasterling et al. 1998	Attention and memory dysfunction in PTSD	Cross-sectional study		As described in Vasterling et al. 1997, but 43 met study criteria	
Vasterling et al. 2003	Olfactory functioning in GW-era veterans, relationships to war-zone duty, hazards exposures, psychologic distress	Cross-sectional study		Members of south Louisiana National Guard and military reservist units activated for GW duty: n = 105; group recruited as one arm of population described by Brailey et al. 1998: 844 deployed, 326 nondeployed, longitudinal study; n = 319 of these underwent olfactory testing Deployed to war zone: n = 72 Not deployed to war zone: n = 33	

Reference	Eligible population	Type of study or methods	Date(s) of enrollment	Subgroup (n = eligible subjects)	Contacted or Located (% of eligible)	Responded or Enrolled (Response Rate)	Comments
Southwick et al. 1993	Members of two National Guard reserve units, medical company and military police company (n = 240)	Longitudinal study; interviews at two times, eligible if completed psychosocial evaluations	1 month and 6 months after return from gulf	GWVs 160 at 1st meeting (66.7% of total number)	119 completed questionnaire at 1st meeting (74.4%)	84 completed questionnaires 1 month and 6 months after return (70.6%; 52.5% of total at 1st meeting)	No formal assessment of selection bias Investigated where nonresponders were, but no other information was provided

Reference	Eligible population	Type of study or methods	Date(s) of enrollment	Subgroup (n = eligible subjects)	Contacted or Located (% of eligible)	Responded or Enrolled (Response Rate)	Comments
Stretch et al. 1995	Active-duty and reserve personnel assigned to all Army, Navy, Air Force and Marine units in Hawaii and Pennsylvania (n = 16,167)	Cross-sectional survey distributed to units; Hawaii National Guard did not distribute (n = 2,000)	Enrollment time not available	Total 16,167 Deployed Nondeployed	14,167 (87.6%)	4,334 (30.6%; 26.8% of eligible) 1,739 (1,524 to gulf, others elsewhere) 2,512	No formal assessment of selection bias (speculated on reasons for nonresponses)

Derivatives from Stretch et al. 1995

Reference	Purpose	Study design	Population (where appropriate)			Comments
			Eligible	Located	Enrolled (Response Rate)	
Stretch et al. 1996a	Psychologic health	Cross-sectional survey	Population in Stretch et al. 1995			
Stretch et al. 1996b	PTSD symptoms	Cross-sectional survey	Population in Stretch et al. 1995			

Reference	Eligible population	Type of study or methods	Date(s) of enrollment	Subgroup (n = eligible subjects)	Contacted or Located (% of eligible)	Responded or Enrolled (Response Rate)	Comments
Kang and Bullman 1996	Gulf War veterans (n = 695,516) and other veterans (n = 746,291) identified by Defense Manpower Data Center	Retrospective mortality study	Followup through 1993				Also compared with general population

Reference	Eligible population	Type of study or methods	Date(s) of enrollment	Subgroup (n = eligible subjects)	Contacted or Located (% of eligible)	Responded or Enrolled (Response Rate)	Comments
Kang and Bullman 2001	GWVs who arrived in the gulf before March 1, 1991 (n = 621,902) and NGVs (n = 746,248)	Retrospective mortality study	Followup began at exit from PG for GWVs and from 5/1/1991 for comparison-group veterans Followup ended 12/31/1997				

Reference	Eligible population	Type of study or methods	Date(s) of enrollment	Subgroup (n = eligible subjects)	Contacted or Located (% of eligible)	Responded or Enrolled (Response Rate)	Comments
Bullman et al. 2005	GWVs deployed to PG— 8/1990-3/1991 (n = 351,041)	Retrospective mortality study; exposure to Khamisiyah defined by 2000 plume model	Followup through 2000	100,487 considered exposed to Khamisiyah plume			

NOTE: The committee reported in this table the figures given in each study publication, *except response rates*. For uniformity, the committee calculated response rates as the number of study participants divided by the number of participants who were located (rather than the number of eligible participants). AFQT = Armed Forces Qualification Test; ANG = Air National Guard; CFS = chronic fatigue syndrome; DOD = Department of Defense; GW = Gulf War;

GWS = Gulf War syndrome; GWV = Gulf War veteran; HSC = Health Symptom Checklist; ICF = idiopathic chronic fatigue; MCS = multiple chemical sensitivity; MOD = Ministry of Defence (UK); NDV = nondeployed veteran; NGV = non-Gulf War veteran; ODSS = Operation Desert Shield / Operation Desert Storm; ODTP = Oregon Dual Task Procedures; PG = Persian Gulf; PTSD = posttraumatic stress disorder; PTSS = posttraumatic stress symptomatology; SF-12 = Medical Outcome Study Short Form 12; SF-36 = Medical Outcome Study Short Form 36; SF36-PF = Medical Outcome Study Short Form 36 Physical Functioning; UN = United Nations.

REFERENCES

Amato AA, McVey A, Cha C, Matthews EC, Jackson CE, Kleingunther R, Worley L, Cornman E, Kagan-Hallet K. 1997. Evaluation of neuromuscular symptoms in veterans of the Persian Gulf War. *Neurology* 48(1):4-12.

Anger WK, Storzbach D, Binder LM, et al. 1999. Neurobehavioral deficits in Persian Gulf veterans: Evidence from a population-based study. *Journal of the International Neuropsychological Society* 5(3):203-212.

Axelrod BN, Milner IB. 1997. Neuropsychological findings in a sample of Operation Desert Storm veterans. *Journal of Neuropsychiatry and Clinical Neurosciences* 9(1):23-28.

Barrett DH, Doebbeling CC, Schwartz DA, Voelker MD, Falter KH, Woolson RF, Doebbeling BN. 2002. Posttraumatic stress disorder and self-reported physical health status among U.S. Military personnel serving during the Gulf War period: A population-based study. *Psychosomatics* 43(3):195-205.

Benotsch EG, Brailey K, Vasterling JJ, Uddo M, Constans JI, Sutker PB. 2000. War zone stress, personal and environmental resources, and PTSD symptoms in Gulf War veterans: A longitudinal perspective. *Journal of Abnormal Psychology* 109(2):205-213.

Binder LM, Storzbach D, Anger WK, Campbell KA, Rohlman DS. 1999. Subjective cognitive complaints, affective distress, and objective cognitive performance in Persian Gulf War veterans. *Archives of Clinical Neuropsychology* 14(6):531-536.

Binder LM, Storzbach D, Campbell KA, Rohlman DS, Anger WK, Members of the Portland Environmental Hazards Research Center. 2001. Neurobehavioral deficits associated with chronic fatigue syndrome in veterans with Gulf War unexplained illnesses. *Journal of the International Neuropsychological Society* 7(7):835-839.

Black DW, Doebbeling BN, Voelker MD, Clarke WR, Woolson RF, Barrett DH, Schwartz DA. 1999. Quality of life and health-services utilization in a population-based sample of military personnel reporting multiple chemical sensitivities. *Journal of Occupational and Environmental Medicine* 41(10):928-933.

Black DW, Doebbeling BN, Voelker MD, Clarke WR, Woolson RF, Barrett DH, Schwartz DA. 2000. Multiple chemical sensitivity syndrome: Symptom prevalence and risk factors in a military population. *Archives of Internal Medicine* 160(8):1169-1176.

Black DW, Carney CP, Forman-Hoffman VL, Letuchy E, Peloso P, Woolson RF, Doebbeling BN. 2004a. Depression in veterans of the first gulf war and comparable military controls. *Annals of Clinical Psychiatry* 16(2):53-61.

Black DW, Carney CP, Peloso PM, Woolson RF, Schwartz DA, Voelker MD, Barrett DH, Doebbeling BN. 2004b. Gulf War veterans with anxiety: Prevalence, comorbidity, and risk factor. *Epidemiology* 15(2):135-142.

Bourdette DN, McCauley LA, Barkhuizen A, Johnston W, Wynn M, Joos SK, Storzbach D, Shuell T, Sticker D. 2001. Symptom factor analysis, clinical findings, and functional status in a population-based case control study of Gulf War unexplained illness. *Journal of Occupational and Environmental Medicine* 43(12):1026-1040.

Brailey K, Vasterling JJ, Sutker PB. 1998. Psychological Aftermath of Participation in the Persian Gulf War. Lundberg A, Editor. *The Environment and Mental Health: A Guide for Clinicians*. London: Lawrence Erlbaum Associates. Pp. 83-101.

Bullman TA, Mahan CM, Kang HK, Page WF. 2005. Mortality in US Army Gulf War Veterans Exposed to 1991 Khamisiyah Chemical Munitions Destruction. *American Journal of Public Health* 95(8):1382-1388.

Carney CP, Sampson TR, Voelker M, Woolson R, Thorne P, Doebbeling BN. 2003. Women in the Gulf War: Combat experience, exposures, and subsequent health care use. *Military Medicine* 168(8):654-661.

Chalder T, Hotopf M, Unwin C, Hull L, Ismail K, David A, Wessely S. 2001. Prevalence of Gulf war veterans who believe they have Gulf war syndrome: Questionnaire study. *British Medical Journal* 323(7311):473-476.

Cherry N, Creed F, Silman A, Dunn G, Baxter D, Smedley J, Taylor S, Macfarlane GJ. 2001a. Health and exposures of United Kingdom Gulf war veterans. Part I: The pattern and extent of ill health. *Occupational and Environmental Medicine* 58(5):291-298.

Cherry N, Creed F, Silman A, Dunn G, Baxter D, Smedley J, Taylor S, Macfarlane GJ. 2001b. Health and exposures of United Kingdom Gulf war veterans. Part II: The relation of health to exposure. *Occupational and Environmental Medicine* 58(5):299-306.

David AS, Farrin L, Hull L, Unwin C, Wessely S, Wykes T. 2002. Cognitive functioning and disturbances of mood in UK veterans of the Persian Gulf War: A comparative study. *Psychological Medicine* 32(8):1357-1370.

Davis LE, Eisen SA, Murphy FM, Alpern R, Parks BJ, Blanchard M, Reda DJ, King MK, Mithen FA, Kang HK. 2004. Clinical and laboratory assessment of distal peripheral nerves in Gulf War veterans and spouses. *Neurology* 63(6):1070-1077.

Dlugosz LJ, Hocter WJ, Kaiser KS, Knoke JD, Heller JM, Hamid NA, Reed RJ, Kendler KS, Gray GC. 1999. Risk factors for mental disorder hospitalization after the Persian Gulf War: U.S. Armed Forces, June 1, 1991-September 30, 1993. *Journal of Clinical Epidemiology* 52(12):1267-1278.

Doebbeling BN, Clarke WR, Watson D, Torner JC, Woolson RF, Voelker MD, Barrett DH, Schwartz DA. 2000. Is there a Persian Gulf War syndrome? Evidence from a large population-based survey of veterans and nondeployed controls. *American Journal of Medicine* 108(9):695-704.

Doyle P, Maconochie N, Davies G, Maconochie I, Pelerin M, Prior S, Lewis S. 2004. Miscarriage, stillbirth and congenital malformation in the offspring of UK veterans of the first Gulf war. *International Journal of Epidemiology* 33(1):74-86.

Eisen SA, Kang HK, Murphy FM, Blanchard MS, Reda DJ, Henderson WG, Toomey R, Jackson LW, Alpern R, Parks BJ, Klimas N, Hall C, Pak HS, Hunter J, Karlinsky J, Battistone MJ, Lyons MJ. 2005. Gulf War veterans' health: Medical evaluation of a US cohort. *Annals of Internal Medicine* 142(11):881-890.

Everitt B, Ismail K, David AS, Wessely S. 2002. Searching for a Gulf War syndrome using cluster analysis. *Psychological Medicine* 32(8):1371-1378.

Forbes AB, McKenzie DP, Mackinnon AJ, Kelsall HL, McFarlane AC, Ikin JF, Glass DC, Sim MR. 2004. The health of Australian veterans of the 1991 Gulf War: Factor analysis of self-reported symptoms. *Occupational and Environmental Medicine* 61(12):1014-1020.

Ford JD, Campbell KA, Storzbach D, Binder LM, Anger WK, Rohlman DS. 2001. Posttraumatic stress symptomatology is associated with unexplained illness attributed to Persian Gulf War military service. *Psychosomatic Medicine* 63(5):842-849.

Fukuda K, Straus SE, Hickie I, Sharpe MC, Dobbins JG, Komaroff A. 1994. The chronic fatigue syndrome: A comprehensive approach to its definition and study. International Chronic Fatigue Syndrome Study Group. *Annals of Internal Medicine* 121(12):953-959.

Fukuda K, Nisenbaum R, Stewart G, Thompson WW, Robin L, Washko RM, Noah DL, Barrett DH, Randall B, Herwaldt BL, Mawle AC, Reeves WC. 1998. Chronic multisymptom illness affecting Air Force veterans of the Gulf War. *Journal of the American Medical Association* 280(11):981-988.

Goldstein G, Beers SR, Morrow LA, Shemansky WJ, Steinhauer SR. 1996. A preliminary neuropsychological study of Persian Gulf veterans. *Journal of the International Neuropsychological Society* 2(4):368-371.

Goss Gilroy Inc. 1998. *Health Study of Canadian Forces Personnel Involved in the 1991 Conflict in the Persian Gulf.* Ottawa, Canada: Goss Gilroy Inc. Department of National Defence.

Gray GC, Coate BD, Anderson CM, Kang HK, Berg SW, Wignall FS, Knoke JD, Barrett-Connor E. 1996. The postwar hospitalization experience of US veterans of the Persian Gulf War. *New England Journal of Medicine* 335(20):1505-1513.

Gray GC, Kaiser KS, Hawksworth AW, Hall FW, Barrett-Connor E. 1999a. Increased postwar symptoms and psychological morbidity among US Navy Gulf War veterans. *American Journal of Tropical Medicine and Hygiene* 60(5):758-766.

Gray GC, Smith TC, Knoke JD, Heller JM. 1999b. The postwar hospitalization experience of Gulf War Veterans possibly exposed to chemical munitions destruction at Khamisiyah, Iraq. *American Journal of Epidemiology* 150(5):532-540.

Gray GC, Reed RJ, Kaiser KS, Smith TC, Gastanaga VM. 2002. Self-reported symptoms and medical conditions among 11,868 Gulf War-era veterans: The Seabee Health Study. *American Journal of Epidemiology* 155(11):1033-1044.

Haley RW, Kurt TL. 1997. Self-reported exposure to neurotoxic chemical combinations in the Gulf War. A cross-sectional epidemiologic study. *Journal of the American Medical Association* 277(3):231-237.

Haley RW, Hom J, Roland PS, Bryan WW, Van Ness PC, Bonte FJ, Devous MD Sr, Mathews D, Fleckenstein JL, Wians FH Jr, Wolfe GI, Kurt TL. 1997a. Evaluation of neurologic function in Gulf War veterans. A blinded case-control study. *Journal of the American Medical Association* 277(3):223-230.

Haley RW, Kurt TL, Hom J. 1997b. Is there a Gulf War Syndrome? Searching for syndromes by factor analysis of symptoms. *Journal of the American Medical Association* 277(3):215-222.

Haley RW, Billecke S, La Du BN. 1999. Association of low PON1 type Q (type A) arylesterase activity with neurologic symptom complexes in Gulf War veterans. *Toxicology and Applied Pharmacology* 157(3):227-233.

Haley RW, Marshall WW, McDonald GG, Daugherty MA, Petty F, Fleckenstein JL. 2000a. Brain abnormalities in Gulf War syndrome: Evaluation with 1H MR spectroscopy. *Radiology* 215(3):807-817.

Haley RW, Fleckenstein JL, Marshall WW, McDonald GG, Kramer GL, Petty F. 2000b. Effect of basal ganglia injury on central dopamine activity in Gulf War syndrome: Correlation of proton magnetic resonance spectroscopy and plasma homovanillic acid levels. *Archives of Neurology* 57(9):1280-1285.

Haley RW, Luk GD, Petty F. 2001. Use of structural equation modeling to test the construct validity of a case definition of Gulf War syndrome: Invariance over developmental and validation samples, service branches and publicity. *Psychiatry Research* 102(2):175-200.

Haley RW, Vongpatanasin W, Wolfe GI, Bryan WW, Armitage R, Hoffmann RF, Petty F, Callahan TS, Charuvastra E, Shell WE, Marshall WW, Victor RG. 2004. Blunted circadian variation in autonomic regulation of sinus node function in veterans with Gulf War syndrome. *American Journal of Medicine* 117(7):469-478.

Hom J, Haley RW, Kurt TL. 1997. Neuropsychological correlates of Gulf War syndrome. *Archives of Clinical Neuropsychology* 12(6):531-544.

Hotopf M, David A, Hull L, Ismail K, Unwin C, Wessely S. 2000. Role of vaccinations as risk factors for ill health in veterans of the Gulf war: Cross sectional study. *British Medical Journal* 320(7246):1363-1367.

Hotopf M, David AS, Hull L, Nikalaou V, Unwin C, Wessely S. 2003a. Gulf war illness — Better, worse, or just the same? A cohort study. *British Medical Journal* 327(7428):1370-1372.

Hotopf M, Mackness MI, Nikolaou V, Collier DA, Curtis C, David A, Durrington P, Hull L, Ismail K, Peakman M, Unwin C, Wessely S, Mackness B. 2003b. Paraoxonase in Persian Gulf War veterans. *Journal of Occupational and Environmental Medicine* 45(7):668-675.

Ikin JF, Sim MR, Creamer MC, Forbes AB, McKenzie DP, Kelsall HL, Glass DC, McFarlane AC, Abramson MJ, Ittak P, Dwyer T, Blizzard L, Delaney KR, Horsley KWA, Harrex WK, Schwarz H. 2004. War-related psychological stressors and risk of psychological disorders in Australian veterans of the 1991 Gulf War. *British Journal of Psychiatry* 185:116-126.

Ikin JF, McKenzie DP, Creamer MC, et al. 2005. War zone stress without direct combat: The Australian naval experience of the Gulf War. *Journal of Traumatic Stress* 18(3):193-204.

IOM (Institute of Medicine). 1999a. *Gulf War Veterans: Measuring Health*. Washington, DC: National Academy Press.

IOM. 1999b. *Strategies to Protect the Health of Deployed US Forces: Medical Surveillance, Record Keeping, and Risk Reduction*. Washington, DC: National Academy Press.

Iowa Persian Gulf Study Group. 1997. Self-reported illness and health status among Gulf War veterans: A population-based study. *Journal of the American Medical Association* 277(3):238-245.

Ishoy T, Suadicani P, Guldager B, Appleyard M, Gyntelberg F. 1999a. Risk factors for gastrointestinal symptoms. The Danish Gulf War Study. *Danish Medical Bulletin* 46(5):420-423.

Ishoy T, Suadicani P, Guldager B, Appleyard M, Hein HO, Gyntelberg F. 1999b. State of health after deployment in the Persian Gulf. The Danish Gulf War Study. *Danish Medical Bulletin* 46(5):416-419.

Ishoy T, Andersson AM, Suadicani P, Guldager B, Appleyard M, Gyntelberg F, Skakkebaek NE, Danish Gulf War Study. 2001a. Major reproductive health characteristics in male Gulf War Veterans. The Danish Gulf War Study. *Danish Medical Bulletin* 48(1):29-32.

Ishoy T, Suadicani P, Andersson A-M, Guldager B, Appleyard M, Skakkebaek N, Gyntelberg F. 2001b. Prevalence of male sexual problems in the Danish Gulf War Study. *Scandinavian Journal of Sexology* 4(1):43-55.

Ismail K, Everitt B, Blatchley N, Hull L, Unwin C, David A, Wessely S. 1999. Is there a Gulf War syndrome? *Lancet* 353(9148):179-182.

Ismail K, Kent K, Brugha T, Hotopf M, Hull L, Seed P, Palmer I, Reid S, Unwin C, David AS, Wessely S. 2002. The mental health of UK Gulf war veterans: Phase 2 of a two phase cohort study. *British Medical Journal* 325(7364):576.

Jamal GA, Hansen S, Apartopoulos F, Peden A. 1996. The "Gulf War syndrome". Is there evidence of dysfunction in the nervous system? *Journal of Neurology, Neurosurgery and Psychiatry* 60(4):449-451.

Joseph SC. 1997. A comprehensive clinical evaluation of 20,000 Persian Gulf War veterans. *Military Medicine* 162(3):149-155.

Kang HK, Bullman TA. 1996. Mortality among US veterans of the Persian Gulf War. *New England Journal of Medicine* 335(20):1498-1504.

Kang HK, Mahan CM, Lee KY, Magee CA, Murphy FM. 2000. Illnesses among United States veterans of the Gulf War: A population-based survey of 30,000 veterans. *Journal of Occupational and Environmental Medicine* 42(5):491-501.

Kang HK, Bullman TA. 2001. Mortality among US veterans of the Persian Gulf War: 7-year follow-up. *American Journal of Epidemiology* 154(5):399-405.

Kang HK, Mahan CM, Lee KY, Murphy FM, Simmens SJ, Young HA, Levine PH. 2002. Evidence for a deployment-related Gulf War syndrome by factor analysis. *Archives of Environmental Health* 57(1):61-68.

Kang HK, Natelson BH, Mahan CM, Lee KY, Murphy FM. 2003. Post-traumatic stress disorder and chronic fatigue syndrome-like illness among Gulf War veterans: A population-based survey of 30,000 veterans. *American Journal of Epidemiology* 157(2):141-148.

Karlinsky JB, Blanchard M, Alpern R, Eisen SA, Kang H, Murphy FM, Reda DJ. 2004. Late prevalence of respiratory symptoms and pulmonary function abnormalities in Gulf War I Veterans. *Archives of Internal Medicine* 164(22):2488-2491.

Kelsall HL, Sim MR, Forbes AB, Glass DC, McKenzie DP, Ikin JF, Abramson MJ, Blizzard L, Ittak P. 2004a. Symptoms and medical conditions in Australian veterans of the 1991 Gulf War: Relation to immunisations and other Gulf War exposures. *Occupational and Environmental Medicine* 61(12):1006-1013.

Kelsall HL, Sim MR, Forbes AB, McKenzie DP, Glass DC, Ikin JF, Ittak P, Abramson MJ. 2004b. Respiratory health status of Australian veterans of the 1991 Gulf War and the effects of exposure to oil fire smoke and dust storms. *Thorax* 59(10):897-903.

Knoke JD, Gray GC. 1998. Hospitalizations for unexplained illnesses among US veterans of the Persian Gulf War. *Emerging Infectious Diseases* 4(2):211-219.

Knoke JD, Gray GC, Garland FC. 1998. Testicular cancer and Persian Gulf War service. *Epidemiology* 9(6):648-653.

Knoke JD, Smith TC, Gray GC, Kaiser KS, Hawksworth AW. 2000. Factor analysis of self-reported symptoms: Does it identify a Gulf War syndrome? *American Journal of Epidemiology* 152(4):379-388.

Kurt TL. 1998. Epidemiological association in US veterans between Gulf War illness and exposures to anticholinesterases. *Toxicology Letters* 102-103:523-526.

Lange JL, Schwartz DA, Doebbeling BN, Heller JM, Thorne PS. 2002. Exposures to the Kuwait oil fires and their association with asthma and bronchitis among gulf war veterans. *Environmental Health Perspectives* 110(11):1141-1146.

Lindem K, Proctor SP, Heeren T, Krengel M, Vasterling J, Sutker PB, Wolfe J, Keane TM, White RF. 2003a. Neuropsychological performance in Gulf War era veterans: Neuropsychological symptom reporting. *Journal of Psychopathology and Behavioral Assessment* 25(2):121-127.

Lindem K, Heeren T, White RF, Proctor SP, Krengel M, Vasterling J, Sutker PB, Wolfe J, Keane TM. 2003b. Neuropsychological performance in Gulf War era veterans: Traumatic stress symptomatology and exposure to chemical-biological warfare agents. *Journal of Psychopathology and Behavioral Assessment* 25(2):105-119.

Maconochie N, Doyle P, Davies G, Lewis S, Pelerin M, Prior S, Sampson P. 2003. The study of reproductive outcome and the health of offspring of UK veterans of the Gulf war: Methods and description of the study population. *BMC Public Health* 3(1):4.

McCauley LA, Joos SK, Spencer PS, Lasarev M, Shuell T. 1999a. Strategies to assess validity of self-reported exposures during the Persian Gulf War. *Environmental Research* 81(3):195-205.

McCauley LA, Joos SK, Lasarev MR, Storzbach D, Bourdette DN. 1999b. Gulf War unexplained illnesses: Persistence and unexplained nature of self-reported symptoms. *Environmental Research* 81(3):215-223.

McCauley LA, Joos SK, Barkhuizen A, Shuell T, Tyree WA, Bourdette DN. 2002a. Chronic fatigue in a population-based study of Gulf War veterans. *Archives of Environmental Health* 57(4):340-348.

McCauley LA, Lasarev M, Sticker D, Rischitelli DG, Spencer PS. 2002b. Illness experience of Gulf War veterans possibly exposed to chemical warfare agents. *American Journal of Preventive Medicine* 23(3):200-206.

McKenzie DP, Ikin JF, McFarlane AC, Creamer M, Forbes AB, Kelsall HL, Glass DC, Ittak P, Sim MR. 2004. Psychological health of Australian veterans of the 1991 Gulf War: An assessment using the SF-12, GHQ-12 and PCL-S. *Psychological Medicine* 34(8):1419-1430.

Morgan CA 3rd, Hill S, Fox P, Kingham P, Southwick SM. 1999. Anniversary reactions in Gulf War veterans: A follow-up inquiry 6 years after the war. *American Journal of Psychiatry* 156(7):1075-1079.

Murphy FM, Kang H, Dalager NA, Lee KY, Allen RE, Mather SH, Kizer KW. 1999. The health status of Gulf War veterans: Lessons learned from the Department of Veterans Affairs Health Registry. *Military Medicine* 164(5):327-331.

Nisenbaum R, Reyes M, Mawle AC, Reeves WC. 1998. Factor analysis of unexplained severe fatigue and interrelated symptoms: Overlap with criteria for chronic fatigue syndrome. *American Journal of Epidemiology* 148(1):72-77.

Nisenbaum R, Barrett DH, Reyes M, Reeves WC. 2000. Deployment stressors and a chronic multisymptom illness among Gulf War veterans. *Journal of Nervous and Mental Disease* 188(5):259-266.

Nisenbaum R, Ismail K, Wessely S, Unwin C, Hull L, Reeves WC. 2004. Dichotomous factor analysis of symptoms reported by UK and US veterans of the 1991 Gulf War. *Population Health Metrics* 2(1):8.

NRC (National Research Council). 2000a. *Strategies to Protect the Health of Deployed US Forces: Analytical Framework for Assessing Risks*. Washington, DC: National Academy Press.

NRC. 2000b. *Strategies to Protect the Health of Deployed US Forces: Detecting, Characterizing, and Documenting Exposures*. Washington, DC: National Academy Press.

NRC. 2000c. *Strategies to Protect the Health of Deployed US Forces: Force Protection and Decontamination*. Washington, DC: National Academy Press.

Pierce PF. 1997. Physical and emotional health of Gulf War veteran women. *Aviation Space and Environmental Medicine* 68(4):317-321.

Proctor SP, Heeren T, White RF, Wolfe J, Borgos MS, Davis JD, Pepper L, Clapp R, Sutker PB, Vasterling JJ, Ozonoff D. 1998. Health status of Persian Gulf War veterans: Self-reported symptoms, environmental exposures and the effect of stress. *International Journal of Epidemiology* 27(6):1000-1010.

Proctor SP, Heaton KJ, White RF, Wolfe J. 2001. Chemical sensitivity and chronic fatigue in Gulf War veterans: A brief report. *Journal of Occupational and Environmental Medicine* 43(3):259-264.

Reid S, Hotopf M, Hull L, Ismail K, Unwin C, Wessely S. 2001. Multiple chemical sensitivity and chronic fatigue syndrome in British Gulf War veterans. *American Journal of Epidemiology* 153(6):604-609.

Reid S, Hotopf M, Hull L, Ismail K, Unwin C, Wessely S. 2002. Reported chemical sensitivities in a health survey of United Kingdom military personnel. *Occupational and Environmental Medicine* 59(3):196-198.

Roland PS, Haley RW, Yellin W, Owens K, Shoup AG. 2000. Vestibular dysfunction in Gulf War syndrome. *Otolaryngology–Head and Neck Surgery* 122(3):319-329.

Roy MJ, Koslowe PA, Kroenke K, Magruder C. 1998. Signs, symptoms, and ill-defined conditions in Persian Gulf War veterans: Findings from the Comprehensive Clinical Evaluation Program. *Psychosomatic Medicine* 60(6):663-668.

Simmons R, Maconochie N, Doyle P. 2004. Self-reported ill health in male UK Gulf War veterans: A retrospective cohort study. *BMC Public Health* 4(1):27.

Simon GE, Daniell W, Stockbridge H, Claypoole K, Rosenstock L. 1993. Immunologic, psychological, and neuropsychological factors in multiple chemical sensitivity. A controlled study. *Annals of Internal Medicine* 119(2):97-103.

Smith TC, Gray GC, Weir JC, Heller JM, Ryan MA. 2003. Gulf War veterans and Iraqi nerve agents at Khamisiyah: Postwar hospitalization data revisited. *American Journal of Epidemiology* 158(5):457-467.

Southwick SM, Morgan A, Nagy LM, Bremner D, Nicolaou AL, Johnson DR, Rosenheck R, Charney DS. 1993. Trauma-related symptoms in veterans of Operation Desert Storm: A preliminary report. *American Journal of Psychiatry* 150(10):1524-1528.

Southwick SM, Morgan CA 3rd, Darnell A, Bremner D, Nicolaou AL, Nagy LM, Charney DS. 1995. Trauma-related symptoms in veterans of Operation Desert Storm: A 2-year follow-up. *American Journal of Psychiatry* 152(8):1150-1155.

Spencer PS, McCauley LA, Joos SK, Lasarev MR, Schuell T, Bourdette D, Barkhuizen A, Johnston W, Storzbach D, Wynn M, Grewenow R. 1998. U.S. Gulf War Veterans: Service periods in theater, differential exposures, and persistent unexplained illness. *Toxicology Letters* 102-103:515-521.

Spencer PS, McCauley LA, Lapidus JA, Lasarev M, Joos SK, Storzbach D. 2001. Self-reported exposures and their association with unexplained illness in a population-based case-control study of Gulf War veterans. *Journal of Occupational and Environmental Medicine* 43(12):1041-1056.

Steele L. 2000. Prevalence and patterns of Gulf War illness in Kansas veterans: Association of symptoms with characteristics of person, place, and time of military service. *American Journal of Epidemiology* 152(10):992-1002.

Storzbach D, Campbell KA, Binder LM, McCauley L, Anger WK, Rohlman DS, Kovera CA. 2000. Psychological differences between veterans with and without Gulf War unexplained symptoms. *Psychosomatic Medicine* 62(5):726-735.

Storzbach D, Rohlman DS, Anger WK, Binder LM, Campbell KA. 2001. Neurobehavioral deficits in Persian Gulf veterans: Additional evidence from a population-based study. *Environmental Research* 85(1):1-13.

Stretch RH, Bliese PD, Marlowe DH, Wright KM, Knudson KH, Hoover CH. 1995. Physical health symptomatology of Gulf War-era service personnel from the states of Pennsylvania and Hawaii. *Military Medicine* 160(3):131-136.

Stretch RH, Bliese PD, Marlowe DH, Wright KM, Knudson KH, Hoover CH. 1996a. Psychological health of Gulf War-era military personnel. *Military Medicine* 161(5):257-261.

Stretch RH, Marlowe DH, Wright KM, Bliese PD, Knudson KH, Hoover CH. 1996b. Post-traumatic stress disorder symptoms among Gulf War veterans. *Military Medicine* 161(7):407-410.

Suadicani P, Ishoy T, Guldager B, Appleyard M, Gyntelberg F. 1999. Determinants of long-term neuropsychological symptoms. The Danish Gulf War Study. *Danish Medical Bulletin* 46(5):423-427.

Sutker PB, Uddo M, Brailey K, Allain AN. 1993. War-zone trauma and stress-related symptoms in Operation Desert Shield/Storm (ODS) returnees. *Journal of Social Issues* 49(4):33-50.

Sutker PB, Uddo M, Brailey K, Allain AN, Errera P. 1994a. Psychological symptoms and psychiatric diagnoses in Operation Desert Storm troops serving graves registration duty. *Journal of Traumatic Stress* 7(2):159-171.

Sutker PB, Uddo M, Brailey K, Vasterling JJ, Errera P. 1994b. Psychopathology in war-zone deployed and nondeployed Operation Desert Storm troops assigned graves registration duties. *Journal of Abnormal Psychology* 103(2):383-390.

Sutker PB, Davis JM, Uddo M, Ditta SR. 1995a. Assessment of psychological distress in Persian Gulf troops: Ethnicity and gender comparisons. *Journal of Personality Assessment* 64(3):415-427.

Sutker PB, Davis JM, Uddo M, Ditta SR. 1995b. War zone stress, personal resources, and PTSD in Persian Gulf War returnees. *Journal of Abnormal Psychology* 104(3):444-452.

Sutker PB, Corrigan SA, Sundgaard-Riise K, Uddo M, Allain AN. 2002. Exposure to war trauma, war-related PTSD, and psychological impact of subsequent hurricane. *Journal of Psychopathology and Behavioral Assessment* 24(1):25-37.

Thompson KE, Vasterling JJ, Benotsch EG, Brailey K, Constans J, Uddo M, Sutker PB. 2004. Early symptom predictors of chronic distress in Gulf War veterans. *Journal of Nervous and Mental Disease* 192(2):146-152.

United Kingdom Ministry of Defence. 2000. *Background to the Use of Medical Countermeasures to Protect British Forces During the Gulf War (Operation Granby).* [Online]. Available: http://www.mod.uk/issues/gulfwar/info/medical/ukchemical.htm [accessed September 26, 2003].

Unwin C, Blatchley N, Coker W, Ferry S, Hotopf M, Hull L, Ismail K, Palmer I, David A, Wessely S. 1999. Health of UK servicemen who served in Persian Gulf War. *Lancet* 353(9148):169-178.

Unwin C, Hotopf M, Hull L, Ismail K, David A, Wessely S. 2002. Women in the Persian Gulf: Lack of gender differences in long-term health effects of service in United Kingdom Armed Forces in the 1991 Persian Gulf War. *Military Medicine* 167(5):406-413.

Vasterling JJ, Brailey K, Constans JI, Borges A, et al. 1997. Assessment of intellectual resources in Gulf War veterans: Relationship to PTSD. *Assessment* 4(1):51-59.

Vasterling JJ, Brailey K, Constans JI, Sutker PB. 1998. Attention and memory dysfunction in posttraumatic stress disorder. *Neuropsychology* 12(1):125-133.

Vasterling JJ, Brailey K, Tomlin H, Rice J, Sutker PB. 2003. Olfactory functioning in Gulf War-era veterans: Relationships to war-zone duty, self-reported hazards exposures, and psychological distress. *Journal of the International Neuropsychological Society* 9(3):407-418.

Wagner AW, Wolfe J, Rotnitsky A, Proctor SP, Erickson DJ. 2000. An investigation of the impact of posttraumatic stress disorder on physical health. *Journal of Traumatic Stress* 13(1):41-55.

White RF, Proctor SP, Heeren T, Wolfe J, Krengel M, Vasterling J, Lindem K, Heaton KJ, Sutker P, Ozonoff DM. 2001. Neuropsychological function in Gulf War veterans: Relationships to self-reported toxicant exposures. *American Journal of Industrial Medicine* 40(1):42-54.

Wolfe F, Smythe HA, Yunus MB, Bennett RM, Bombardier C, Goldenberg DL, Tugwell P, Campbell SM, Abeles M, Clark P, et al. 1990. The American College of Rheumatology 1990 criteria for the classification of fibromyalgia. Report of the Multicenter Criteria Committee. *Arthritis and Rheumatism* 33(2):160-172.

Wolfe J, Proctor SP, Erickson DJ, Heeren T, Friedman MJ, Huang MT, Sutker PB, Vasterling JJ, White RF. 1999a. Relationship of psychiatric status to Gulf War veterans' health problems. *Psychosomatic Medicine* 61(4):532-540.

Wolfe J, Erickson DJ, Sharkansky EJ, King DW, King LA. 1999b. Course and predictors of posttraumatic stress disorder among Gulf War veterans: A prospective analysis. *Journal of Consulting and Clinical Psychology* 67(4):520-528.

Wolfe J, Proctor SP, Erickson DJ, Hu H. 2002. Risk factors for multisymptom illness in US Army veterans of the Gulf War. *Journal of Occupational and Environmental Medicine* 44(3):271-281.

Zwerling C, Torner JC, Clarke WR, Voelker MD, Doebbeling BN, Barrett DH, Merchant JA, Woolson RF, Schwartz DA. 2000. Self-reported postwar injuries among Gulf War veterans. *Public Health Reports* 115(4):346-349.

5

HEALTH OUTCOMES

In this chapter, the committee evaluates the evidence and draws conclusions about long-term health outcomes associated with serving in the Persian Gulf War. The studies reviewed in this chapter generally compare Gulf War veterans with veterans who, during the same period, were either deployed to the Gulf War or were deployed elsewhere. This chapter draws on information from many of the studies that were described in Chapter 4. The committee presents the health outcomes in order of their ICD-10 codes.[1] The committee did not examine health outcomes related to or resulting from infectious and parasitic diseases because another IOM committee[2] is examining those outcomes; its report will be released in fall 2006.

For each health outcome presented here, the committee first identifies the primary studies and then the secondary studies, as defined by the committee's criteria (see Chapter 3). Because many cohort studies in this chapter examine multiple outcomes, a study might be referred to in more than one place. The same study might be deemed a primary study for several health outcomes or a primary study for one outcome and a secondary study for another outcome. The key determinant is how well the study's health outcomes are defined and measured. For example, a study that was well designed for assessing a neurobehavioral effect might not be as well designed for assessing peripheral neuropathy. In general, only primary studies appear in the tables accompanying the discussions of health outcomes.

With rare exceptions, the chapter excludes studies of participants in Gulf War registries established by the Department of Veterans Affairs (VA) and the Department of Defense (DOD). Registry participants are not representative of all Gulf War veterans in that they are self-selected veterans who come to receive care. The VA and DOD registries were not intended to be representative of the entire group of Gulf War veterans.

CANCER
(ICD-10 C00-D48)

Over a million people are diagnosed with cancer each year in the United States. About one of every two American men and one of every three American women will have cancer at

---

[1] The International Statistical Classification of Diseases and Related Health Problems (ICD) provides a detailed description of known diseases and injuries. Every disease (or group of related diseases) is given a unique code. It is periodically revised and the tenth edition is known as the ICD-10.

[2] The Committee on Gulf War and Health: Infectious Diseases.

some point during their lifetime. Cancer can develop at any age, but about 77% of all cancers are diagnosed in people 55 of age and older. Military personnel during the Gulf War were had a mean age of 28 years, thus it is likely too early for the development of most cancers in Gulf War veterans. Cancer strikes Americans of all racial and ethnic groups, and the rate at which new cancers occur (the incidence) varies from group to group (ACS 2006).

The mortality and hospitalization studies reviewed later in this chapter do not definitively identify overall increases in cancer among Gulf War veterans. However, there are other studies that have examined the possibility of brain and testicular cancer. Those cancer studies are evaluated in this section. Because few of the studies dealt with specific cancers, this section groups primary and secondary studies together.

Primary and Secondary Studies

Brain Cancer

Brain cancer is relatively rare. The annual age-adjusted incidence of brain cancer in the United States is 6.4 cases per 100,000 men and women. The median age at diagnosis of brain cancer is 55 years; about 22% of cases are diagnosed between the ages of 20-44 years old (Ries et al. 2005). There is one published study of brain cancer in Gulf War veterans.

Bullman and colleagues assessed cause-specific mortality among the 100,487 Gulf War veterans identified with the 2000 sarin plume model (described in Chapter 2) as having been subjected to nerve-agent exposure from the March 1991 demolition of weapons at Khamisiyah (Bullman et al. 2005). Compared with 224,980 Gulf War veterans similarly deployed but considered unexposed to the plume, there was an increased risk of brain cancer deaths with followup through December 31, 2000 (relative risk [RR] 1.94, 95% confidence interval [CI] 1.12-3.34). There was also a suggestion of a dose-response relationship, with the risk increasing from those who were unexposed to those exposed for 1 day to those exposed for 2 days. Specific subtypes of brain cancer were not considered. Because brain cancer is considered to have a latent period of 10-20 years, and the study included less than 9 years of followup, the results should be interpreted with caution. Further followup is necessary to draw any conclusions about the risk of brain cancer among Gulf War veterans.

Testicular Cancer

Two articles have focused specifically on testicular cancer. Testicular cancer is relatively uncommon in the United States. The annual age-adjusted incidence is 5.3 cases per 100,000 men. However, it is one of the few cancers whose usual age of onset is in the same range as the age of the Gulf War veterans, about 20 to 44 years old (Ries et al. 2005). In general, little is known about environmental risk factors for testicular cancer.

Knoke and colleagues (1998) examined testicular cancer among US servicemen on active duty during the time of the Gulf War (August 8, 1990, through July 31, 1991) and who remained on active duty at the end of the deployment period. Eligible servicemen included 517,223 people deployed to the gulf and 1,291,323 nondeployed. The authors identified cases of all first-hospital admissions, in US military hospitals worldwide, for a principal diagnosis of testicular cancer from the period of July 31, 1991 through April 1, 1996. Cases were identified by examining the DOD hospitalization database through April 1, 1997. A total of 505 cases were ascertained: 134 among the deployed and 371 among the nondeployed. In Cox proportional-hazards models adjusted for race and ethnicity, age, and occupation, no association with deployment status was observed (RR 1.05, 95% CI 0.86-1.29). The deployed did have an increased risk in the early months after the end of the deployment period; it persisted, but did not increase, for about 3 years. The initial increased risk was originally reported in a study of all hospitalizations in the cohort (Gray et al. 1996). However, by the end of the followup period (1996), the cumulative probability of hospitalization of the two groups was the same (0.034% for deployed and 0.035% for nondeployed). There was no interaction between covariates and deployment status. The authors also assessed the association of testicular cancer with specific occupations. The highest RRs were observed for electronic-equipment repair (RR 1.56, 95% CI 1.23-2.00), construction-related trades (RR 1.42, 95% CI 0.93-2.17), and electrical or mechanical repair (RR 1.26, 95% CI 1.01-1.58).

To assess whether the transient increase in risk among the deployed servicemen was related to factors associated with service (for example, the healthy warrior effect), the authors repeated the analysis beginning with followup on January 1, 1990. The rate of hospitalization among the deployed was lower during the prewar and deployment period, increased after the deployment period, and then decreased again. That suggests a healthy warrior effect. The overall relative risk for deployment in this analysis (which used the earlier followup date) was 0.89 (95% CI 0.75-1.06).

In conclusion, Knoke and colleagues did not observe an association of Gulf War service with a risk of testicular cancer (RR 1.05, 95% CI 0.86-1.29) during almost 5 years of followup. The followup period was short for a cancer assessment, but it did include the age-range (22-31 years) when the disease might appear. No specific Gulf War exposures were assessed although risk by occupational group was calculated.

There was some evidence of an association of testicular cancer with Gulf War deployment in a pilot cancer-registry-based study. Levine et al. (2005) matched a cohort of 697,000 Gulf War veterans (all personnel on active duty, in the reserves, and in the National Guard deployed to the Persian Gulf) and 746,248 non-Persian Gulf-region veterans (a stratified random sample of military personnel serving at the time of the conflict but not deployed) with the central cancer registries of New Jersey and the District of Columbia. Between 1991 and 1999, testicular cancer cases were identified in 17 deployed and 11 nondeployed for a crude proportional incidence rate (PIR) of 3.05 (95% CI 1.47-6.35). After adjustment for state of residence, deployment status, race, and age, the PIR was reduced to 2.33 (95% CI 0.95-5.70). The excess in the number of cases peaked 4-5 years after deployment, as opposed to the findings in the Knoke et al. study, where the excess was seen in the first few months after the soldiers returned home. The numbers of cancers included in this study are small, and no definitive conclusions can be made until additional registries are added to the overall study.

All Cancers

Macfarlane and colleagues (2003) assessed all first diagnoses of malignant cancer in a cohort of UK armed-services personnel. The deployed group consisted of all military personnel who served in the Persian Gulf in the period September 1990-June 1991 (n = 51,721). The comparison group was randomly selected from members of the armed services who were in service on January 1, 1991 but not deployed in the Persian Gulf, and was stratified to match the Persian Gulf cohort on age, sex, service branch, rank, and level of fitness for active service (n = 50,755). Followup was from April 1, 1991 until diagnosis of cancer, emigration, death, or July 31, 2002, whichever was earlier. Cancers were identified through the National Health Service Central Register. During followup, 270 incident cases of cancer among the Gulf War veterans and 269 among the nondeployed group were identified; the RR—after adjustment for sex, age group, service branch, and rank—was 0.99 (95% CI 0.83-1.17). Thus, there was no evidence of an association of Gulf War Service with site-specific cancers. In subgroups of cohort members who participated in morbidity surveys that yielded more information on potential risk factors (28,518 Gulf War veterans and 20,829 nondeployed veterans), the RR was 1.11 (95% CI 0.86-1.44). That result did not change after adjustment for smoking or alcohol use, and there was no evidence of associations with exposure to pesticides; multiple vaccinations against anthrax, plague, and pertussis; or reported exposure to depleted uranium. See Table 5.1 for a summary of the cancer studies.

Summary and Conclusion

There is no consistent evidence of a higher overall incidence of cancer in Gulf War veterans than in nondeployed veterans. However, many veterans are young for cancer diagnosis and, for most cancers, the followup period after the Gulf War is probably too short to expect the onset of cancer.

The incidence of and mortality from cancer in general, and brain and testicular cancer in particular, have been assessed in cohort studies. An association of brain-cancer mortality with possible nerve-agent exposure (as modeled by DOD's exposure model of 2000) was observed in one study (Bullman et al. 2005). As discussed in more detail in Chapter 2, there are many uncertainties in the exposure model. Further followup is warranted to see whether the association with brain cancer holds up with time. Results for testicular cancer were mixed: one study concluded that there was no evidence of an excess risk, and another, a small registry-based study, suggested that there may be an increased risk. Although the results are inconsistent, the committee believes that followup is warranted to see whether such an association exists when more time has passed, as it is still early for the development of most cancers in Gulf War veterans.

TABLE 5.1 Cancer Outcomes

Reference	Design	Population	Outcomes	Results	Adjustments	Comments or Limitations
Brain cancer						
Bullman et al. 2005 (population from same source as Kang and Bullman 1996; Kang and Bullman 2001)	Cohort mortality study	US GWVs, grouped on basis of exposure to Khamisiyah chemical-munitions destruction (sarin gas) determined by 2000 plume model Exposed (n = 100,487) Unexposed (n = 224,980) Unknown (n = 25,574)	Cause-specific mortality ascertained from BIRLS and NDI, followup from date left Gulf through 12/31/2000	Exposed (cases = 25) vs unexposed (cases = 27) Adj RR 1.94 (95% CI 1.12-3.34); GWVs: Exposed 2+ days: RR 3.26 (95% CI 1.33-7.96) Exposed 1 day: RR 1.72 (95% CI 0.95-3.10) 12.2 deaths/100,000 for each day exposure (95% CI 4.8-19.7) vs general population: Exposed 2+ days: SMR 2.13 (95% CI 0.78-4.63) Unexposed SMR 0.71 (95% CI 0.46-1.04) [similar results when limited to medical-record confirmed cases]	Age at entry, race, sex, unit component and rank	Latent period too soon (risk increases with time since exposure); multiple comparisons; death certificate diagnosis
Testicular cancer						
Levine et al. 2005	Population-based survey—pilot study	US, all personnel (including reserves) deployed to Gulf War (GWVs) and random sample of NDVs; GWVs (n = 697,000) NDVs (n = 746,248)	Cancers diagnosed 1991-1999 and registered by DC Cancer Registry or NJ State Cancer Registry	GWVs (cases = 17) vs NDVs (cases = 11) (358 males with cancer) Crude PIR 3.05 (95% CI 1.47-6.35) Adj PIR 2.33 (95% CI 0.95-5.70) SIR (compared with SEER) for GWVs 1.42, NDVs 0.94; GWVs peaked 1995-1996, NDVs constant	Age, state of residence, deployment status, race	
Knoke et al. 1998 (followup of	Cohort study	US, all regular, active-duty male service members	First diagnosis of testicular cancer at US military	GWVs (cases = 134) vs NDVs (cases = 371) RR 1.05 (95% CI 0.86-1.29)	Race or ethnicity, age, occupation	Short followup time, but right age range;

Reference	Design	Population	Outcomes	Results	Adjustments	Comments or Limitations
Gray et al. 1996		GWVs (n = 517,223) NDVs (n = 1,291,323)	hospitals worldwide 7/31/1991- 4/1/1996			no specific exposures evaluated; military hospitals only
Gray et al. 1996	Cohort study	US, regular, active-duty military personnel deployed in gulf and randomly selected nondeployed personnel GWVs (n = 579,931) NDVs (n = 700,000)	Cause-specific hospitalizations in military hospitals 10/1/1988- 7/31/1990, 8/1/1991- 12/31/1991, 1992, 1/1/1993- 9/30/1993	1991 followup period GWVs (cases = 29) vs NDVs (cases = 14) RR 2.12 (95% CI 1.11-4.02) Later followup periods No excess risk	Age, sex	Active duty personnel only; short followup
All cancers						
Macfarlane et al. 2003 (followup of Macfarlane et al. 2000)	Cohort study	UK armed service personnel GWVs (n = 51,721) NDVs (n = 50,755); Subgroup participated in morbidity study with information on risk factors and exposures (GWVs = 28,518, NDVs = 20,829)	First diagnosis of cancer 4/1/1991- 7/31/2002 identified through NHSCR	GWVs (cases = 270) vs NDVs (cases = 269) Main study: RR 0.99 (95% CI 0.83-1.17) Morbidity study subgroup: RR 1.12 (95% CI 0.86-1.45)	Main analysis: sex, age group, service branch, rank; Morbidity study: smoking, alcohol	Short followup; low age; grouped all cancer sites because of small numbers
Kang and Bullman 2001 (also in Gray et al. 1996— followup thru 1993)	Cohort mortality study	US, all military personnel in Gulf before 3/1/1991 and random sample of nondeployed GWV (n = 621,902) NDV (n = 746,248)	Cause specific mortality ascertained from BIRLS, death certificates and NDI; followup from date left gulf through 12/31/1997	Males: GWVs (cases = 477) vs controls (cases = 860): adjusted RR 0.90 (95% CI 0.81-1.01) Females: GWVs (cases = 49) vs controls (cases = 103): adjusted RR 1.11 (95% CI 0.78-1.57)	Age, race, branch of service, unit component, marital status	Short latency; low age range; Death certificates

NOTE: BIRLS = Beneficiary Identification Records Locator System; GWV = Gulf War veteran; NDI = National Death Index; NDV = nondeployed veteran; NHSCR = National Health Service Central Register; PIR = proportional incidence ratio; SEER = Surveillance Epidemiology and End Results; SIR = standardized incidence ratio; SMR = standardized mortality ratio.

MENTAL AND BEHAVIORAL DISORDERS
(ICD-10 F00-F99)

War is a known health risk for psychiatric conditions (Pizarro et al. 2006; Wessely 2005). The description of the extent and type of psychiatric affliction and its course has depended on the development of modern psychiatric diagnostic systems and epidemiologic methods. The development of a structured diagnostic system and diagnostic instruments has facilitated the diagnosis of behavioral disorders. Moreover, the prevalence of psychiatric disorders in epidemiologic samples drawn from the general population has become available (Kessler et al. 2005) and provides baseline data with which to compare data from specific inquiries. Thus, after the Persian Gulf War, many methodologic and scientific details were in place to support a sound assessment of the psychologic consequences of war. The Persian Gulf War was highly unusual in that the air war lasted 40 days and the ground war concluded in 5 days, so there was a limited theater and set of conditions amenable in many respects to scientific study. In fact, each of the large cohort studies of Gulf War veterans, described in Chapter 4, included items pertaining to mental health. Nested within them was analysis of mental health characteristics based on direct interview techniques or validated symptom scales.

Types of psychiatric ill health that could be associated with the Gulf War, particularly posttraumatic stress disorder (PTSD), were predicted on the basis of their descriptions from previous wars (O'Toole et al. 1996; Roy-Byrne et al. 2004). As background, psychiatric disorders in the general population are common, as well as disabling and chronic (Department of Health and Human Services 1999). Diagnosable psychiatric disorders are found in about 20% of the US population, but their prevalence in military populations is lower, largely as a result of the healthy-warrior effect. Psychiatric disorders can be grouped into several large classes, for example, mood disorders (that is, depression and bipolar disorder); anxiety disorders (that is, generalized anxiety disorder, obsessive-compulsive disorder, panic disorder, post-traumatic stress disorder, and social phobia); and substance use disorders (for example, abuse of drugs and/or alcohol).

The specification of characteristics of mental diagnoses has made research on their incidence and prevalence possible (Tasman et al. 2003). Depression, a type of mood disorder, is characterized by lifelong vulnerability to episodes of depressed mood and loss of interest and pleasure in daily activities. Some symptoms of clinical depression include sleeping too little or too much, reduced appetite and weight loss or increased appetite and weight gain, restlessness, irritability, difficulty concentrating, feeling guilty, hopeless or worthless, and thoughts of suicide or death. Depression is categorized as major depressive disorder (MDD) or, when it accompanies mania, as bipolar disorder. PTSD is a subtype of anxiety disorder; it occurs after exposure to a traumatic event and is diagnosed when a person manifests severe distress on recollection of the event, avoids the situation, and suffers symptoms of anxiety in daily life. Substance abuse is defined as a maladaptive pattern of substance use (there are many types of abused substances) that results in a failure to fulfill major social roles (such as work or family-care performance), that involves use of the substance despite physical hazards and in association with legal consequences, and that involves use despite deleterious social and interpersonal consequences.

The prevalence of those disorders in the general population is addressed in the US National Comorbidity Survey Replication, a nationally representative face-to-face household survey conducted in the period February 2001-April 2003 (Kessler et al. 2005). The data show that the prevalence estimates for all anxiety disorders were 28.8% (lifetime) and 18.1% (in the last 2 months); for all mood disorders, 20.8% (lifetime) and 9.5% (in the last 12 months); and for substance use disorders, 14.6% (lifetime) and 3.8% (in the last 12 months). Those prevalence estimates are generally higher than those in deployed veterans exposed to war and much higher than in the control veteran populations. It is expected, however, that the military population is healthier than a broad epidemiologic population-based sample, given their age, general health, and military screening.

Primary Studies

In general, each of the large epidemiologic studies of Gulf War veterans' health included items pertaining to mental health. Moreover, there was often a nested case-control study of mental health characteristics in the primary epidemiologic cohort studies that used direct interview techniques. The studies reviewed here drew their data from the direct-interview studies from the large Gulf War cohort studies reviewed in Chapter 4, and often used validated instruments to complement the interview.

Black et al. (2004b) reanalyzed the population-based, telephone interviews from the Iowa cohort of 4,886 randomly selected veterans (military and reserve), deployed and nondeployed (Iowa Persian Gulf Study Group 1997). The initial cohort study had uncovered higher than anticipated levels of anxiety; therefore, this analysis of the interview data looked more carefully into the features of anxiety in that population. The original cohort was interviewed by telephone using the Primary Care Evaluation of Mental Disorders (PRIME-MD). The Post Traumatic Stress Disorder Checklist-Military (PCL-M) was used for the diagnosis of PTSD (Blanchard et al. 1996; Dobie et al. 2002; Forbes et al. 2001; Weathers and Ford 1996); the CAGE[3] was used to estimate alcoholism, and structured questions identified medical conditions and military preparedness. The study identified over a twofold increase in the prevalence of generalized anxiety disorder, panic disorder, PTSD, and any anxiety disorder in veterans deployed to the Gulf War vs the nondeployed. Nearly 6% of deployed veterans met criteria for having any anxiety disorder, compared with nearly 3% of nondeployed veterans (odds ratio [OR] 2.3, 95% CI 1.5-3.5). Participation in combat was an important independent risk factor for the development of anxiety disorders, particularly PTSD (OR 2.1, 95% CI 1.7-4.2). Anxious Gulf War veterans were more likely to have had a pre-existing psychiatric condition, to have taken medications associated with a psychiatric diagnosis, or to have had a previous psychiatric hospitalization. Anxiety conditions occurred with several psychiatric and some medical conditions.

In a separate analysis of the same population based on the telephone survey data, Barrett et al. (2002) used the PCL-M to examine PTSD in the Gulf War veterans. Each of 17 items was rated on a 1-5 severity scale and the scores were summed. A score of 50 or more defined PTSD. Symptoms of physical health were solicited and recognition assessment of medical problems was obtained. PTSD-positive veterans had a mean score of 58.7, whereas those without PTSD had a mean score of 19.7. The PTSD score was significantly associated with the magnitude of other physical illness and negative emotional characteristics.

---

[3] CAGE = acronym for four questions: Have you ever thought you should **C**ut down on your drinking? Have you ever felt **A**nnoyed by others' criticism of your drinking? Have you ever felt **G**uilty about your drinking? Do you have a morning **E**ye-opener?

In a nested case-comparison study, Black et al. (2004a) conducted face-to-face interviews with 602 veterans in 1999-2002. They used the Structured Clinical Interview for DSM-IV (SCID) with a random group of veterans drawn from strata of the PRIME-MD-interviewed group who reported one or more of the following symptom-based conditions during their previous interview: depression (major or minor depression), widespread chronic pain (established criteria for generalized, severe, and chronic pain), and cognitive dysfunction (amnesia or cognitive impairment of a moderate and prolonged intensity). Veterans were stratified by each symptom combination (one, two, or all) and by deployed or non-deployed status. Controls did not have any of those conditions and might have been deployed or not deployed. The veterans were selected randomly for interview from each stratum to optimize the match between cases and controls; interviewers were trained in the use of the diagnostic instruments.

Personality disorders were assessed with the SNAP (Schedule for Nonadaptive and Adaptive Personality). Level of functioning was assessed using the SF-36 (36-Item Short-Form Health Survey). The Whiteley Index was used to determine hypochondriasis. The study found a 32% rate of lifetime depression diagnosis (all types) and that was the same in deployed and nondeployed veterans. There were few diagnostic differences between the depressed deployed and the depressed nondeployed veterans, except for PTSD (27.3% in deployed, 5.0% in nondeployed), anxiety disorders (51.5% deployed and 25.0% non-deployed) and any disorder (68.2% deployed and 51.7% non-deployed). The deployed depressed veterans were more likely to have a diagnosis of any substance-use disorder (69.7% in deployed vs 51.7% in non-deployed). What was most surprising about the direct interview analysis was that there was so little difference between the deployed and the nondeployed veterans in aspects of depression (36.6% vs 30.3%). Most differences were found in anxiety disorder (51.5% vs 25.0%) as noted above.

Kang et al. (2003) conducted a population-based stratified random sample of 15,000 US Gulf War troops compared to a similar sample of troops deployed elsewhere. Phase 1 was a mail survey and phase 2 was a telephone-based survey of PTSD symptoms and chronic fatigue (CFS) symptoms. In the interview cohort, 12.1% of Gulf War veterans and 4.3% of other veterans had symptoms of PTSD, with an adjusted OR of 3.1 (95% CI 2.7-3.4) for PTSD in the Gulf War group; and 6% of the Gulf War veterans and 1.2% of the other veterans (OR 4.8) had CFS symptoms. It was interesting to note that PTSD symptoms showed a monotonic relationship to intensity of war stress, whereas the CFS symptoms did not show any relationship to war stress. The rates of PTSD tracked rates of stressors closely. Deployment, but not war stress, was associated with CFS symptoms.

Wolfe et al. (1999a; 1999b), and Proctor et al. (1998) examined cohorts of veterans randomly sampled and stratified from the Fort Devens and New Orleans-deployed Gulf War veterans, as well as a cohort deployed to Germany (a noncombat area). The Gulf War-deployed veterans from Fort Devens were followed longitudinally from the day of their arrival home from the gulf (time 1) to about 2 years later (time 2) with a 78% participation rate. The Fort Devens cohort was mainly male, Caucasian, and National Guard; rates of PTSD measured at time 1 were 3%. From those cohorts, stratified random samples were selected for closer study with direct interview (220 of the Fort Devens cohort, 73 of the New Orleans cohort, and 48 of the Germany-deployed). The researchers used questionnaires (the 52-item expanded Health Symptoms Checklist [HSC] and the Expanded Combat Exposure Scale), a neuropsychologic test battery, an environmental interview, and psychiatric diagnostic instruments (the Clinician-Administered

PTSD Scale [CAPS] or the Mississippi Scale for Combat-Related PTSD) (Proctor et al. 1998). Current PTSD (time 2) was diagnosed in 8.1% of the Fort Devens group, 7.6% of the New Orleans group, and none of the Germany group. Health status and function were lower in the Fort Devens cohort. The three most prevalent symptoms in the Fort Devens group were "forgetfulness," "fatigue," and "unsatisfactory sleep."

Wolfe et al. (1999b) also recruited cases from the Fort Devens and Germany cohorts with a stratified random-sampling strategy (148 from the Fort Devens group, 73 from the New Orleans group, and 48 from the Germany group). They used the Laufer Combat Scale to assess exposure to combat situations and the Mississippi Scale for Combat-Related PTSD to assess PTSD. The deployed Fort Devens group had higher levels of current and lifetime PTSD and current and lifetime MDD than the Germany group; little else regarding psychiatric function was different between the groups. Compared with the PTSD prevalence in the general population (7.8%) (Kessler et al. 1995), the Germany group (controls) had much lower rates of PTSD. However, the low prevalence estimates in the controls increases from zero to 5-8% when the veterans are deployed to active war situations. A strength of this study is that it is characterized by direct interview.

A different study of the cohort examined above (Wolfe et al. 1999a) looked at the course and predictors of PTSD and found that there was a higher rate of PTSD at time 2 (8%) than at time 1 (3%), indicating the development of new cases. Responders at time 2 were more likely to be younger, belong to racial minorities, and be deployed; however, the absence of differences in PTSD rates due to those characteristics indicates a lack of selection bias at time 2. Women were significantly more likely to have PTSD (OR 3.2 at time 1; OR 2.3 at time 2), although their numbers were very low at each assessment.

Brailey et al. (1998) studied Gulf War veterans on their return from service (an average of 9 months after their return) with a face-to-face debriefing and psychologic assessment, comparing Gulf War-deployed (n = 876) with nondeployed veterans (n = 396 mobilized but not deployed), including National Guard and reserve troops. A subset of 349 received a followup assessment an average of 16 months later. They used standard psychiatric rating scales for their assessments including: the Beck Depression Inventory (BDI), the State Anger, State Anxiety, the Brief Symptom Inventory (BSI) Depression, BSI Anxiety, BSI Hostility, and the Health Symptom Checklist (HSC). The deployed veterans had higher scores than the nondeployed on the BDI, the State Anger, the BSI Anxiety, and the HSC. When the Gulf War-deployed veterans were reassessed on average of 16 months later, they showed increases on all scales, including the BDI, the State Anger, the BSI Anxiety, the BSI Hostility, HSC, and on both PTSD scales (the 17-item DSM-III R PTSD Checklist and the Mississippi Scale for Desert Storm War Zone Personnel). They showed increased rates of depression (6.9% to 13.8%), PTSD (2.3% to 10.6%), and hostility (4.9% to 13.8%). The authors correlated war stress with those symptoms and found that the higher the war-zone stress, the more severe the depressive and anxiety symptoms. Troops who were assigned to high-risk activities, such as grave registration, showed a high prevalence of PTSD (46%), depression (25%), and substance abuse (13%).

Goss Gilroy et al. (1998) assessed all 3,113 Canadian Gulf War veterans deployed to the war zone and a comparison group of nondeployed veterans with a mail questionnaire; the methodologic details are in Chapter 4. Using the PCL-M, the investigators found that symptoms of PTSD were 2.5 times more prevalent in the deployed than in the nondeployed veterans (OR 2.69, 95% CI 1.7-4.2). Using the PRIME-MD, the investigators found that the deployed had higher prevalences of major depression (OR 3.67, 95% CI 3.0-4.4), chronic dysphoria, and anxiety. Anxiety and depression were more severe in lower income veterans.

The studies of psychologic outcomes in Australian Gulf War veterans were distinguished by inclusion of the entire deployed population and the use of direct assessments. McKenzie et al. (2004) used the SF-12, the PCL-M, and the GHQ-12 (12-item version of the General Health Questionnaire) to assess 1,424 male Gulf War veterans (86.5% Navy) and 1,548 male Australian Defence Force members who were not deployed to the Gulf War. On those self-rating instruments, the Gulf War-deployed had overall poorer psychologic health and more PTSD-like symptoms than control veterans. The psychologic distress increased with age in the comparison group but decreased with age in the Gulf War veterans (that is, the young Gulf War veterans had the worst psychologic ill-health). Moreover, the perceived level of exposure to war stress was associated with both psychologic ill-health and PTSD-like symptoms, although very few experienced direct combat.

Ikin et al. (2004) conducted a comprehensive health assessment of the same (Australian) cohort, including an interview-administered psychologic health assessment with the Composite International Diagnostic Interview (CIDI), a structured interview of demonstrated reliability and validity. The CIDI data allowed them to make an estimate of pre-Gulf War disorder, post-Gulf War disorder, and current (last 12 months) disorder. Those interview data were used with postal questionnaire data to form a complete workup of 1,381 Gulf War veterans, and 1,377 comparison veterans. Both the veterans and the controls completed both the health assessment and the postal questionnaire. The two groups were demographically similar, although the Gulf War veterans were significantly younger, more likely to have been in the Navy, and less highly ranked than the comparison veterans. The two veteran groups were similar in prevalence of prewar psychiatric disorders. However, the Gulf War veterans were more likely than the comparison group to have developed any disorder after the war (31% vs 21%). The greatest risks were for the anxiety disorders (for example, PTSD: OR 3.9, 95% CI 2.3-6.5), major depression (OR 1.6, 95% CI 1.3-2.0), and alcohol dependence/abuse (OR 1.5, 95% CI 1.2-2.0); the rates of somatic disorders were low in both groups. In addition, the Gulf War group was 2-5 times more likely to have anxiety, PTSD, obsessive-compulsive disorder (OCD), social phobia, or panic, than the comparison group in the preceding 12 months. On average, the Gulf War veterans had twice as many current psychiatric disorders as the comparison veterans. The strengths of this study were the large sample, the comparable control group, the use of well-validated psychologic interviews, and the analyzed participation bias, which was estimated to be low.

A study of DOD postwar hospitalizations for mental disorders (June, 1991-September 30, 1993) using 10 categories from the International Classification of Diseases, 9th Revision, Clinical Modification, 6th Edition (ICD-9-CM) was conducted by Dlugosz and colleagues (1999). It compared all active-duty personnel during the Gulf War era (n = 1,984,996) with those who did not serve. It also sought to identify risk factors for hospitalization. Nearly half the postwar hospitalizations were for alcohol-related disorders. Gulf War veterans were at greater risk for hospitalizations than nondeployed veterans due to drug-related disorders (RR 1.29, 95% CI 1.10-1.52) and acute reactions to stress (RR 1.45, 95% CI 1.08-1.94). Adjustments were made for age, sex, and military service branch. Although the database of ICD-9 codes does not allow determination of whether stress reactions expressly included PTSD, the authors noted that if posttraumatic stress was diagnosed, it would probably have been coded as an unspecified acute reaction to stress (ICD-9 code 308.9). Alcohol-related diagnoses were not increased. Exposure to the ground war in Iraq was associated with a greater risk of alcohol-related hospitalizations in men (RR 1.13, 95% CI 1.04-1.23). Serving as support for the ground war without being in direct combat was associated with a greater risk of drug-related hospitalizations in men (RR 1.42, 95% CI 1.03-1.96) and women (RR 3.61, 95% CI 1.70-7.66). The limitation of this study is that it examined only hospitalizations and thus was not representative of most psychiatric disorders which require outpatient treatment rather than hospitalization. It also did not include veterans who left the military after the Gulf War.

Table 5.2 summarizes the results of the primary studies on psychiatric outcomes.

Secondary Studies

Findings on many other major cohorts of Gulf War veterans support what has been found in primary studies (Gray et al. 2002; McCauley et al. 2002). The most important limitation was their reliance on self-reports of "physician-diagnosed disorders" rather than measurement of symptoms with validated questionnaires or face-to-face interviews. In the UK cohort studied by Unwin et al. (1999), investigators asked some questions taken from the Mississippi Scale for Combat-Related PTSD but did not administer the entire questionnaire. They found that "post-traumatic stress reaction" was about 2-3 times more likely in deployed than in two nondeployed groups. The magnitude of the increase is consistent with that seen in the primary studies. Several other secondary studies have found an association between serving in the Gulf War and psychiatric disorders (Holmes et al. 1998; Magruder et al. 2005; Simmons et al. 2004; Steele 2000; Stretch et al. 1996a; Stretch et al. 1996b; Sutker et al. 1995).

Summary and Conclusion

Two well-designed studies using interview-based assessments have found that several psychiatric disorders, notably PTSD and depression, are 2-3 times more likely in Gulf War-deployed than in nondeployed veterans (Black et al. 2004b; Wolfe et al. 1999b). Direct interviews are labor-intensive; so many other studies administered validated symptom questionnaires. The findings were remarkably similar, that is, an overall increase in magnitude, by a factor of 2-3, of psychiatric disorders. When war exposure was assessed with symptoms, studies characteristically showed higher rates, particularly of PTSD, in veterans who had more traumatic war experiences than in those with lower levels of exposure. In other words, studies found a dose-response relationship between the degree of traumatic war exposure and PTSD. Nevertheless, deployment to a war zone without direct combat exposure is a traumatic war exposure, considering that one well-designed study found deployment without combat to increase the risk of psychiatric disorders by about 60% (Ikin et al. 2004).

Other risk factors were war preparedness, enlisted status (possibly correlated with war preparedness), smoking, and previous psychiatric diagnosis. Two studies indicated that severe PTSD symptoms would worsen over time, and so suggested that careful assessment of the longitudinal course of postwar psychiatric conditions is needed.

It is confirmatory to see that the primary studies, regardless of their techniques of ascertainment or their target population, reported almost identical conclusions regarding the psychiatric outcomes of Gulf War deployment for veterans, that is, that depression, substance abuse or dependence, and anxiety disorders, especially PTSD, were increased in Gulf War veterans after deployment, and that symptom severity was associated with the level of war stress.

TABLE 5.2 Psychiatric Disorders

Study	Design	Population	Outcomes	Results	Adjustments	Comments
Black et al. 2004b, Iowa Persian Gulf Study Group 1997	Population-based interview study, by telephone; stratified random sample with proportional allocation	1,896 deployed vs 1,799 nondeployed veterans listing Iowa as home state at time of enlistment	PRIME-MD (major depression, panic disorder, GAD) PCL-M, combat exposure assessed in basic demographic questionnaire. CAGE questionnaire (alcohol abuse)	Panic disorder (OR 2.2, 95% CI 1.2-3.8); GAD (OR 2.5, 95% CI 1.5-4.1); PTSD (OR 2.5, 95% CI 1.2-5.0); any anxiety disorder (OR 2.3, 95% CI 1.5-3.5)	Age, sex, race, branch of military, rank, military status, prior mental-health condition	Large, population-based sample
Barrett et al. 2002	Population-based survey; completed telephone survey about their health status	3,682 Gulf War veterans and control subjects	PCL-M	Gulf War-deployed were twice as likely to screen positive for PTSD than those not deployed; current smoking stautus was associated with PTSD status; PTSD associated with higher number of medical symptoms than those without PTSD and lower levels of functioning and quality of life		Brief PTSD screen used; used 50 as the cut-off score with the PCL-M; low number of subjects who screened positive for PTSD; the sample from Iowa might not be representative of all US military personnel
Black et al. 2004a	Nested case-comparison; face-to-face interviews	602 veterans and controls	SCID (face-to-face interviews); SNAP; SF-36; Whiteley Index	PTSD (27% vs 5% in deployed vs controls); anxiety disorders (52% vs 25%); any disorder (68% vs 52%)	Validated PTSD checklist against SCID (70.4% sensitivity and 86.2% specificity of questionnaire)	

Study	Design	Population	Outcomes	Results	Adjustments	Comments
Kang et al. 2003	Cross-sectional; population-based stratified random sample of Gulf War deployed compared with deployed elsewhere	11,441 deployed vs 9,476 nondeployed	Mail survey and telephone-based survey of PTSD symptoms	PTSD (OR 3.1, 95% CI 2.7-3.4)	Sex, age, marital status, rank, and unit component	Nationally representative sample, questionnaire only
Wolfe et al. 1999a Wolfe et al. 1999b Proctor et al. 1998	Cross-sectional survey and interviews from larger cohorts followed longitudinally	220 Ft. Devens vs 73 New Orleans vs 48 Germany; New Orleans and Germany cohorts only studied at time 2	Health Symptom Checklist, Mississippi PTSD Scale (times 1 and 2), SCID, CAPS (clinician diagnostic interviews, time 2 only)	Risk factors for PTSD were being female; having high combat exposure; current PTSD, depression, dysthymia; more prevalent in Ft. Devens vs Germany; prevalence of PTSD increased from time 1 (3%) to time 2 (8%) in Ft. Devens	Sex, reported health symptoms	Small sample deployed to Germany
Brailey et al. 1998	Longitudinal; psychologic interviews 9 months after war, and subgroup followup at 16 months; Louisiana National Guard and Reserve troops (Marine, Army, Air Force, Navy)	876 deployed (349 at time 2) vs 396 nondeployed	BDI, State Anger; State Anxiety; the BSI Depression; BSI Anxiety; BSI Hostility, the HSC, PTSD Checklist and the Mississippi Scale	Depression increased over time in deployed veterans from time 1 (6.9%) to time 2 (13.8%); PTSD increased from time 1 (2.3%) to time 2 (10.6%)	Age, education	Large attrition by time 2
Goss Gilroy Inc. 1998	Cross-sectional survey; all deployed	3,113 deployed vs 3,439 nondeployed	PCL-M and the PRIME-MD	PTSD (OR 2.69, 95% CI 1.7-4.2); major depression (OR 3.67, 95% CI 3.0-4.4); alcohol abuse (NS)	PTSD: income, service; Major depression: rank, income	Entire cohort surveyed, actual alcohol findings not reported
McKenzie et al. 2004	Cross sectional survey of all Australian deployed	1,424 male Australian Gulf War veterans vs 1,548 male Australian	SF-12; PCL-M, and GHQ-12	PTSD: OR 2.0, 95% CI 1.5-2.9	Service, branch, rank, age category, education, marital	86.5% Navy in Gulf War veterans

Study	Design	Population	Outcomes	Results	Adjustments	Comments
Ikin et al. 2004	Cross sectional survey of all Australian deployed veterans	1,381 GWVs vs 1,377 comparison veterans Defence Force nondeployed veterans	CIDI	Prevalence of any Disorder: 31% in GWVs vs 21% in comparison group; PTSD: OR 3.9, 95% CI 2.3-6.5); major depression: OR 1.6, 95% CI 1.3-2.0	Service type, rank, age, education, marital status	GWVs younger, more likely in the Navy, and lower ranked than comparison group
Dlugosz et al. 1999	Post-war hospitalizations June 1991-September 1993	Active-duty men and women (1,775,236) (209,760) June 1991-September 1993; Gulf War veterans vs non-Gulf War veterans	ICD-9 CM categories for 10 mental disorders	Gulf War-deployed had increased risk of hospitalizations due to: acute reactions to stress (RR 1.45, 95% CI 1.08-1.94); drug-related disorders (RR 1.29, 95% CI 1.10-1.52)	Age, sex, service-branch adjusted rates; modeling adjusted for age, race, length and branch of service, medical catchment area, prewar mental-health hospitalizations, dates of service in Gulf War, length of deployment in Gulf War, duty occupation, sex, marital status, rank	Active duty only; no assessment of outpatient treatment

NOTE: BDI = Beck Depression Inventory; BSI = Brief Symptom Inventory; CAGE = acronym for four questions: Have you ever thought you should Cut down on your drinking? Have you ever felt Annoyed by others' criticism of your drinking? Have you ever felt Guilty about your drinking? Do you have a morning Eye opener?; CAPS = Clinician Administered PTSD Scale; CIDI = Composite International Diagnositic Interview; ICD = International Classification of Diseases; GAD = generalized anxiety disorder; GHQ-12 = 12-Item General Health Questionnaire; HSC = 52-Item Expanded Health Symptoms Checklist; PCL-M = Post Traumatic Stress Disorder Checklist-Military Version; PRIME-MD = Primary Care Evaluation of Mental Disorders-MD; SCID = Structured Clinical Interview for DSM-IV; SNAP = Schedule for Nonadaptive and Adaptive Personality; SF-12 = 12-Item Short Form Health Survey; SF-36 = 36-Item Short Form Health Survey.

NEUROBEHAVIORAL AND NEUROCOGNITIVE OUTCOMES
(ICD-10 F00-F99)

After the Persian Gulf War, veterans often reported memory and concentration difficulty (for example, CDC 1995). The nature of the symptoms, coupled with concern about potential exposure to chemical warfare agents that affect the nervous system, led to studies that used objective neurobehavioral tests. This section focuses on neurobehavioral performance as measured by tests of cognition and, in some cases, sensory integrity or motor speed and coordination. It begins with background on the tests used to evaluate neurocognitive and neurobehavioral capabilities. The confounding factors that can affect performance on these tests also are described. The evaluative portions of the section are divided according to the two major questions addressed by this research in Gulf War veterans: (1) Do veterans who were deployed to the Persian Gulf War differ in neurobehavioral performance tests from veterans who were not so deployed? (2) Do deployed veterans reporting symptoms that meet various case definitions of possible "Gulf War Syndrome" differ in neurobehavioral performance tests from veterans who report having no symptoms?

For the purposes of this section, the committee defines primary studies as high-quality studies that used neurobehavioral tests that had previously been used to detect adverse effects in population-based research on occupational groups. Most secondary studies also used those tests but had methodologic limitations. The secondary studies were reviewed and are included in the discussion because they evaluated the same functional domains, such as attention and memory, and in some cases used the same neurobehavioral tests as did primary studies; they therefore provide valuable supplementary information that helps to increase or decrease confidence in the conclusions drawn from the primary studies. Confidence in a secondary study is substantially reduced if its statistical analysis did not adjust for confounders or if individually administered neurobehavioral tests were given by examiners not blinded to the status of cases and controls. Nevertheless, secondary studies are included in the tables and text whether or not they support the results of primary studies; this departure from most other sections of this report is necessary because of the ill-defined nature of nervous system symptomatology and the need to consider the basis of veterans' most prevalent complaints exhaustively.

Neurobehavioral Tests and Confounding Factors

Neurobehavioral tests have been used for over a century to measure human performance, initially to evaluate performance in schoolchildren (for example, Matarazzo 1972). In the 1960s, neurobehavioral tests came into use to study populations occupationally exposed to neurotoxic agents, and this type of research continues and is still evolving (Lucchini et al. 2005). Neurobehavioral testing is conducted specifically to determine whether cognitive, sensory, or motor performance has been affected by chemical exposure. Most, if not all, of the tests identified in this section have been used for that purpose (Anger 2003), and their demonstrated validity in research increases our confidence in their ability to provide data that will assist researchers in answering questions about illnesses in Gulf War veterans.

The neurobehavioral or neurocognitive tests used in these studies are affected by a variety of factors that can confound their results: age, education, sex, and native intellectual ability (Lezak et al. 2004). Demographic factors (age, education, and sex) are typically obtained through self-reports, but native intellectual ability is difficult to measure in a convincing manner. Three approaches have been used to characterize it: the Armed Forces Qualifying Test (AFQT), which is administered on enlistment in the US military (e.g., Storzbach et al. 2001); the information subscale of the Wechsler Adult Intelligence Scale (WAIS), which measures knowledge attained in life (e.g., Proctor et al. 2003); and the National Adult Reading Test (NART) (e.g., David et al. 2002). The WAIS and NART were administered after the veterans returned from the Persian Gulf War, so they yielded estimates based on current performance; the AFQT is the one truly premilitary measure. Each of those tests correlated with measures of overall intelligence, such as the full WAIS, but is not believed to be affected by exposures, such as to neurotoxic chemicals (that idea has been put forward as one possible cause of symptoms attributed to "Gulf War syndrome"). In addition, it is important that the examiners who administer the neurobehavioral or neurocognitive tests be blind to the condition or status of the veterans and the control population. Blinding is of less concern if the tests are administered on a computer.

Studies That Respond to Question 1 (Outcomes in Gulf War-Deployed Veterans vs Veterans Deployed Elsewhere or Not Deployed)

The committee identified two primary and five secondary studies that compared deployed veterans with those deployed elsewhere or not deployed (Table 5.3). David et al. (2002) compared the neurobehavioral test performance of 209 UK soldiers deployed to the Persian Gulf, 54 UK Bosnia peacekeeping soldiers, and 78 UK Gulf War-Era nondeployed soldiers. A broad array of neurobehavioral tests were administered to all participants and the results were analyzed, although evaluation is limited by the lack of standard deviations of the mean test scores. No differences were found among the groups after correction for age, education, intelligence (according to the NART) and the Beck Depression Inventory (BDI) score (Table 5.3).

Proctor et al. (2003) studied 143 Gulf War veterans and 72 nondeployed veterans of the Danish military. A broad array of neurobehavioral tests were administered to participants and the results were analyzed. Proctor et al. (2003), too, did not find any differences in the overall analysis of neurobehavioral test performance (Table 5.3).

Three secondary studies addressed whether deployed veterans differed from nondeployed veterans (Axelrod and Milner 1997; Vasterling et al. 2003; White et al. 2001). Only one of the three (Axelrod and Milner 1997) found reliable differences in neurobehavioral test performance between the groups after correction for age and education (Table 5.3). The study by White et al. (2001) was considered a secondary study for this outcome because the Gulf War group combined two demographically heterogeneous samples—one from Fort Devens and the other from New Orleans—and because the comparison population, Germany-deployed veterans, was small.

In its evaluation of those studies, the committee was concerned that the investigators' analyses might have masked differences. For example, David et al. (2002) adjusted the results for depression because it is found to coexist with cognitive measures (e.g., Brown et al. 1994). That adjustment could have made it impossible to detect cognitive differences. David et al. (2002), White et al. (2001) and Vasterling et al. (2003) used an overconservative Bonferroni statistical adjustment to correct for multiple comparisons (Sterne and Davey Smith 2001), which also might have masked differences. The committee, therefore, estimated the effect size (d) (Cohen 1992) of the corrected and precorrection significant test differences (or trends) and searched for a pattern or consistency of results among the studies. The percentage of the neurobehavioral tests given that were reported as significant is listed in column 6; it varied from

0% to 30% (mean 21.2%). The tests on which performance deficits in deployed veterans were reported as significant by the report authors are listed in **boldface** type in column 5 of Table 5.3. Tests that were significant before but not after correction (and thus are not considered significant by the article's authors) are included in column 5, where they are printed in regular type. The effect size (d) for each test difference is listed if it could be calculated. Although the results are subject to interpretation, a consistent pattern of neurobehavioral deficits does not emerge from an inspection of column 5 of Table 5.3. Specifically, deficits in higher cognitive function were seen in pre-Bonferroni correction analyses in the David et al. (David et al. 2002) and Proctor et al. (2003) studies and not in the secondary studies. Other cognitive measures were deficient in one primary study and two secondary studies (Axelrod and Milner 1997; Proctor et al. 2003; White et al. 2001), but the measures were different. Although deficits in the same or similar measures of motor coordination (pegboard) were seen in three studies, including the David et al. (2002) study, it is difficult to infer that wartime events might have changed coordination. The percentage of precorrection significance varied from 0 to 31% across the five studies (see last column of Table 5.3), and overall, 23% of the tests had at least one significant measure. However, the effect sizes ranged from 0.2 to 0.4 (column 5 of Table 5.3); all of which are small according to Cohen's criteria (1992).

TABLE 5.3 Comparisons of Gulf War-Deployed Veterans with Those Not Deployed or Deployed Elsewhere

Reference	Population	Adjustments	Significant Findings and Statistical Adjustment	Significant Test Deficits	Number of Neurobehavioral Tests and % Significance
Primary Studies					
David et al. 2002	209 GWVs vs 54 Bosnia-deployed vs 78 nondeployed	Age, education, intelligence, depression	No; ANCOVA & Bonferroni	**Purdue pegboard;** Verbal IQ; Performance IQ; block design; logical memory; (No SDs, so no d)	16 (31% sig)
Proctor et al. 2003	143 GWVs vs 72 nondeployed (Denmark)	Age	No; MANCOVA	CVLT (d = 0.3 and 0.32); Wisconsin card sort (d = 0.3 and 0.3)	7 (29% sig)
Secondary Studies					
White et al. 2001	293 GWVs vs 50 Germany-deployed	Age, sex, education	No; Bonferroni	Continuous performance test (d = 0.2); PASAT (d = 0.3); trailmaking A errors (d = 0.3); Purdue pegboard (d = 0.4)	13 (31% sig)
Vasterling et al. 2003	72 GWVs vs 33 nondeployed	Age, rank	No; Bonferroni		7 (0% sig)
Axelrod and Milner 1997	44 GWVs vs standard or adjusted T scores (see Table 5.7 for specifics)	Age (Stroop), education (both tests)	Yes; no adjustment	**Stroop word, color, color-word; grooved pegboard** (no standard comparators, so no d)	13 (15% sig)

NOTE: ANCOVA = Analysis of Covariance; CVLT = Continuous Verbal Learning Test; d = effect size; GWV = Gulf War veteran; MANCOVA = Multivariate Analysis of Covariance; PASAT = Paced Auditory Serial Arithmetic Test. Measures considered significant after final statistical analyses are shown in **boldface**. Measures significant before final statistical comparison (Bonferroni) and thus not considered significant by the authors are listed in regular print in column 5.

Studies That Respond to Question 2 (Symptomatic vs Nonsymptomatic Veterans)

Table 5.4 lists the symptoms used by each primary and secondary study to categorize participants as symptomatic. Two primary studies and six secondary studies (Table 5.5) compared symptomatic veterans with those who did not report being symptomatic.

TABLE 5.4 Symptoms Serving as Basis for Categorizing Participants into Symptomatic and Not Symptomatic

Study	David et al. 2002 (Primary Study)	Storzbach et al. 2001 (Primary Study)	Axelrod and Milner 1997	Goldstein et al. 1996	Bunegin et al. 2001	Lange et al. 2001	Hom et al. 1997	Sillanpaa et al. 1997
Basis for Inclusion in Symptomatic Group:	Any One	Any One	Probably Any One	Any One	Chemical Intolerance and Any One	One	Highest Factor Scores On	Factor in Regression
Cognitive deficiencies (unspecified)			X					
Memory loss	X	X		X	X		(X)	X
Confusion		X			X		(X)	
Attention or concentration problems	X	X			X		(X)	X
Mood swings	X	X						
Somnolence		X		X				
Gastrointestinal distress		X		X				
Fatigue	X	X		X		X	(X)	
Muscle and joint pain	X	X		X			(X)	
Skin or mucous membrane lesions		X		X				
Chemical intolerance					X			
Headache, sensory dysfunction, dizziness, depression, tumor, weight gain, sinus problems, or newborn problems				X				
Main factors: impaired cognition (attention, memory, reasoning); confusion-ataxia (disturbances in thinking, orientation, balance, impotence), arthromyoneuropathy (joint and muscle pain, muscle fatigue, lifting difficulty, extremity paresthesia)							X	

NOTE: (X) = symptom listed in publication as example of group of symptoms that was basis for categorization, but report did not list full range of specific symptoms.

Table 5.5 below reveals that each case definition is slightly different. However, the core symptoms are cognitive and are the same or would be assumed to be the same (Axelrod and Milner 1997; Hom et al. 1997) in all but the Lange et al. study (2001) in which fatigue was the exclusive symptom. The results are summarized in Table 5.5.

Of the two primary studies, David et al. (2002) categorized the veterans as either ill (n = 151) or well (n = 188) on the basis of their SF-36 physical functioning scale scores. They then recategorized them more stringently on the basis of the Center for Disease Control and Prevention working definition of Fukuda et al.(1998) for reporting (n = 65) or not reporting (n = 33) Gulf War-related symptoms (Table 5.5, column 1), analyzing each separately. (These are additional analyses from the same David et al. (2002) study identified above as comparing veterans on the basis of their deployment status.) David et al. (2002) did not find overall differences after adjusting for age, education, estimated intelligence (NART), and depression (BDI score); but they did report some cognitive test differences before making the final Bonferroni correction for multiple comparisons (column 5 of Table 5.5).

Storzbach et al. (2001) compared 239 deployed veterans who reported having at least one symptom associated with the "Gulf War syndrome" (Table 5.5) with 112 nonsymptomatic deployed veterans in a case-control study. They reported poorer performance on three neurobehavioral tests (Oregon Dual Task Procedure [ODTP] errors and latency, Digit Span Backward, and Simple Reaction Time) after adjustment for age, education, AFQT scores, and a Bonferroni correction for multiple comparisons (Table 5.5). In the analysis, Storzbach et al. identified a group of "slow" responders (12% of the symptomatic group) who had very slow latencies in choosing answers on the memory component of the ODTP, although they made few errors. Anger et al. (1999) reported on the first 100 participants in whom this slow group was identified; see Table 5.5. The slow ODTP subgroup proved to have been responsible for the statistically significant differences between cases and controls in the larger group: removing them from the analysis virtually eliminated the differences. The effect size between the slow responders and the controls on the ODTP was very high ($d = 2.9$). The effect sizes for the performance differences of the slow ODTP group (n = 30) vs the non-symptomatic (n = 112) ranged from 0.7 to 2.9 (mean $d = 1.5$) (Table 5.5, column 5).

All other studies that compared symptomatic veterans with nonsymptomatic veterans are listed as secondary studies. They are outlined in the lower six rows of Table 5.5 and are described in more detail in Table 5.7. As can be seen in Table 5.5, some categorized veterans as symptomatic on the basis of subsets of the overall symptom constellation, and some defined symptomatic in a narrower way. Some of the studies were not well described, had design flaws, asked limited questions, and studied small samples (n = 8-48). Nonetheless, as noted above, they used similar or the same neurobehavioral tests and thereby allow an evaluation of the consistency or pattern of results.

Five of the six secondary studies reported performance differences between symptomatic cases and controls or between cases and standard scores. As can be seen in Table 5.5, only two secondary studies (Axelrod and Milner 1997; Sillanpaa et al. 1997) adjusted for age and education, and only one (Lange et al. 2001) adjusted for multiple comparisons. The study that did not report a difference (Sillanpaa et al. 1997) did adjust for age, education, and an estimate of exposure (not clearly described). Each study had additional weaknesses. Axelrod and Milner (1997) used standard scores instead of controls and did not mention whether examiners were blinded. The basis of subject selection was not clear in Goldstein et al. (1996), and the issue of blinding of examiners was not addressed in the publication. The basis of subject selection was not clear in Bunegin et al. (2001), nor was examiner blinding addressed, although this is not a serious problem for the computer-based tests used in this study. In Hom et al. (1997), the basis of selection of nonsymptomatic controls was not described, that is, whether it was random or targeted. Lange et al. (2001) studied veterans with fatiguing illness but did not describe the case definition, and blinding of the examiners was not addressed although individually administered tests were used. Sillanpaa et al. (1997) used standard scores instead of a control group. Notwithstanding those limitations in the secondary studies, they found effects consistent with those found in the primary studies.

TABLE 5.5 Comparisons of Symptomatic and Nonsymptomatic Veterans

Reference	Population	Adjustment	Significant and Statistical Adjustment	Significant Test Deficits (Effect Size if Calculable)	Number of Neurologic Tests
Primary Studies					
David et al. 2002	151 symptomatic vs 188 nonsymptomatic	Age, education, intelligence, depression	No; ANCOVA, Bonferroni	Performance IQ, digit symbol; trail-making B, trail-making A, sustained attention reaction time (No d; no SDs in Table 3 of study)	16 (25% sig)
David et al. 2002	65 CDC cases (per Fukuda 1998 definition) vs 33 well	Age, education, intelligence	No; ANCOVA, Bonferroni	WAIS vocabulary (d = 0.6), digit symbol (d = 0.8), PASAT [near 0.05] (d = 0.7), Stroop [near 0.05] (d = 0.8)	16 (25% sig)
Storzbach et al. 2001	239 GWVs symptomatic vs 112 nonsymptomatic; analyses of 30 "slow ODTP" vs 112 nonsymptomatic	Age, education, intelligence	Yes; ANOVA, regression, Bonferroni	**Symbol Digit** (d = 1.0), **ODTP errors** (d = 1.1) and **latency** (d = 2.9); **Selective Attention** (d = 0.7); **digit span forward** (d = 0.9); **backward** (d = 2.5); **simple reaction time** (d = 1.3)	6 (83% sig)
Secondary Studies					
Axelrod and Milner 1997	17 GWVs symptomatic vs 27 non symptomatic	Age, education	Yes; t-test; no adjustment	**Semantic fluency** (d = 0.7), **Stroop word** (d = 0.8), **color** (d = 0.8), **color-word** (d = 0.7)	13 (15% sig)
Goldstein et al. 1996	21 GWVs with ½ symptomatic vs 38 locals		Yes; MANOVA; no adjustment	**Impairment index** (d = 0.7) (individual < 0.05) COWAT (d = 0.6); continuous performance/trial 3 (d = 0.6)	11 (18% sig)
Bunegin et al. 2001	8 GWVs with symptoms vs 8 without symptoms		Yes; ANOVA; no adjustment	**Pattern memory** (d = 0.9), **digit span forward** (d = 0.8) and **backward** (d = 1.0); **switching attention** (d = 0.7)	6 (50% sig)
Hom et al. 1997	26 GWVs symptomatic vs 20 non-deployed or nonsymptomatic		Yes; 1-tail t-test and Mann-Whitney U; no adjustment	**Halstead category test** (d = 0.7); **WAIS comprehension** (d = 0.6); **WAIS vocabulary** (d = 0.6); **bilateral simultaneous tactile-left** (d = 0.6); **finger recognition-right** (d = 0.5) and **left-right confusion** (d = 0.8); **trail-making B** (d = 0.7); **grip strength dominant** (d = 1.0) and **nondominant** (d = 0.7)	22 (36% sig)
Lange et al. 2001	48 GWVs with fatigue vs 39 without		Yes; MANOVA by domain, Bonferroni	**Category test** (d = 0.5); **trail-making B-A** (d = 0.8); **complex** (d = 0.6) and **simple** (d = 0.8) **reaction time**; **digit span backward** (d = 0.5); **PASAT** (d = 0.6)	10 (60% sig)

Reference	Population	Adjustment	Significant and Statistical Adjustment	Significant Test Deficits (Effect Size if Calculable)	Number of Neurologic Tests
Sillanpaa et al. 1997	16 with memory and attitude symptoms vs 13 with memory or attention symptoms vs 19 nonsymptomatic	Age, education, exposure estimates	No; MANOVA; no adjustment		7 (WAIS = 1) (0% sig)

NOTE: ANCOVA = Analysis of Covariance; CDC = Centers for Disease Control and Prevention; COWAT = Controlled Oral Word Association Test; GWV = Gulf War veteran; MANOVA = Multivariate Analysis of Variance; ODTP = Oregon Dual Task Procedure; PASAT = Paced Auditory Serial Addition Test; sig = significant; WAIS = Wechsler Adult Intelligence Scale. Measures the authors reported as significant after all statistical comparisons are in **boldface**. Measures that are not considered significant (column 5) after final statistical adjustments are in regular print.

Taking the primary and secondary studies together (see Table 5.5), differences or trends were seen in numerous measures in at least six and probably seven studies, but not in Sillanpaa et al. (1997). The percentage of neurobehavioral tests that were reported as significant out of the total number of tests administered is listed in column 6 of Table 5.5; the percentage is 0% to 83% (mean = 35.9%). The mean effect size of all the measures in the table is 0.86, which is "large" according to Cohen (1992). Thus, evidence of a consistent and large neurocognitive deficit in returning symptomatic veterans emerges from the disparate studies. That one study (Storzbach et al. 2001) found that a subgroup defined the extreme of the distribution of test scores suggests that a unique experience or exposure may have been responsible for the differences. Hom et al. (1997) may have been studying a similar subgroup in their factor-analytic study of "Gulf War syndrome" in a larger group from which the subgroup was selected.

Related Findings: Malingering and Association of Symptoms with Objective Test Results

Several studies provide ancillary information related to the two main questions. Results of a standard test of malingering (TOMM—Test of Memory Malingering) suggested that veterans were not attempting to fake poor memory performance in the study of White et al. (2001). In fact, all veterans earned a perfect or nearly perfect score on the TOMM (low scores are associated with malingering) (Lindem et al. 2003c). The computer-based ODTP used by Storzbach et al. (2001) was adapted from a standard neuropsychologic test of malingering. The component of the ODTP that was designed to detect poor motivation or malingering (memory errors) led to the exclusion of only two of 351 participants. That suggests that the veterans were not faking poor performance on the tests and supports the validity of the neurobehavioral test results.

Neurobehavioral performance scores were correlated only weakly with self-reported symptoms of poor memory drawn from the SCL-90 test (Binder et al. 1999) or self-reported symptoms of poor overall neuropsychologic performance (Lindem et al. 2003b). Thus, subjective symptoms were not a good predictor of objective test performance. Lindem and colleagues (Lindem et al. 2003a; Lindem et al. 2003b) reported associations between neurobehavioral performance and mood (Lindem et al. 2003b), PTSD symptom severity (Lindem et al. 2003a), and anger and depression measures (Hull et al. 2003).

Those findings suggest that neurobehavioral monitoring or screening tests should be administered to all veterans, not only those who report symptoms of concentration and memory loss, because subjective reports are not accurate predictors of objective lower performance (Binder et al. 1999; Lindem et al. 2003b). Tests that should be considered are those listed multiple times in the tables, particularly tests with the highest effect size (such as PASAT, Stroop, Digit Span, and Trail-making). Having predeployment data on the tests would make them far more useful for individual assessments, and tests that did not reveal differences in veterans should also be included to provide a control for broad-based low performance or intentionally poor individual performance (Storzbach et al. 2001).

Summary and Conclusion

Primary studies found nonsignificant trends of poorer neurobehavioral performance when Gulf War veterans were compared with nondeployed veterans or those deployed to Germany. However, when PTSD (White et al. 2001) or depressed mood (David et al. 2002) was treated as a confounder in the statistical analyses those trends disappeared. The results were adjusted for depression because it is often found to coexist with PTSD. That adjustment could have made it impossible to detect cognitive differences.

One study concluded that Gulf War veterans who report symptoms associated with the Gulf conflict performed more poorly on neurobehavioral tests than veterans who did not report symptoms (Storzbach et al. 2001); another study found substantial neurobehavioral deficits in deployed veterans but had intentionally recruited veterans who experienced a high prevalence of post-Gulf War illness (Hom et al. 1997). That study, however, failed to adjust for key confounders and for the large number of statistical comparisons in their study, raising doubt about the validity of their findings.

With regard to malingering, the component of the ODTP that was designed to detect poor motivation or malingering (memory errors) led to the exclusion of only two of 351 participants (Storzbach et al. 2001). That suggests that the veterans were not faking poor performance on the tests and supports the validity of the neurobehavioral test results.

In conclusion, primary studies of deployed Gulf War veterans vs veterans not deployed to the Gulf do not demonstrate differences in cognitive and motor measures as determined through neurobehavioral testing. However, returning Gulf War veterans who had at least one symptom commonly reported by Gulf War veterans (fatigue, memory loss, confusion, inability to concentrate, mood swings, somnolence, gastrointestinal distress, muscle or joint pain, or skin or mucous membrane complaints) demonstrated poorer performance on cognitive tests than returning veterans who did not report such symptoms. Results of primary and secondary studies are summarized in Tables 5.6 and 5.7.

TABLE 5.6 Neurobehavioral or Neurocognitive Outcomes: Primary Studies

Reference	Population	Outcome Measures	Results	Statistics and Adjustment	Comments or Limitations
David et al. 2002 (derivative of Unwin et al. 1999) Design: case-control	341 randomly selected male deployed; Bosnia peacekeeping; Gulf War-Era non-deployed UK soldiers who were either ill or well (from the first decile of the SF-36 physical functioning subscale)	WAIS-R scaled scores: Vocabulary Digit span Arithmetic Similarities Picture arrangement Block design Object assembly Digit symbol PASAT Sustained attention to response task Stroop Trailmaking A and B WMS: Logical memory (Immediate and delayed recall) Verbal paired associates (Immediate and delayed recall) Camden recognition memory test Purdue pegboard Individually administered tests, blinded examiners	Gulf War-deployed had significantly lower scores on 5 cognitive tests after demographic confounder and LSD corrections: Digit symbol Trail-making Stroop PASAT Verbal associates Final Bonferroni adjustments for multiple comparisons and BDI revealed the only difference was in the Purdue pegboard	ANCOVA, adjusted for education, age, NART, BDI Multiple comparison adjustment: LSD procedure and Bonferroni	Careful treatment of potential confounders such us depression mood, IQ and education. However, it is difficult to determine the total number (n) in the original Unwin et al. (1999) study or this one. Also, the definition of the controls in this study is confusing
Proctor et al. 2003 (derivative of Ishoy et al. 1999b) Design: cross-sectional	143 male Gulf War-deployed Danish military veterans vs 72 male non-deployed military forces randomly selected from 84% and 58% of total number of Danish armed forces deployed and nondeployed, respectively, at the time of the Gulf War	WAIS-R scaled scores: Block design Information CVLT WMS Visual reproductions NES continuous-performance test Trail-making Wisconsin card sorting test Purdue pegboard TOMM Individually administered tests except test in computer-based	Authors' conclusion: no overall differences in neuropsychologic domains. Significant tests within domains: CVLT Wisconsin Card Sort	MANCOVA by neuropsychologic domain, adjusted for age	Rationale of initial mailing is unclear (see Ishoy et al. 1999, in Chapter 4)

Reference	Population	Outcome Measures	Results	Statistics and Adjustment	Comments or Limitations
Storzbach et al. 2001 (derivative of McCauley et al. 1999) Design: case-control	239 randomly selected Gulf War male and female deployed veterans with symptoms vs 112 deployed with no symptoms; case = one of memory loss, confusion, inability to concentrate, mood swings, somnolence, gastrointestinal distress, fatigue, muscle and joint pain, skin or mucous membrane lesions lasting 1 month or longer, starting during or after service in gulf, and present during 3 months before questionnaire received	neurobehavioral system (NES); blinded examiners Symbol digit Serial digit learning ODTP Selective attention test Digit span Simple reaction time BARS computer-based testing system Blinded examiners	Cases significantly worse than controls on: Digit span backward Simple reaction time ODTP Errors Latency (including a slow group of 13% of sample with scores > 2 SD slower than control mean latency) PCA showed the slow ODTP (Slow Case in 1999) were responsible for group differences in neurobehavioral performance; 2 of 354 excluded for possible poor motivation because of excess errors in ODTP	ANCOVA, adjusted for age, sex and AFQT, but effect was small so t-tests were used; Bonferroni correction for multiple comparisons	

NOTE: AFQT = Armed Forces Qualifying Test; ANCOVA = analysis of covariance; BARS = Behavioral Assessment and Research System; BDI = Beck Depression Inventory; CVLT = California Verbal Learning Test; LSD = least significant difference; MANCOVA = multivariate analysis of covariance; NART = National Adult Reading Test; NES = Neurobehavioral Evaluation System; ODTP = Oregon Dual Task Procedure; PASAT = Paced Auditory Serial Addition Test; PCA = principal components analysis; TOMM = Test of Memory Malingering; WAIS = Wechsler Adult Intelligence Scale; WMS = Wechsler Memory Scale.

TABLE 5.7 Neurobehavioral or Neurocognitive Outcomes: Secondary Studies

Reference	Population	Outcome Measures	Results	Statistics and Adjustment	Comments and Limitations
White et al. 2001 (derivative of Proctor et al. 1998) Design: cross-sectional	343 veterans (293 randomly selected deployed who returned to Massachusetts and Louisiana, and 50 from entire National Guard reserves in Maine sent to Germany during the Gulf War)	WAIS-R scores: Information Block design WMS-R: Digit span Paired associate learning Visual reproduction CVLT NES CPT Trail-making A and B Wisconsin card-sorting test PASAT Finger tapping Purdue pegboard TOMM Individually administered tests except test in computer-based NES Blinding not addressed	No differences between the domains; trends of deficits in Gulf War-deployed groups seen on continuous performance, PASAT, Wisconsin Card Sorting Test, trail-making A errors, Purdue pegboard; no evidence of malingering on TOMM	Multivariate regression controlling for age, sex, education, sampling design; Bonferroni-adjusted differences within neuropsychologic domains	Small sample of 50 controls from Maine reserve unit; combined demographically heterogeneous groups to form deployed group
Vasterling et al. 2003 (derivative of Brailey et al. 1998) Design: cross-sectional	72 randomly selected male and female south Louisiana National Guard and military reserve unit deployed vs 33 male and female activated but not deployed veterans (estimated 33% response rate) from larger Brailey et al. study	UPSIT Digit span Rey Auditory Verbal Learning Test Continuous Visual Memory Test Wisconsin card-sorting test Purdue pegboard WAIS-R information Individually administered tests Examiner blinding not addressed	No differences on neurobehavioral tests	ANCOVA with age and rank as co-covariates; Bonferroni correction for multiple comparisons within neuropsychologic domains	Small sample, large difference in sample size between control and deployed group, lack of adjustment for multiple comparisons

Reference	Population	Outcome Measures	Results	Statistics and Adjustment	Comments and Limitations
Axelrod and Milner 1997 Design: case vs standard scores	44 male deployed veterans of Michigan Army National Guard reserve unit vs standard or T scores; subjects were volunteers from 78 veterans in unit; data grouped by (1) self-report of cognitive symptoms or (2) none since Gulf War	Rey Auditory Verbal Learning Test^R Reitan-Indiana Aphasia Screen^{AE} WAIS-R (score)^S Controlled Oral Word Association Test^R Semantic (Category) Fluency^R Peabody Individual Achievement Test$^{R, S}$ Wisconsin card-sort test^{AE} Stroop Color and Word Test^A Trail-making test^{AE} WMS$^{R, AE}$ Finger tapping test^{AE} Grooved pegboard^{AE} Grip strength^{AE} ^{A} Age-adjusted T scores ^{AE} Age- and education-adjusted T scores ^{R} Raw scores ^{S} Standard scores	Deployed veterans had deficits on Stroop^A, Grooved Pegboard^{AE}; symptomatic veterans had deficits on Stroop word, color & color-word naming^A, semantic fluency^R	t-test; no adjustment for demographic confounders, but analysis indicated groups did not differ on these issues; no adjustment for multiple comparisons; individual tests were compared with scores according to legend in column 3	Small sample; lack of group differences on demographic confounders does not eliminate need for adjustments; no controls; standard scores do not adjust for occupational group characteristics or local differences, such as in education

145

Reference	Population	Outcome Measures	Results	Statistics and Adjustment	Comments and Limitations
Goldstein et al. 1996 Design: cross-sectional	21 male and female deployed with health-related complaint vs 38 male and female demographically matched community controls (part of study of chemical exposure); basis for selection (such as, random) not addressed	WAIS-R: Information Similarities Digit span Digit symbol Picture completion Block design Verbal associative learning task Incidental memory task Symbol-digit learning task Trail-making Grooved pegboard CPT Brown-Peterson distractor-type memory task Individually administered tests, examiner blinding not addressed	Impairment index comparison revealed significant deficits in deployed with health-related complaints vs controls (no complaints), but SCL-90-R covariates attenuated other differences	MANOVA; no adjustments for confounders or multiple comparisons	Small sample; selection basis for participants not clear
Bunegin et al. 2001 Design: case-control	Eight deployed male veterans with one or more neurobehavioral symptoms and eight asymptomatic male controls who had no negative responses to odors; participants recruited from newspaper ads and from San Antonio veterans hospital	While exposed to acetone or not: Horizontal addition Pattern memory Switching attention Digit span NES computer-based tests Examiner blinding not addressed	Symptomatic veterans had less correct responses on digit span, pattern memory, switching attention (side direction side) than controls; (acetone exposure had no effect on performance)	RMANOVA; no differences in groups on age, but no adjustment for age or other confounders or multiple comparisons	Very small sample; selection basis for participants not clear; community participants were respondents to newspaper ads

Reference	Population	Outcome Measures	Results	Statistics and Adjustment	Comments and Limitations
Hom et al. 1997 (derivative of Haley et al. 1997b) Design: case-control	26 male deployed veterans with highest attention, memory, reasoning "syndrome" scores (syndrome based on (Haley et al. 1997b) factor analysis of 249 Naval Mobile Construction Batallion veterans) vs 10 deployed with no health problems plus 10 nondeployed from same battalion (the 20 controls were selected to be demographically similar to "syndrome" group in age and education and were treated as single group for analysis)	WAIS-R: Information Comprehension Digit span Arithmetic Similarities Vocabulary Picture arrangement Picture completion Block design Object assembly Halstead category test Tactual performance tests A & B Speech-sounds perception test Finger Oscillation test Trail-making A & B Reitan-Indiana aphasia screening examination Reitan-Klove sensory-perception and lateral dominance examinations Reitan word-finding test WMS-Russell Revision Verbal and Figural immediate and delayed recall Wide range achievement test 3 Individually administered tests Blinded examiners	One-tail t tests showed deficits in deployed in Halstead category test, trailmaking B, grip measures, WAIS comprehension, WAIS vocabulary bilateral, simultaneous tactile-left, finger recognition-right, left-right confusion	One-tail t tests for neurobehavioral tests; two-tail tests for psychologic tests; no adjustment for demographic confounders or multiple comparisons	Small sample; many neurobehavioral measures without statistical adjustment for multiple comparisons; used one-tail tests for those comparisons

Reference	Population	Outcome Measures	Results	Statistics and Adjustment	Comments and Limitations
Lange et al. 2001 Design: case-control, secondary study	48 deployed veterans who reported fatiguing illness vs 39 veterans who did not	NES Simple Reaction Time NES Complex Reaction Time PASAT WAIS-R: Digit span Block design CVLT Rey-Osterrieth complex figure test Trail-making A & B Judgment of line orientation test Grooved pegboard test Individually administered tests except tests in computer-based NES; blinding not addressed	Significant deficits in deployed fatigued on simple and complex reaction time, digit span backward, PASAT, trail-making B, category test; followed by multiple regression fitted with psychopathology variables MDD, anxiety disorders, PTSD as independent variables and cognitive measures as dependent variables, revealing that MDD predicted simple and complex reaction-time performance, and MDD and anxiety disorders predicted PASAT performance	MANOVA followed by t tests adjusted for multiple comparisons by Bonferroni	Source of subjects unclear (may not be randomly selected); usual limitation of a small sample, but statistical analysis is sound, research questions are clearly spelled out
Sillanpaa et al. 1997 Design: case vs standard scores (rather than a control group), secondary study	82 deployed Army Reserve Military Police referred for neuropsychologic analyses vs normative data; somatic complaints on MMPI, SCL-90, and neurologic symptom checklist were basis for defining participants as symptomatic or not	WAIS-R-full scale IQ Rey auditory verbal learning test Wisconsin card-sorting test NES CPT Grip strength Grooved pegboard Vibratory sense-neurologic screen Fingertip number writing perception Individually administered tests, except tests in computer-based NES, blinded examiners	No differences	MANOVA; no confounder adjustments, although all explored and education was significant; no adjustment for multiple comparisons	No control group; standard scores do not adjust for occupational group characteristics or local differences, such as in education

Reference	Population	Outcome Measures	Results	Statistics and Adjustment	Comments and Limitations
Lindem et al. 2003c (derivative of Proctor et al. 1998) Design: cross-sectional, secondary study	77 veterans (58 randomly selected deployed from Louisiana, and 19 from National Guard reserves in Maine deployed to Germany during Gulf War) divided into those with high and low scores on TOMM; their neurobehavioral test performance was compared	WAIS-R full scale IQ Rey Auditory Verbal Learning Test Wisconsin card-sorting test NES CPT Grip strength Grooved pegboard Vibratory sense ("neurologic screen") Fingertip number-writing perception Individually administered tests except tests in computer-based NES; blinded examiners	21 veterans scoring high on TOMM had significantly lower scores on some neuropsychologic tests compared to 56 scoring lower; overall, most veterans' scores indicated they were well motivated	MANCOVA by domain; also adjusted for multiple comparisons by requiring significant univariate and grouping variables; adjusted for age and education	Small sample limits conclusions
Binder et al. 1999 (derivative of McCauley et al. 1999) Design: case-control, secondary study	100 randomly selected northwest US male and female deployed veterans grouped as symptomatic (cases) vs asymptomatic (controls) (first 100 of sample from Storzbach et al. 2000)	Correlated self-ratings of cognition (8 SCL-90 R items indicative of cognitive complaints) and affective stress (Beck Depression and Anxiety Inventories) with neurobehavioral performance: Symbol digit ODTP Selective attention task Digit span Simple reaction time BARS computer-based testing system; blinded examiners	Correlations between neurobehavioral performance and subjective cognitive complaints on SCL-90 R were 0.28 or less; affective complaints correlated 0.58 with subjective cognitive complaints; thus, self-ratings of cognitive deficits are poorly correlated with objective measures of cognitive performance	R, plus q statistic that showed cognitive symptoms (SCL-90 R) had a stronger relationship with affective distress (Beck scales) than with cognitive test performance	This suggests that epidemiologic studies that use self-reported symptoms are hard to interpret and at least must adjust for affective distress (such as depression)

150

Reference	Population	Outcome Measures	Results	Statistics and Adjustment	Comments and Limitations
Lindem et al. 2003b (derivative of Proctor et al. 1998) Design: case-control analysis of cross-sectional study, secondary study	240 veterans (193 randomly selected deployed who returned to Massachusetts and Louisiana and 47 from entire National Guard reserves in Maine sent to Germany during Gulf War; veterans were grouped by degree of symptom reporting into high, medium, and low on HSC	WAIS-R: Information Block design WMS-R: Digit span Paired Associate Learning Visual reproduction CVLT NES CPT Trail-making A & B Wisconsin card-sorting test PASAT Finger tapping Purdue pegboard TOMM Individually administered tests except test in computer-based NES; blinding not addressed	Subjective complaint (on HSC) not associated with objective neurobehavioral performance deficits; veterans deployed to gulf reported more severe neuropsychologic symptoms than those deployed to Germany	MANCOVA by domain; also adjusted for multiple comparisons by requiring significant univariate and grouping variables; adjusted for age, education, WAIS-R information, deployment status, medical conditions, PTSD severity, depression	Sound statistical analysis
Lindem et al. 2003a (derivative of Proctor et al. 1998) Design: cross-sectional, secondary study	225 veterans (178 randomly selected deployed who returned to Massachusetts and Louisiana and 47 from entire National Guard reserves in Maine sent to Germany during Gulf War); PTSD symptom severity (based on CAPS) was related to objective measures of neuropsychologic performance	WAIS-R: Information Block design WMS-R: Digit span Paired Associate Learning Visual reproduction CVLT NES CPT Trail-making A & B Wisconsin card-sorting test PASAT Finger tapping Purdue pegboard TOMM Individually administered tests except test in computer-based NES; blinding not addressed	PTSD symptom severity (based on CAPS) was correlated with neuropsychologic performance on: WAIS-R information, CPT, tapping, Purdue pegboard, CVLT	Partial correlational analyses controlled for age, education, WAIS-R information, deployment status, depression as measured by the Structured Clinical Interview for DSM-III-R (SCID); R^2 measures obviate multiple comparison adjustment	

Reference	Population	Outcome Measures	Results	Statistics and Adjustment	Comments and Limitations

Hull et al. 2003 (derivative of Unwin et al. 1999)
 	
Design: case-control, secondary study
	136 males from three groups: deployed UK soldiers, nondeployed UK soldiers, UK soldiers deployed to Bosnia–divided into symptomatic cases (from lowest decile of SF-36 physical functioning subscale) and nonsymptomatic cases	WAIS-R – scaled scores:
		  Vocabulary
		  Digit span
		  Arithmetic
		  Similarities
		  Picture arrangement
		  Block design
		  Object sssembly
		  Digit symbol
		WMS-R:
		  Logical memory--Immediate and delayed recall
		  Verbal Paired Associates-- Immediate and delayed recall
		Letter-number sequencing task
		PASAT
		Sustained Attention to Response Task
		Stroop
		Trail-making A & B
		Camden Recognition Memory Tests
		Purdue pegboard
		Individually administered tests; blinded examiners
			Moderate (0.41) relationships between anger and depression with neuropsychologic measures	Structural equation modeling	Small sample for structural equation modeling

Reference	Population	Outcome Measures	Results	Statistics and Adjustment	Comments and Limitations
Anger et al. 1999 (derivative of McCauley et al. 1999) (also see Storzbach et al. 2001 in primary studies, Table 5.6 above) Design: case-control, secondary study	101 randomly selected northwest deployed male and female veterans grouped (first 101 scheduled for study) as symptomatic (cases, n = 66) vs asymptomatic (controls, n = 35); cases: symptoms lasting 1 month or longer, began during or after service in gulf, present during 3 months before questionnaire received	Symbol digit Serial digit learning ODTP Selective Attention Test Digit span Simple reaction time BARS computer-based testing system; blinded examiners	Cases significantly worse on ODTP after Bonferroni correction, including "slow group" of 13% of the sample with very low scores; cases inferior on all other neurobehavioral tests, and three were significant before Bonferroni correction; performance test results and ODTP performance suggested good motivation	t tests; Bonferroni adjustment for multiple comparisons	Report of "slow case" group after first 100 participants; also described in full sample by Storzbach et al. (2000); a limitation is that slow group has only 13 veterans, so it is hard to identify important determinants of slow cases

NOTE: ANCOVA = analysis of covariance; BARS = Behavioral Assessment and Research System; CAPS = Clinician Administered PTSD Scale; CPT = Continuous Performance Test; CVLT = California Verbal Learning Test; HSC = Health Symptom Checklist; MANOVA = multivariate analysis of variance; MDD = Major Depressive Disorder; MMPI = Minnesota Multiphasic Personality Inventory; NES = Neurobehavioral Evaluation System; ODTP = Oregon Dual Task Procedure; PASAT = Paced Auditory Serial Addition Test; R = Revised; RMANOVA = Repeated Measures Analysis of Variance; SCL = Symptom Check List; TOMM = Test of Memory Malingering; UPSIT = University of Pennsylvania Smell Identification Test; WAIS = Wechsler Adult Intelligence Scale; WMS = Wechsler Memory Scale.

DISEASES OF THE NERVOUS SYSTEM
(ICD-10 G00-G99)

The neurologic outcomes reviewed in this section are amyotrophic lateral sclerosis (ALS), peripheral neuropathy, and other neurologic outcomes.

Amyotrophic Lateral Sclerosis

ALS is a neuromuscular disorder; it is often referred to as Lou Gehrig's disease and might also be called motor neuron disease or Charcot's disease. It affects approximately 20,000 to 30,000 people in the United States (NINDS 2006; The ALS Association 2006). ALS affects all races and ethnic backgrounds and the risk is higher in men than women of the same age (Annegers et al. 1991). The disease is often relentlessly progressive and almost always fatal. The rate of progression is quite variable from patient to patient.

ALS causes degeneration of the motor neurons in the cerebral motor cortex (called upper motor neurons) and the brain stem and spinal cord (called lower motor neurons) (Rowland 2000). The motor neurons are nerve cells that provide communication between the highest levels of the nervous system and the voluntary muscles of the body (NINDS 2006). When the upper motor neurons degenerate, their connections to the lower motor neurons and spinal interneurons are disrupted. That disruption leads to weakness of muscles in a characteristic distribution and the development of spasticity. Lower motor neuron degeneration disrupts nerve contact to the muscles resulting in muscle atrophy. Spontaneous muscle activity, called fasciculation, occurs. Eventually, affected people are unable to move their arms and legs and cannot speak or swallow. When the connection is disrupted between the neurons and the muscles responsible for breathing, patients either die from respiratory failure or require mechanical ventilation to continue to breathe. The majority of persons with ALS die from respiratory failure within 5 years from the onset of symptoms. To be diagnosed with ALS, patients must have signs and symptoms of both upper and lower motor neuron damage that cannot be attributed to other causes (such as progressive muscular atrophies and varieties of peroneal muscular atrophy, Kennedy's syndrome, or multifocal motor mononeuropathy).

Five to ten percent of ALS cases are familial (inherited) and the remainder are sporadic (Rowland 2000; Siddique et al. 1999). Only one parent needs to carry the mutant gene for ALS to occur in about half of the children in cases of familial ALS (NINDS 2006). The specific gene mutations causing the majority of familial ALS cases are unknown. However, about 20% of familial cases are believed to be caused by a mutation in a gene that encodes the enzyme superoxide dismutase 1 (NINDS 2006).

The cause of sporadic ALS is unknown. Despite a number of epidemiologic studies examining occupation (for example, Italian professional soccer players, farmers, and electricians), physical trauma, strenuous physical activity, lifestyle factors (for example, diet, body mass index, cigarette use, and alcohol consumption), ethnic group, and socioeconomic status, there are no consistent findings (Armon 2003; Armon 2004a; Chio et al. 2005; Rowland 2000; Valenti et al. 2005). It has been suggested that the risk of ALS may decrease with decreasing latitude (north-south gradient of risk) (McGuire et al. 1996a).

Primary Studies

Horner and colleagues (2003) conducted a nationwide, epidemiologic case-ascertainment study to determine if Gulf War veterans have elevated rates of ALS. They used active and passive methods of case ascertainment. Active methods included screening of inpatient, outpatient, and pharmacy medical databases of VA or DOD. Passive methods included establishment of a toll-free telephone number, solicitations through relevant Internet sites, and mass mailings of study brochures to practicing neurologists in the VA and to members of the American Academy of Neurology. The ALS diagnosis was verified by medical-record review.

Among nearly 2.5 million eligible military personnel, nearly 700,000 had been deployed to the Gulf War. In the deployed population, there were 107 cases of ALS identified, for an overall occurrence of 0.43 per 100,000 persons per year. Most of the cases were found with active ascertainment methods. An increased risk of ALS was found among all deployed personnel (RR 1.92, 95% CI 1.29-2.84). The attributable risk (that is, the risk difference or excess risk) associated with Gulf War deployment was 18% (95% CI 4.9-29.4%). The foremost limitation of the study was potential underascertainment of cases, particularly among nondeployed veterans, because nondeployed veterans had less incentive to participate. Because of the rarity of ALS, underascertainment of a few cases, particularly if it is greater among the nondeployed, can substantially exaggerate results. Underascertainment of nondeployed cases would make the risk among the deployed appear higher by comparison. Contributing to this concern is the finding that the incidence of ALS in deployed veterans was actually lower than that of an age-matched sample from Washington state (McGuire et al. 1996b). Another study limitation was failure to consider smoking as a possible confounding factor in the study design. Smoking may be a risk factor for the development of ALS (Nelson et al. 2000). Those study limitations were raised in a letter to the editor by an ALS researcher (Armon 2004b). In response, the study authors pointed out the low likelihood that smoking rates would be different among deployed and nondeployed veterans. They also undertook a secondary analysis to address concerns about differential case ascertainment among deployed vs nondeployed veterans.

The secondary analysis (Coffman et al. 2005) estimated the occurrence of ALS with capture-recapture statistical methods. The study used three methods of capture-recapture: log-linear models, sample coverage, and ecologic models. The investigators found modest underascertainment of cases in nondeployed military personnel and little underascertainment in the deployed. Correcting the rates for underascertainment, the investigators still found a higher age-adjusted risk of ALS among the deployed than among the nondeployed (RR 1.77, lower bound 1.21, with log-linear model). The results confirmed the original findings of Horner et al. (2003) of an increase in ALS among deployed veterans. Their estimates also address the criticism that the incidence among the deployed is lower than that in the study of Washington State. See Table 5.8 for a summary of the primary ALS studies discussed above.

Secondary Studies

Haley (2003) found an excess incidence of ALS among deployed veterans in comparison with the expected incidence based on US vital statistics. His analysis spanned 1991-1998. In the first half of that period the increased incidence was not apparent; from 1995 to 1998, the incidence more than doubled (standardized mortality ratio [SMR] 2.27, 95% CI 1.27-3.88). Although the study used passive and active means of case ascertainment similar to those of Horner et al., it differed in several key respects: it restricted cases to those below the age of 45 years (instead of all ages); it used 8 years of followup (instead of 10 years), and it used as a comparison population the age-adjusted rates from US mortality statistics (instead of age-adjusted rates in nondeployed veterans). The major criticism of the study is its use of mortality statistics for the general population to estimate the "expected" incidence (Armon 2004b). That may have underestimated the expected rates for the comparison population, thereby making the SMRs appear higher than they actually were.

Several US and UK mortality studies have not found an excess risk of ALS, but they did not have sufficiently long followup or sufficiently detailed methods (Defence Analytical Services Agency 2005; Kang and Bullman 1996; Macfarlane et al. 2000). A hospitalization study (Smith et al. 2000) also found no difference in relative risk of ALS (RR 1.66, 95% CI 0.62-4.44), but the authors acknowledge that they had too few cases to make valid comparisons between deployed and nondeployed veterans. The study was also limited by inclusion of only active-duty military personnel and only 6 years of followup. Nicolson and colleagues, studying eight Gulf War veterans with ALS and two other comparison populations, found that ill Gulf War veterans had the highest frequency of systemic mycoplasm infections (Nicolson et al. 2002). Although the authors suggest that mycoplasma might be involved in the pathogenesis or progression of ALS, insufficient information was given regarding the selection of cases and controls to determine whether they are representative.

Summary and Conclusion

Two primary studies and one secondary study found that deployed veterans appear to be at increased risk for ALS. The primary study, by Horner et al., which had the possibility of underascertainment of cases in the nondeployed population, was confirmed by a secondary analysis by Coffman et al. That analysis, using capture-recapture analysis, confirmed the nearly doubled risk. A secondary study by Haley, using general population estimates as the comparison group, found a slightly higher relative risk but within the same general range. Other US and UK mortality studies and a hospitalization study have not found excess risk of ALS. The committee concludes that the further followup is warranted.

TABLE 5.8 Amyotrophic Lateral Sclerosis (ALS)

Reference	Design	Population	Outcome	Results	Adjustment	Comments or Limitations
Horner et al. 2003	Retrospective cohort	All active, Gulf War–deployed military personnel (1990-1991), compared with non-Gulf War–deployed veterans	ALS	All deployed forces, significant increased risk of ALS RR = 1.92% 95% CL -1.29-2.84	Age-adjusted average, annual 10-year incidence; attributable risk	Case ascertainment method through screening of VA and DOD medical databases and benefit files (and TriCare) by ICD-9 code for ALS or riluzole use; toll-free telephone enrollment; Internet notices; mass mailings to neurologists, VA centers, and veteran service organizations
Coffman et al. 2005	Capture-recapture reanalysis of Horner et al. cohort	See Horner et al. 2003	ALS	Found no under ascertainment of ALS cases among deployed	Log-linear models; sample coverage; ecologic models	Possible slight undercounts not likely to substantively affect results

NOTE: DOD = Department of Defense; VA = Department of Veterans Affairs.

Peripheral Neuropathy and Other Neurologic Outcomes

Peripheral Neuropathy

This section reviews studies of peripheral neuropathy, polyneuropathy, or neuromuscular symptoms, as identified by the investigators' conducting the studies. Peripheral neuropathy, broadly defined, is a disease of the peripheral nerve tissues (that is, nerve fibers ensheathed by Schwann cells, including nerve roots), which transmit information from the brain and spinal cord to other parts of the body.

Numerous types of peripheral neuropathy have been characterized, each with its own set of symptoms, patterns of development, and prognosis. Peripheral neuropathy can be classified by a variety of factors, such as the population of nerves fibers affected (for example, motor, sensory, or autonomic). Additionally, neuropathy can be classified by the time course (acute, subacute and chronic, remitting, or relapsing); and by pathology (axonal, demyelinating, or other). Peripheral neuropathy might be inherited (for example, resulting from inborn errors in the genetic code or mutations) or acquired (for example, from physical injury, tumors, toxins, autoimmune responses, nutritional deficiencies, alcoholism, vascular and metabolic disorders, or infections from conditions such as leprosy, HIV, herpes simplex and zoster, or hepatitis associated) (National Institute of Neurological Disorders and Stroke 2006). Polyneuropathy is a neurologic disorder characterized by progressive weakness and impaired sensory function in arms and legs. The committee notes that an objective inquiry of peripheral neuropathy depends on clinical recognition of absent ankle reflexes; distal symmetric leg and foot weakness and atrophy; sensation loss in toes and feet; and abnormalities in nerve conduction. The committee also regards studies with objective and quantitative measures, such as those with nerve-conduction tests, to be optimal.

The best population-based questionnaire study for assessment of peripheral neuropathy is that of Cherry and colleagues (Cherry et al. 2001a), who studied UK troops deployed to the Gulf War. Almost 35% of Gulf War veterans who reported handling pesticides for more than a month indicated numbness or tingling on mannequin diagrams compared with 13.6% of veterans who did not report handling pesticides. However, although the study was well-designed and suggested a dose-response relationship, it was limited by recall bias, lack of clinical evaluations, and the absence of nerve-conduction studies. Self-reporting of peripheral neuropathy symptoms has poor diagnostic accuracy (Franse et al. 2000). Because of those limitations, the committee defined primary studies as requiring, at a minimum, medical evaluations and objective nerve testing.

Primary Studies

In the medical evaluation component of the large, population-based cohort assembled by VA, Davis and colleagues (2004) reported on the presence of distal symmetric polyneuropathy in deployed and nondeployed veterans. That condition was evaluated through history, physical examination, and standardized electrophysiologic assessment of motor and sensory nerves in 1,061 deployed veterans and 1,128 nondeployed veterans. Spouses of deployed and nondeployed veterans were also studied, as was a population of 240 Khamisiyah-exposed deployed veterans. Exposure to potential nerve agents was assessed with one of the first DOD models of atmospheric dispersion (Winkenwerder 2002). Blood studies were performed to rule out metabolic causes of neuropathy. Although the study provided results on distal symmetric polyneuropathy as distal sensory or motor neuropathy identified on the basis of the neurologic examination, nerve conduction study, or both, the committee favored distal symmetric polyneuropathy identified with a nerve conduction study as the best, most reliable measure of peripheral neuropathy. No significant differences between adjusted population prevalence of distal symmetric polyneuropathy in deployed and nondeployed veterans were found (OR 0.65, 95% CI 0.33-1.28). There also were no differences on physical examination or self-reported peripheral neuropathy, although at the time of examination, deployed veterans reported more numbness and tingling than did nondeployed veterans. The veterans exposed to the Khamisiyah ammunition depot explosion did not differ significantly from nonexposed deployed veterans (OR 1.04, 95% CI 0.25-4.37). The prevalence of distal symmetric polyneuropathy in the spouses of deployed and non-deployed veterans also did not differ; however, the measure of distal symmetric polyneuropathy was obtained through self-reports as opposed to medical evaluation or nerve conduction study. One limitation of the study is potential participation bias: only 53% of deployed veterans and 39% of nondeployed veterans invited to participate were actually examined.

Neuromuscular symptoms of UK veterans were evaluated with objective testing of peripheral nerves, skeletal muscles, or neuromuscular junctions in a case-control study (Rose et al. 2004; Sharief et al. 2002). Ill veterans (with more than four neuromuscular symptoms and lower functioning according to the SF-36) were compared with healthy deployed veterans, 13 symptomatic Bosnian veterans, and 22 symptomatic Gulf-War-era controls. All groups had been randomly selected from 8,195 male military personnel. In the first publication, veterans underwent nerve-conduction studies, quantitative sensory and autonomic testing, and concentric needle and single-fiber electromyography. In the second, they underwent quantitative myometry through the ischemic forearm exercise test, the subanaerobic bicycle exercise test, and a muscle biopsy. The studies revealed no statistically significant differences between deployed and nondeployed veterans who had symptoms of Gulf War illness. The sole exception was the greater effort required for the bicycle exercise test with increased lactate production; this finding could reflect mitochondrial damage or inactivity resulting from ill health. See Table 5.9 for a summary of the peripheral neuropathy findings.

Secondary Studies

Other studies supporting the absence of findings include those of Eisen et al. (2005), Joseph et al. (2004), Rivera-Zayas et al. (2001), and Amato et al. (1997). Eisen et al. (2005) appears to have reported in less detail on the peripheral neuropathy findings in the same cohort as previously reported by Davis and colleagues (2004) but the precise relationship between the two publications is not clear. The Joseph et al. study, a retrospective review of electrophysiologic testing of 56 Gulf War veterans and 120 nondeployed veterans, showed no objective evidence of a higher incidence of neuromuscular disease in deployed veterans than in nondeployed veterans. In the Rivera-Zayas et al. study, 12 of 162 Gulf War veterans tested electrophysiologically with positive questionnaires for neuropathy had normal results except for two subjects who had carpal tunnel syndrome. Amato et al. showed that in 20 Gulf War veterans who had severe muscle fatigue, weakness, and myalgias, nerve conduction studies, repetitive nerve stimulation, quantitative and single-fiber electromyography, and muscle biopsies were inconclusive.

Other Neurological Outcomes

Haley and colleagues performed detailed neurologic assessments in several case-control studies of the original cohort of Seabee reservists. The cases were veterans who had met criteria for factor-derived syndromes. Under the hypothesis that those veterans were ill from neurotoxic exposures, especially to organophosphates, the assessments covered broad neurologic function

(Haley et al. 1997a), autonomic function (Haley et al. 2004), vestibular function (Roland et al. 2000), basal ganglia injury (Haley et al. 2000a; 2000b), and paraoxonase (PON) genotype and serum concentrations (Haley et al. 1999). Separate groups of investigators also studied PON genotype or activity (Hotopf et al. 2003; Mackness et al. 1997). A case-control study of neuropsychologic functioning (Hom et al. 1997) is discussed elsewhere in this chapter.

The committee regarded those case-control studies as secondary studies primarily because of their lack of generalizability and strong potential for selection bias. Although their study design was characterized as nested case-control, the studies of Haley et al. were not true nested case-control studies. Cases were, appropriately, selected from the original cohort, but controls were not. Ten of the 20 controls were from 150 newly discovered members of the battalion who had not been deployed. Those 10 were not from the original cohort and there is no indication that they were tested to determine whether they should be treated as cases. The selection of those controls raises the possibility of selection bias. With regard to the other concern, lack of generalizability, the authors selected as cases the most severely affected veterans—that is, those who scored highest on factor analysis-derived syndromes—rather than a random sample of those who met a particular case definition.

Summary and Conclusion

One large, well-designed study conducted by VA did not find evidence of excess peripheral neuropathy in Gulf War veterans. The study used the most thorough and objective case definition. Several other secondary studies supported a conclusion of no excess risk. Some studies (such as that of Cherry et al. (2001a)) did report higher rates of peripheral neuropathy, but they used self-reports, which the committee did not accept as a reliable measure of peripheral neuropathy. Furthermore, because researchers use different case definitions of peripheral neuropathies, there are problems of ascertainment, which makes comparisons among groups difficult.

The committee finds no increase in the prevalence of peripheral neuropathy in deployed vs nondeployed veterans, as defined by history, physical examination, and electrophysiologic studies. Regarding other neurologic outcomes, Haley and colleagues found evidence of basal ganglia injury and other abnormalities with detailed neurologic assessments, in several case-control studies. The committee regarded the studies as secondary because of their lack of generalizability and their strong potential for selection bias.

TABLE 5.9 Peripheral Neuropathy

Reference	Design	Population	Outcomes	Results	Adjustments	Comment or Limitations
Davis et al. 2004	Cross-sectional, prevalence, medical evaluation	1,061 deployed veterans vs 1,128 nondeployed veterans; 240 Khamisiyah-exposed deployed veterans vs 807 non-Khamisiyah-exposed deployed veterans; 482 spouses of deployed vs 527 spouses of nondeployed veterans as controls	Distal symmetric polyneuropathy identified by nerve-conduction study[a]	Deployed vs. nondeployed veterans: OR 0.65 (95% CI 0.33-1.28); Khamisiyah-exposed deployed veterans vs non-Khamisiyah-exposed deployed veterans: OR 1.04 (95% CI 0.25-4.37); no difference between spouses of deployed vs spouses of nondeployed veterans	Excludes coexisting conditions[b]	Low participation rate: 53% in deployed veterans, 39% in nondeployed veterans
Rose et al. 2004, Sharief et al. 2002	Case-control	49 symptomatic deployed veterans vs 26 healthy deployed veterans, 13 symptomatic Bosnia deployed veterans, 22 symptomatic Gulf War-era veterans	Nerve-conduction studies, quantitative sensory and autonomic testing, concentric needle and single-fiber electromyography, ischemic forearm exercise test, subanaerobic bicycle exercise test, muscle biopsy	No significant differences except bicycle exercise test more effortful and produced higher plasma lactate		One positive finding could reflect mitochondrial damage or inactivity resulting from ill health

[a] Although study defined distal symmetric polyneuropathy as distal sensory or motor neuropathy identified on basis of neurologic examination, nerve conduction study, or both, committee defined it by nerve-conduction study alone.
[b] Alcohol dependence, diabetes mellitus, renal insufficiency, hypothyroidism, AIDS/HIV, collagen vascular disease, and neurotoxic medications.

CHRONIC FATIGUE SYNDROME

Chronic fatigue syndrome (CFS) is marked by severe and persistent fatigue with a cluster of other symptoms. Fatiguing syndromes were chronicled 100 years ago and have long been the focus of considerable controversy in the medical establishment (Straus 1991; Wessely 1998). The study of unexplained fatiguing illnesses was greatly facilitated and legitimized in the last decade with the development of a case definition sponsored by the Centers for Disease Control and Prevention (CDC). That case definition has helped clinicians and scientists to recognize and classify CFS (Table 5.10). CDC's case definition, first published in 1988 and revised in 1994 (see below), requires fatigue and related impairment in function, and the occurrence of four of eight other defining symptoms over at least 6 months (Fukuda et al. 1994; Holmes et al. 1988). Of the eight symptoms, the most commonly reported are headaches, post-exertional malaise, impaired cognition, and muscle pain (Buchwald and Garrity 1994).

The etiology of CFS is unknown, and no widely accepted laboratory tests or pathologic physical signs are widely accepted (Epstein 1995). Several biologic correlates of the syndrome have emerged, including dysregulation of the hypothalamic-pituitary-adrenal axis, immune activation, and other measures (Goshorn 1998), but they might be present in only a minority of patients; and those findings are not specific to CFS. Although infectious agents may trigger some cases of CFS, a complex, multifactorial etiology that incorporates biologic, psychologic, and social factors is likely (Wessely 1998). The degree of disability associated with CFS is striking, with high rates of unemployment (Bombardier and Buchwald 1996; Buchwald et al. 1996) and poor quality of life related to health (Hardt et al. 2001; Komaroff et al. 1996).

Before defining a primary and secondary study, it should be noted that CFS is a diagnosis of exclusion. The CDC criteria require that three elements be completed as part of a comprehensive evaluation. The first element, determining whether the symptom criteria for CFS are present, requires that a person be queried specifically about length and severity of fatigue and about eight ancillary symptoms. The second element, determining whether other medical conditions are present, mandates a complete physical examination, a battery of specified laboratory tests, and a medical history. The third element, assessing exclusionary psychiatric conditions, requires an interview by a trained professional to obtain diagnostic information.

Thus, in this report, a primary study for CFS is one in which CFS has been diagnosed. A secondary study is one in which a CFS-like condition has been documented. Both primary and secondary studies needed to include a suitable control group so that findings could be interpreted. Other studies that estimated the prevalence of symptoms of "chronic fatigue" (Gray et al. 1999a; Unwin et al. 1999), or multisymptom illness (Fukuda et al. 1998), are not considered further in this section. Likewise, studies that used scalar measures of disability and poor quality of life related to health (Reid et al. 2001) as surrogates for the CDC criteria are not included. Finally, self-reports of CFS (Unwin et al. 1999) and self-reports of a physician diagnosis of CFS (Gray et al. 2002) were not included among the secondary studies, because diagnostic data obtained that way are highly inaccurate. For example, in the Eisen et al. (2005) study, which the committee considered to be the only primary study, only two or three of 38 deployed and eight nondeployed veterans who self-reported CFS received a formal diagnosis after a comprehensive examination. Others, using a method of classifying a case of CFS based on cutoff scores on a fatigue scale and a functional status instrument, found that only 11% of veterans reporting a diagnosis of CFS met operational CFS study criteria.

TABLE 5.10 Case Definition of Chronic Fatigue Syndrome

Reference	Case Definition
CDC Criteria	Presence of both the following: Clinically evaluated, unexplained, persistent or relapsing chronic fatigue that is of new or definite onset (that is, has not been lifelong), is not the result of ongoing exertion, is not substantially alleviated by rest, and results in substantial reduction in levels of occupational, educational, social, or personal activities. Clinical evaluation includes medical history, physical examination, laboratory studies, and psychiatric assessment. Concurrent occurrence of four or more of the following, which must have persisted or recurred during 6 or more consecutive months of illness and must not have predated the fatigue: self-reported impairment of short-term memory or concentration severe enough to cause substantial reduction in levels of occupational, educational, social, or personal activities; sore throat; tender cervical or axillary lymph nodes; muscle pain; multijoint pain without joint swelling or redness; headaches of a new type, pattern, or severity; unrefreshing sleep; and postexertional malaise lasting more than 24 hours.

SOURCE: Fukuda et al. 1994.

Primary Studies

Only one primary study was identified (see Table 5.12). CFS was one of 12 primary health outcome measures studied by Eisen and colleagues (2005), who conducted medical evaluations in phase III of the nationally representative, population-based VA study. In the period 1999-2001, 1,061 of 11,441 deployed and 1,128 of 9,476 nondeployed veterans selected were evaluated. Those veterans had participated in the phase I survey study conducted in 1995 (Kang et al. 2000). The veterans were randomly selected, and the researchers were blinded to their deployment status. The diagnosis of CFS was based on in-person interviews, examinations, and the strict application of the CDC criteria (Fukuda et al. 1994). One study limitation was that deployed veterans were significantly younger, less educated, less likely to be married, and of lower income than nondeployed veterans, although the analysis adjusted for those factors. Another limitation was that, despite three recruitment waves, participation in the 2005 study was low: only 53% of Gulf War veterans and 39% of nondeployed veterans participated. However, to examine biases associated with nonparticipation, the authors obtained sociodemographic and self-reported health data previously collected in 1991 on participants and nonparticipants from the DOD Manpower Data Center and the 1995 VA study (Kang et al. 2000). Because both deployed and nondeployed participants were more likely than nonparticipants to report CFS, this disparity was adjusted for in the analysis of prevalence. The authors concluded that the population prevalence of CFS was higher in deployed than in nondeployed veterans: 1.6% vs 0.1% (OR 40.6, 95% CI 10.2-161.15). Study strengths are its large, population-based design, stratified sampling method, analysis of participation bias, comprehensive examination, and use of computer-based algorithms by researchers who were blinded to deployment status.

Secondary Studies

An earlier questionnaire study, conducted during phase I of the VA study, surveyed 11,441 deployed and 9,476 nondeployed veterans (Kang et al. 2003). Several items in the 48-item symptom questionnaire served as the basis for meeting the case definition for CFS. After exclusion of veterans who self-reported exclusionary medical conditions that could explain their fatigue, 4.9% of deployed and 1.2% of nondeployed veterans (OR 4.2, 95% CI 3.3-5.5) met the case definition. The investigators found that CFS was not related to the severity of combat stressors. The latter was assessed according to responses to questions on wearing chemical protective gear or hearing chemical alarms, being involved in direct combat duty, or witnessing any deaths. The study was limited by its reliance on solely self-reported symptoms without a physical or laboratory examination and on self-reported physician-diagnosed conditions. Those shortcomings resulted in a higher rate of CFS-like illness than was observed when the same cohorts were sampled and underwent more rigorous medical evaluations as in Eisen et al. (2005).

Proctor and colleagues (2001) conducted in-person interviews of 180 Army veterans selected from the larger Fort Devens cohort to determine the prevalence of CFS. The deployed veterans were compared with 46 members of an air ambulance company deployed to Germany during the Gulf War. The prevalence was determined only according to the symptom criteria specified by the CDC case definition (Fukuda et al. 1994). With that approach, the rate was higher in the Gulf-deployed than the Germany-deployed group (7.5% vs 0%, p = 0.02). When additional information from self-reported medical or psychiatric conditions (such as substance abuse and bipolar disorder) and clinical psychiatric interviews was considered, the prevalence in Gulf veterans decreased to 2%, which was no longer statistically significant. The study demonstrated the importance of performing psychiatric assessments, but it was limited by the relatively small sample and the lack of medical or laboratory evaluations.

Canada deployed more than 3,000 sea, land, and air troops to the gulf region; they participated in a naval blockade and were responsible for one-fourth of enemy interceptions in the gulf. A survey of the entire Canadian Gulf War forces found that deployed veterans were at least 5 times as likely as nondeployed veterans to report symptoms of CFS (OR 5.27, 95% CI 3.95-7.03) (Goss Gilroy Inc. 1998). Veterans were not interviewed or examined, and all data were obtained from self-reports. The CFS-like illness was based on responses to questions derived from the CDC criteria and a score above zero on the Chalder fatigue scale.[4] With only minor modifications, the items used in this study were the same as those used by the Iowa Persian Gulf Study Group (1997). The study was limited by the lack of in-person interviews and examinations and by the nontraditional assessment of CFS.

The Iowa study (1997) surveyed 1,896 deployed and 1,799 nondeployed veterans who listed Iowa as their home state at the time of enlistment. The presence of a CFS-like condition was based on a combination of symptoms used in the CDC criteria (Fukuda et al. 1994) and scores on a fatigue scale (Chalder et al. 1993). The investigators found that the prevalence differed by 1.4% (95% CI 0.9-2.0) after adjusting for age, sex, race, branch of military, and rank. Study limitations were the use of self-reports of symptoms on a questionnaire and the lack of medical evaluations. Although rigorously conducted and analyzed, the study suffers shortcomings similar to those of the Canadian study.

Summary and Conclusion

Because the diagnosis of CFS depends entirely on symptoms, not on physical or laboratory findings, the prevalence is highly variable from study to study. In addition, some of the secondary studies reviewed were not limited to CFS but included fatigue or CFS-like illnesses. One primary study demonstrated a higher prevalence of CFS in deployed than in nondeployed veterans although the absolute difference in risk was very small (1.6% vs 0.1%).

---

[4] The Chalder fatigue scale is widely used to measure physical and mental fatigue in CFS patients.

Secondary studies showed higher prevalences of CFS and CFS-like illnesses among veterans deployed to the Persian Gulf than among their counterparts who were not deployed to the gulf or who were deployed elsewhere.

TABLE 5.11 Chronic Fatigue Syndrome (CFS)

Study	Design	Population	Outcomes	Results	Adjustments	Comments or Limitations
Eisen et al. 2005	Population-based; cross-sectional; prevalence; in-person medical and psychiatric evaluations	1,061 deployed and 1,128 nondeployed	CFS based on in-person interviews using CFS criteria of CDC (Fukuda et al. 1994) and exclusionary diagnoses from history, interviews, examinations, laboratory testing	Prevalence: 1.6% vs 0.1% (OR 40.6, 95% CI 10.2-161.15)	Age, sex, race, education, cigarette-smoking, duty type, service branch, rank	Low participation rates, especially among nondeployed; but analysis of nonparticipants and participants reveals that participants, both deployed and nondeployed, are more likely to report symptoms of CFS

DISEASES OF THE CIRCULATORY SYSTEM
(ICD-10 I00-I99)

Cardiovascular disease is a broad, all-encompassing term for any disorder in the heart or the blood vessels. Heart disease is the leading cause of death of both women and men in the United States. In 2002, 696,947 people (51% of them women) died of heart disease, accounting for 29% of all US deaths. The age-adjusted death rate was 241 per 100,000 of population. Several of the studies that examined cardiovascular disease also included reports of diabetes. Type 2 diabetes is the most common form of diabetes and usually appears in adults, often in middle age (over the age of 45 years). Type 2 diabetes is a progressive disease that contributes to nearly 200,000 US deaths per year. It occurs frequently in people who are overweight; are of black, Latino or Hispanic, American Indian, Asian or Pacific Islander descent; or have a family history of the disease.

Primary Studies

There are two primary studies of cardiovascular effects or diabetes; one study examined the effect of possible nerve agent exposure (Smith et al. 2003) and the other examined deployment as the exposure (Eisen et al. 2005). The latter of the two studies is one of few studies that conducted medical evaluations.

The medical evaluation study is the culmination (phase III) of the nationally representative, population-based study sponsored by VA. Eisen and colleagues (2005) examined 12 primary health-outcome measures, two of which are relevant here: hypertension and diabetes. The study evaluated 1,061 Gulf War and 1,128 non-Gulf War veterans who had been randomly selected from 11,441 Gulf War deployed and 9,476 Gulf War-nondeployed veterans who previously participated in the phase I questionnaire conducted in 1995 (Kang et al. 2000). Neither VA study tested for relationships between exposures and health outcomes. Researchers were blind to deployment status. One limitation was that deployed veterans were significantly younger, less educated, less likely to be married, and of lower income, although the analysis attempted to adjust for those factors. Another limitation was that, despite three recruitment waves, the participation rate in the 2005 study was low, with only 53% of Gulf War veterans and 39% of nondeployed Gulf War veterans participating. To determine nonparticipation bias, the study authors obtained previously collected findings on participants and nonparticipants from the DOD Manpower Data Center and gathered sociodemographic and self-reported health findings from the 1995 VA study (Kang et al. 2000). The lack of difference in hypertension and diabetes between deployed participants and nonparticipants and between nondeployed participants and nonparticipants suggested the lack of participation bias. The medical evaluation revealed no statistically significant differences between deployed and nondeployed in the prevalence of hypertension (OR 0.90, 95% CI 0.60-1.33) or diabetes (OR 1.52, 95% CI 0.81-2.85). The nonsignificant statistical increase in diabetes may warrant further investigation.

Two additional DOD hospitalization studies examined hospitalizations in relation to possible exposure to sarin and cyclosarin as a result of demolishing weapons at Khamisiyah, Iraq, in March 1991. The first, by Gray and colleagues (1999b), was superseded by the second, by Smith and colleagues (2003), which used a more advanced exposure model. Those researchers analyzed hospitalizations in 1991-2000 among 431,762 active-duty military deployed to the Gulf War Theater during the time of the Khamisiyah demolition. Exposure status was determined by whether active-duty military were within the plume area defined by meteorologic-dispersion modeling, according to DOD's revised modeling in 2000 (Rostker 2000), and whether they were within a military unit determined, according geographic information systems data, to have been exposed during a 3-day period at concentrations set by CDC for the "general population limit"(GPL), below which no symptoms are expected. The GPL was adjusted because of the briefer duration of troops' potential exposure. Troops were considered exposed if concentrations were above the adjusted GPL of 0.000003 mg/m^3 for sarin and 0.00001 mg/m^3 for cyclosarin.

Investigators studied hospitalization diagnoses from 15 ICD-10 categories, including "circulatory system diseases" and "endocrine, nutritional, and metabolic diseases". On the basis of Cox's proportional hazard modeling, only one category of disease—circulatory system disease—was related to exposure (RR 1.10, 95% CI 1.05 -1.16), after adjustment for other variables in the model. The increased RR was due to one of 10 cardiac diagnoses: cardiac dysrhythmias (RR 1.23, 95% CI 1.04-1.44). While acknowledging that the finding could have been due to chance, the investigators concluded that the excess in dysrhythmia was "small in comparison with potential observational variability, but the findings are provocative and warrant further evaluation." The authors identified study limitations as use of hospitalizations because outpatient data were unavailable, restriction to DOD hospitals (because of the availability of computerized records), and restriction to hospitalizations of those who remained on active duty after the war. It also was not possible to adjust for confounding exposures. A further problem is the uncertainty in the exposure model (see Chapter 2). See Table 5.12 for a summary of the primary papers reviewed for cardiovascular disease and diabetes.

Secondary Studies

Several large-scale epidemiologic studies included self-reported cardiovascular outcomes or diabetes with new onset after the Gulf War. The committee regarded studies that relied on self-reports of cardiac outcomes as secondary studies. None of the studies, except the VA study, verified the diagnosis with physician records.

No postwar differences between deployed and nondeployed were found in two population-based cohort studies. Australian veterans deployed to the Gulf War were as likely as nondeployed veterans to report physician-diagnosed high blood pressure (OR 1.2, 95% CI 0.9-1.6) (Kelsall et al. 2004a). Similarly, Kansas veterans were as likely as their nondeployed counterparts to report physician-diagnosed high blood pressure (OR 1.24, 95% CI 0.82-1.89), heart disease (OR 1.56, CI 0.69-3.56), or diabetes (OR 1.22, 95% CI 0.45-3.30) (Steele 2000). Other studies had mixed findings. The large VA study, with its phase I questionnaire (Kang et al. 2000), found no statistical increase in physician-diagnosed stroke or diabetes in deployed veterans, but did find increases in physician-diagnosed hypertension; the difference in incidence between deployed and nondeployed was 3.84% (95% CI 3.75-3.93). That finding, however, occurred in the same cohort in which later medical evaluations found no statistical increase in hypertension (Eisen et al. 2005). Studying all Seabee commands, Gray and colleagues (2002) found no statistical increase in physician-diagnosed diabetes but did find increased hypertension (OR 1.63, 95% CI 1.36-1.95). McCauley and colleagues (2002) examined the effects of potential nerve-agent exposure on a cohort of veterans from Portland and Washington. The investigators found no statistical differences in self-reporting of physician-diagnosed diabetes, high blood pressure, or heart disease when they divided their cohort into potentially exposed and unexposed, with exposure defined as being within 50 miles of the Khamisiyah demolition. But when dividing their cohort into deployed and nondeployed to the Gulf War, they found more self-reported physician-diagnosed high blood pressure (OR 1.7, 95% CI 1.3-2.4) and heart disease (OR 2.5, 95% CI 1.1-6.6), but no increase in diabetes (OR 1.0, 95% CI 0.5-2.4). Finally, one study of UK veterans (Unwin et al. 1999) found higher prevalence of self-reported high blood pressure: 8.8% of deployed veterans reported high blood pressure compared with 4.3% of Bosnia veterans and 6.6% of Gulf War-era veterans.

Two postwar hospitalization studies did not detect any differences between deployed and nondeployed active-duty personnel regarding circulatory system or endocrine diseases (Gray et al. 1996; Gray et al. 2000). The methods of those studies are described in the section on hospitalizations and are listed in Table 5.13. Those hospitalization studies had several limitations, including lack of generalizability because they covered only active-duty personnel; inability to capture illnesses with longer latency, such as cancer; and inability to measure outpatient treatment, which is more likely to detect diabetes and cardiovascular effects if they occur. Medical evaluation studies, like that by Eisen and colleagues, have greater capacity to detect cardiac outcomes or diabetes.

Fukuda et al. (1998) studied an Air National Guard unit from Pennsylvania in relation to three comparison groups. Their major goal was to develop a case definition of multisymptom illness and to correlate it with clinical and laboratory examinations. Their case definition required one or more chronic symptoms in two of three categories: fatigue, mood and cognition, and musculoskeletal. Some 45% of deployed and 15% of nondeployed veterans met the case definition. Results of clinical examinations were essentially negative in both groups.

Summary and Conclusion

Primary studies found no statistically significant differences in cardiovascular disease between deployed and nondeployed veterans. However, the specific cardiac outcomes under study were limited to hypertension and abnormal blood sugar. The one study that found an increase in cardiovascular disease was limited to hospitalizations and compared deployed veterans who were possibly exposed to the Khamisiyah plume and those who were not exposed. The increase was due entirely to an increase in cardiac dysrhythmia (Smith et al. 2003). The small but significant increase seen in this single study bears watching, though the study did not address the risk in deployed vs nondeployed veterans. In the secondary studies, deployed veterans were generally more likely to self-report hypertension and palpitations, but those reports were not confirmed in medical evaluations. Thus, it does not appear that there is a difference in the prevalence of cardiovascular disease or diabetes between deployed Gulf War veterans and nondeployed.

TABLE 5.12 Cardiovascular Disease or Diabetes

Study	Design	Population	Outcomes	Results	Adjustments	Comments or Limitations
Eisen et al. 2005	Population-based; cross-sectional; prevalence; medical evaluation	1,061 deployed, 1,128 nondeployed	Hypertension = blood pressure > 140/90 mm Hg or history of hypertension and use of antihypertensive medications; diabetes = fasting glucose of 6.99 mmol/L or greater (≥126 mg/dL) or taking hypoglycemic medication	Hypertension: OR 0.90 (95% CI 0.60-1.33) Diabetes: OR 1.52 (95% CI 0.81-2.85)	Age, sex, race, years of education, cigarette-smoking, duty type, service branch, rank	Low participation rates, especially among nondeployed, but analysis of nonparticipants and participants reveals no differences in hypertension or diabetes
Smith et al. 2003	DOD hospitalization study (1991-2000) of those potentially exposed to nerve agent	99,614 active-duty military considered exposed vs 318,458 nonexposed, according to revised DOD exposure model	ICD-10 diagnostic groups plus some individual codes	Circulatory system diseases: RR 1.10 (95% CI 1.05-1.16); Cardiac dysrhythmia: RR 1.23 (95% CI 1.04-1.44); Endocrine, nutritional, and metabolic diseases: RR 1.00 (95% CI 0.94-1.06)	One or more hospitalizations in a specific diagnostic category	Restricted to DOD hospitals; restricted to hospitalizations for only Gulf War veterans who remained on active duty after the war; not possible to adjust for confounding exposures
Gray et al. 1999b	DOD hospitalization study	Three exposure categories, according to revised DOD exposure model: not exposed (n = 224,804), uncertain low dose exposure (n = 75,717), estimated subclinical exposure (n = 48,770)	ICD-10 diagnostic groups plus some individual codes	Circulatory system diseases: RR 1.12 (95% CI 0.93-1.33); Endocrine, nutritional, metabolic diseases: RR 0.75 (95% CI 0.52-1.07)	Sex, age group, prewar hospitalization, race, service type, marital status, pay grade, occupation	Same as for Smith et al. 2003

NOTE: DOD = Department of Defense; VA = Department of Veterans Affairs.

DISEASES OF THE RESPIRATORY SYSTEM
(ICD-10 J00-J99)

This section covers respiratory outcomes according to the major types of studies published in the Gulf War literature: respiratory outcomes in deployed vs nondeployed personnel without regard to specific exposures, and respiratory outcomes in relation to two types of exposures: to oil-well fires and to nerve agents. Table 5.14 contains a summary of all the studies reviewed in this section.

Associations of Respiratory Outcomes with Deployment in the Gulf War Theater

Primary Studies

Two publications reported on respiratory outcomes among participants in the medical evaluation component (phase III) of the large, population-based VA study (Eisen et al. 2005; Karlinsky et al. 2004). Eisen and colleagues (2005) evaluated 1,061 Gulf War and 1,128 non-Gulf War veterans who had been randomly selected from 11,441 Gulf War-deployed and 9,476 Gulf War-nondeployed veterans who previously had participated in a 1995 questionnaire survey (Kang et al. 2000). No statistically significant increase in the prevalence of self-reported asthma, bronchitis, or emphysema was observed among deployed veterans in models that adjusted for smoking and demographic variables (adjusted OR 1.07, 95% CI 0.65-1.77, for the three diseases combined). Obstructive lung disease was defined by the investigators as a history of lung disease (asthma, bronchitis, or emphysema) or pulmonary symptoms (wheezing, dyspnea on exertion, or persistent coughing with phlegm), and either the use of bronchodilators or at least 15% improvement in FEV_1[5] after use of a short-acting bronchodilator. No increase in obstructive lung disease was observed among deployed personnel (adjusted OR 0.91, 95% CI 0.52-1.59). Limitations of the study include potential selection bias owing to low participation rates – 53% and 39% of deployed and nondeployed veterans, respectively.

Karlinsky and colleagues (2004) reported on results of pulmonary function tests (PFTs) on the same VA population that Eisen and colleagues studied. PFT results were classified into five categories: normal pulmonary function, nonreversible airway obstruction, reversible airway obstruction, restrictive lung physiology, and small airway obstruction. The patterns of PFT results were similar in deployed and nondeployed veterans; there were no statistically significant differences. The pattern of PFT results was also reported to be similar in those exposed and those not exposed on the basis of DOD exposure estimates developed in 2002 (see Chapter 2), of exposure to nerve agents through destruction of munitions at the storage site at Khamisiyah in 1991. Prevalences of self-reported pulmonary symptoms were higher in deployed veterans; however, self-reported diagnoses, use of asthma medications, and self-reported physician visits and hospitalizations for pulmonary conditions were similar in deployed and nondeployed. Although no adjustments were made for covariates, demographic variables were similar in the two groups and a history of tobacco-smoking was more common in deployed veterans than in nondeployed veterans (51% vs 44%, p = 0.03). Limitations include the inadequacy of the sampling strategy description to evaluate bias and no explanation of "matching" in the analysis.

---

[5] Forced expiratory volume (FEV_1) measures how much air a person can exhale during the first second of a forced breath.

Kelsall et al. (2004b) reported respiratory outcomes from a cross-sectional study of Australian Gulf War veterans. The gulf-deployed cohort comprised 1,456 participants and the nongulf military cohort included 1,588 participants. Response rates differed greatly between deployed (80.5%) and nondeployed (56.8%). Deployed veterans reported higher prevalences of all respiratory symptoms and some self-reported symptom-based respiratory diagnoses. Lung function measures adjusted for smoking and other covariates were somewhat higher in the deployed group (for example, $FEV_1/FVC\% < 70\%$; OR 0.8, 95% CI 0.5-1.1). FVC^6, but not FEV_1, was associated with self-report of exposure to oil-fire smoke. Although generally well conducted, the study was limited by the potential for selection bias, the lack of doctor-diagnosed respiratory conditions, and, with respect to effects of exposure to oil-fire smoke, the lack of availability of modeled exposure estimates.

Gray and colleagues (1999a) enrolled 1,497 study subjects from 14 Seabee commands in the US Navy, 527 of whom were Gulf War veterans and 970 were nondeployed veterans. Although respiratory symptoms were reported more frequently by the Gulf War veterans, pulmonary function measures adjusted for age, height, race, and smoking status were not associated with Gulf War status (mean FVC: Gulf War 4.96 L, non-Gulf War 4.99 L, p = 0.77; mean FEV_1: Gulf War 4.05 L, non-Gulf War 4.04 L, p = 0.81).

A cross-sectional study of military personnel from Denmark, involved primarily in peacekeeping or humanitarian roles after the end of the Gulf War, also found increased respiratory symptoms among gulf-deployed personnel (n = 686) compared with nondeployed (n = 231) but no statistically significant differences in pulmonary function (FVC percent of expected: Gulf War 100.7, non-Gulf War 100.7; FEV_1 percent of expected: Gulf War 95.6, non-Gulf War 96.4). Smoking patterns were very similar in the two groups (Ishoy et al. 1999b).

Secondary Studies

The overwhelming majority of secondary studies conducted among Gulf War veterans have found that several years after deployment veterans report higher levels of respiratory symptoms and of respiratory illnesses than nondeployed troops whether from the United States (Doebbeling et al. 2000; Gray et al. 2002; Iowa Persian Gulf Study Group 1997; Kang et al. 2000; Karlinsky et al. 2004; Kroenke et al. 1998; Petruccelli et al. 1999; Steele 2000), the UK (Cherry et al. 2001b; Nisenbaum et al. 2004; Simmons et al. 2004; Unwin et al. 1999), Canada (Goss Gilroy Inc. 1998), Australia (Kelsall et al. 2004b), or Denmark (Ishoy et al. 1999b). For example, the findings of the 1997 survey (Goss Gilroy Inc. 1998) mailed to the entire cohort of Canadian Gulf War veterans found an increase in self-reported respiratory disease (OR 1.35, 95% CI 1.16-1.57), bronchitis (OR 2.81, 95% CI 2.22-3.55), and asthma (OR 2.64, 95% CI 1.97-3.55) when adjusted for tobacco-smoking. The study by Eisen et al. (2005), described previously, is an exception in finding few differences among US veterans in respiratory symptoms and self-reported respiratory diagnoses between deployed and nondeployed troops 10 years after the Gulf War. And some of the many studies that have conducted factor analyses on reported symptoms have found respiratory factors (e.g., Cherry et al. 2001a; Cherry et al. 2001b; Hotopf et al. 2004). Of particular interest, some of the UK reports that found differences in respiratory symptoms and self-reported respiratory diagnoses have included comparisons of those deployed to the Gulf War with those deployed to Bosnia (Nisenbaum et al. 2004; Unwin et al. 1999). The Unwin et al. (1999) study found that the risk of self-reported asthma and bronchitis was higher in the Gulf

---

[6] Forced vital capacity (FVC) is the total amount of air exhaled during an FEV test.

War cohort than in the Bosnia cohort (asthma: RR 1.2, 95% CI 0.8-1.6; bronchitis: RR 1.5, 95% CI 1.0-2.3). The Nisenbaum et al. (2004) study found most respiratory symptoms reported at least twice as often by the Gulf War cohort as by the Bosnia cohort.

Associations of Respiratory Outcomes with Specific Exposures Experienced by Gulf War Veterans During Their Deployment

Exposure to Oil-Well Fire Smoke

In February 1991, retreating Iraqi forces set fire to more than 600 oil wells. Fires burned over a 10-month period until November 1991, exposing thousands of US troops to combustion products. Several studies of US Gulf War veterans exposed to oil-well fires stand out from most other Gulf War studies by virtue of their focus on a narrow set of respiratory health outcomes and on a single type of exposure (smoke from oil-well fires) and their exposure assessment on the basis of models of troop-unit movements in relation to air-quality models that incorporated ground-based monitoring data, as well as satellite imagery. The vast majority of Gulf War health studies focused on multiple health outcomes, multiple exposures, and self-reporting of exposures without validation (see Chapter 2). The studies reviewed below examined long-term respiratory effects *after* deployment. Most of them did not distinguish between new cases and exacerbation of pre-existing respiratory illnesses. Asthma, for example, would not have been grounds for exclusion from entry into the military, so cases that predated deployment would be expected to be present in the deployed population. These studies are important because their exposure estimates are probably relatively robust inasmuch as data were actively collected at at least eight sites for many months and were integrated with remote-sensing imagery, and the locations of units on particular days are expected to have been reasonably valid.

Primary Studies

Cowan et al. (2002) conducted a case-control study to identify cases of physician-diagnosed asthma in the DOD registry (n = 873) and controls without asthma (n = 2,464). The DOD registry was established for active-duty Gulf War military who wished to receive a comprehensive physical examination. Cases were defined by using a military physician-assigned diagnosis of asthma (ICD-9-CM [Clinical Modification] codes 493 and 493.91) based on patient complaints, signs and symptoms, and a physical examination. Specified diagnostic criteria were not used. Pulmonary function testing data were not available. Exposure to smoke from oil-well fires was estimated by linking troop locations with modeled oil-fire smoke exposure. National Oceanic and Atmospheric Administration (NOAA) researchers modeled exposure on the basis of meteorologic and ground-station air-monitoring data (Draxler et al. 1994; McQueen and Draxler 1994) with a spatial resolution of 15 km and a temporal resolution of 24 hours (see Chapter 2). DOD personnel records were used to ascertain each study subject's unit and dates of service. Only Army personnel were included in the study, because their location data were more precise. Two exposure measures were used: cumulative smoke exposure (based on the estimated concentrations on all days when each subject was in the Gulf War Theater) and number of days when the subject was exposed at 65 $\mu g/m^3$ or greater.

Self-reported oil-well-fire smoke exposure was associated with a higher risk of asthma (OR 1.56, 95% CI 1.23-1.97). In addition, modeled cumulative oil smoke exposure was related to a greater risk of asthma (OR 1.21, 95% CI 0.97-1.51 for the intermediate-exposure group of up to 1.0mg-day/m^3; and OR 1.40, 95% CI 1.12-1.76 for the high-exposure group of over 1.0 mg-day/m^3) after controlling for sex, age, race or ethnicity, rank, smoking history, and self-reported exposure. When exposure was classified as number of days with exposure at 65 μg/m³ or greater, the risk of asthma also increased. Both exposure metrics showed statistically significant associations with asthma when treated as continuous variables. Smoking appeared to modify the effect: the effect of oil-well-fire smoke exposure was observed among never-smokers and former smokers but not among current smokers. Study strengths include the objective exposure assessment and the use of physician-diagnosed asthma on the basis of clinical evaluations. Limitations include the lack of pulmonary function data and of specified criteria for the diagnosis of asthma and self-selection into the DOD registry, which could have introduced selection bias (for example, if the cohort was enriched with persons who both experienced exposure and had respiratory conditions, the risk estimate could be biased upward). Moreover, the study examined current asthma cases, so a higher incidence of asthma cannot be distinguished from exacerbation or recrudescence of pre-existing disease. The study did not ask about chronic bronchitis or other respiratory effects.

In contrast, the population-based Iowa cohort of 1,560 Gulf War veterans found no statistical association between modeled oil-well-fire exposure and the risk of asthma (Lange et al. 2002). Five years after the war, veterans were asked about their exposures and current symptoms. Exposure was modeled with an approach similar to that of Cowan et al. (2002). Each veteran's exposure was modeled on the basis of the identified unit and its location during the period of oil-well fires (February-October 1991). Cases of asthma were defined by questions aimed at assessing wheezing and chest tightness. Cases of bronchitis were assessed on the basis of self-reported cough and phlegm production. Both questions pertained to symptoms in the preceding month, so it is not possible to determine whether symptoms were chronic. Self-reported exposure to oil-well fires was associated with a greater risk of asthma and bronchitis. However, there was no statistical association between modeled exposure and the risk of asthma or bronchitis in models that controlled for sex, age, race, military rank, smoking history, military service, and level of preparedness. The correlation between self-reported exposure and modeled exposure was modest (0.40-0.48, $p < 0.05$). The authors ascribed the different results for self-reported and objective exposure measurement to recall bias. A study strength is the population-based sampling, which implies that findings can be generalized to all military personnel in the Persian Gulf. A limitation is that the study addressed the outcome of asthma symptoms rather than an asthma diagnosis. And, chronic bronchitis was not defined with the standard epidemiologic definition, so it was impossible to distinguish between acute and chronic symptoms.

In the postwar hospitalization study of 405,142 active-duty Gulf War veterans, Smith et al. (2002) examined the effect of oil-well-fire exposure. Exposure was estimated by using troop location data and estimated oil-smoke concentrations based on the same NOAA modeling used in the Cowan and Lange studies. There was no association between exposure to oil-well fires and the risk of hospitalization for asthma (RR 0.90, 95% CI 0.74-1.10), acute bronchitis (RR 1.09, 95% CI 0.62-1.90), or chronic bronchitis (RR 0.78, 95% CI 0.38-1.57). There was modest increase in the relative risk of emphysema from oil-well fire smoke (RR 1.36, 95% CI 0.62-2.98). Because most adults who have asthma or chronic bronchitis are never hospitalized for the condition, the study would not be expected to have captured most cases. No information was available on tobacco-smoking or other exposures that may be related to respiratory symptoms.

Secondary Studies

Several other studies on smoke from oil-well fires in the Persian Gulf are methodologically less robust. A cohort study of Gulf War veterans evaluated self-reported combustion exposure but examined pulmonary symptoms only as a broad class; asthma and bronchitis were not specifically evaluated (Proctor et al. 1998). A prospective study of 125 British royal engineer bomb disposal units stationed in Kuwait City found no change in forced expiratory flow (FEF 25-75%) after the oil-well fires were set, but asthma and bronchitis were not specifically evaluated (Coombe and Drysdale 1993). Finally, two ecologic studies of Kuwaiti residents found no increase in the rate of asthma hospitalization after the Gulf War (Abul et al. 2001; Al-Khalaf 1998).

Exposure to Nerve Agents

A study by Gray and colleagues (1999b) reported on the occurrence of postwar hospitalization from 1991-1995 for respiratory system disease from three levels of estimated exposure to nerve agents due to destruction of munitions at the storage site at Khamisiyah in 1991. Their exposure estimates were developed by DOD in 1997. A small increase in risk was seen only in the comparison of the highest exposed group (0.09657-0.51436 mg-min/m^3) with the unexposed group (RR 1.26, 95% CI 1.05-1.51). Limitations of the study include the likely exposure misclassification based on late revised DOD estimates, lack of control for tobacco-smoking, lack of a clear dose-response pattern, and low biologic plausibility of effects on the respiratory system.

As described earlier, Karlinsky et al. (2004) found no associations between pulmonary function measures and exposure to nerve agents at Khamisiyah based on the most recent DOD exposure estimates developed in 2000.

Other Exposures

A study of the UK veteran cohort 7 years after the Gulf War (Cherry et al. 2001a) reported a weak but statistically significant ($p < 0.01$) effect of exposure to insect repellent on a respiratory-health index based on self-reports.

A cohort study of US Gulf War veterans evaluated associations of self-reported exposures with a pulmonary symptom score; asthma and bronchitis were not specifically evaluated (Proctor et al. 1998). Exposure to smoke from tent heaters and to smoke from burning human waste were significantly associated with pulmonary symptom score ($p < 0.001$, and $p = 0.015$, respectively).

Summary and Conclusion

Associations of Respiratory Outcomes with Deployment in the Gulf War Theater

Five studies (Eisen et al. 2005; Gray et al. 1999b; Ishoy et al. 1999b; Karlinsky et al. 2004; Kelsall et al. 2004b) representing four distinct cohorts from three countries (United States, Australia, and Denmark) examined associations of Gulf War deployment with pulmonary function measures or respiratory disease diagnoses based in part on such measures. In none of the studies were such associations found. Each study suffers from various methodologic weaknesses, but the uniformity of findings is striking, especially given that the same studies found Gulf War deployment status to be significantly associated with self-reports of respiratory symptoms among three of the four cohorts. Indeed, the overwhelming majority of studies conducted among Gulf War veterans—whether from the United States (Doebbeling et al. 2000; Gray et al. 1999b; Gray et al. 2002; Iowa Persian Gulf Study Group 1997; Kang et al. 2000; Karlinsky et al. 2004; Kroenke et al. 1998; Petruccelli et al. 1999; Steele 2000), the UK (Cherry et al. 2001b; Nisenbaum et al. 2004; Simmons et al. 2004; Unwin et al. 1999), Canada (Goss Gilroy Inc. 1998), Australia (Kelsall et al. 2004b), or Denmark (Ishoy et al. 1999b)—have found that, several years after deployment, deployed veterans report higher levels of respiratory symptoms and of self-reported respiratory illnesses than nondeployed troops. Of particular interest is the UK cohort study reported in Nisenbaum et al. (2004) and Unwin et al. (1999) which found substantially higher prevalences of respiratory symptoms and self-reports of respiratory disease among those deployed in the Gulf War than among those deployed in another war theater, Bosnia.

Associations of Respiratory Symptoms, Signs, and Illnesses with Specific Exposures Experienced by Gulf War Veterans During Their Deployment

The study of Gulf War veterans of Cowan et al. (2002), which used objective exposure measure and methods, found associations between oil-well-fire smoke and doctor-assigned diagnosis of asthma in veterans. Limitations of the study include the lack of pulmonary function data and of specified criteria for the diagnosis of asthma and the self-selection into the DOD registry. Exposures were well estimated and high but brief, and the exposed population was healthier than in most studies of combustion products. At least aggravation of asthma appears biologically plausible as effect of this exposure, and causation of asthma less certain (the Cowan study did not include questions that would have enabled differentiation of aggravation from causation). The other key Gulf War study of oil-well-fire smoke, based on the Iowa cohort (Lange et al. 2002), which found no relationship between the same objective exposure and respiratory health outcomes, had the advantage of avoiding the potential selection biases of the Cowan et al. study. However, its definitions of respiratory diseases were based entirely on self-reports of symptoms and cannot be viewed as adequate. The study of Smith et al. (2002) found no significant associations between the same objective measures of exposure to smoke from oil-well fires and hospitalization for asthma, acute bronchitis, chronic bronchitis, or emphysema. Limitations of the study include the lack of information on tobacco-smoking, and that most adults in the study age range are seldom hospitalized for those diagnoses which imply that most cases would not be expected to be captured.

The study by Gray and collaborators (1999b) found a small increase in postwar hospitalization for respiratory system disease associated with modeled exposure to nerve agents at Khamisiyah. Limitations of the study include the likely exposure misclassification based on later revised DOD exposure estimates, lack of control for tobacco-smoking, lack of a clear dose-response pattern, and low biologic plausibility of effects on the respiratory system. Karlinsky et al. (2004) found no associations between pulmonary function measures and exposure to nerve agents at Khamisiyah based on the improved DOD exposure estimates developed in 2002; the lack of finding casts further doubt on the validity of the findings of the Gray et al. study.

In conclusion, as is the case for a number of other organ systems, respiratory symptoms and self-reported diseases are strongly associated with Gulf War deployment in most studies addressing this question and used comparison groups of nondeployed veterans. However, the findings of no statistical association of objective pulmonary function measures with Gulf War deployment, in the four cohorts in which this has been investigated, leaves the clinical interpretation of the increased symptoms and self-reported diseases uncertain.

With respect to associations of specific exposures in the Gulf War Theater with pulmonary outcomes, the positive study by Cowan et al. of objective measures of oil-well fire smoke and doctor-assigned respiratory diagnoses is methodologically the strongest to have addressed the question. A well-conceived study that examined associations of pulmonary function measures, and specific-criteria-based physician-diagnosed respiratory diseases and used the same objective measures of exposure to smoke from oil-well fire would be useful. With respect to nerve agents at Khamisiyah, no study that used valid objective estimates of exposure has found statistically significant associations with pulmonary function measures or physician-diagnosed respiratory disease (see Table 5.13 for a summary of respiratory outcomes).

TABLE 5.13 Respiratory Outcomes

Study	Design	Population	Outcomes	Results	Adjustments	Comments or Limitations
Eisen et al. 2005	Population-based, cross-sectional, prevalence, medical evaluation	1,061 US deployed vs 1,128 non-deployed	Self-reported asthma, bronchitis, or emphysema; obstructive lung disease (history of disease or symptoms plus use of bronchodilators or 15% improvement in FEV_1 after bronchodilator use)	Asthma, bronchitis, or emphysema: OR 1.07 (95% CI 0.65-1.77) Obstructive lung disease: OR 0.91 (95% CI 0.52-1.59)	Age, sex, race, years of education, cigarette-smoking, duty type, service branch, rank	Low participation rates, especially among nondeployed
Karlinsky et al. 2004	Cross-sectional, medical evaluation	1,036 US deployed vs 1,103 non-deployed	PFT results classified into five categories: normal, nonreversible obstruction, reversible obstruction, restrictive, small-airways obstruction	No association of PFT-based classifications with deployment status, nor with exposure to nerve agents at Khamisiyah based on 2002 DOD exposure models		No adjustment for smoking or other confounders; description of sampling strategy inadequate to evaluate bias; no explanation of "matching" or control of matching in analysis
Kelsall et al. 2004b	Cross-sectional, medical evaluation	1,456 Australian deployed vs 1,588 nondeployed	Asthma; bronchitis; $FEV_1/FVC\% <70\%$	Asthma: OR 1.2 (95% CI 0.8-1.8); Bronchitis: OR 1.9 (95% CI 1.2-3.1); $FEV_1/FVC\% <70\%$: OR 0.8 (95% CI 0.5-1.1); FVC, but not FEV_1, associated with self-report of oil-well-fire exposure	Service type, rank, age, education, marital status	Generally well done; substantial potential for selection bias (response rates: deployed 81%, comparison 57%); no use of modeled oil-fire exposures
Gray et al. 1999a	Cross-sectional, medical evaluation	527 Gulf War veterans vs 970 nondeployed from 14 US Navy Seabees commands	Cough; shortness of breath; FVC (L); FEV_1 (L)	Cough : OR 1.8 (95% CI 1.2-2.8) Shortness of breath: OR 4.0 (95% CI 2.2-7.3)	Age, height, race, smoking status	No use of modeled oil-fire exposures

Study	Design	Population	Outcomes	Results	Adjustments	Comments or Limitations
Ishoy et al. 1999b	Cross-sectional, population-based, medical evaluation	686 peace-keeping Danish deployed to Gulf War Theater vs 231 nondeployed controls	Shortness of breath; FVC; FEV$_1$; peak flow	FVC (L): 4.96 vs 4.99, p = 0.77 FEV$_1$ (L): 4.05 vs 4.04, p = 0.81 14 % vs 3.5% Percent of predicted: 100.7 vs 100.7, NS 95.6 vs 96.4, NS 94.0 vs 92.8, NS	None	Appropriate population-based controls but differential participation: 84% deployed vs 58% nondeployed; smoking histories similar in deployed and nondeployed

Studies of respiratory outcomes specifically associated with modeled oil-well-fire exposure

Study	Design	Population	Outcomes	Results	Adjustments	Comments or Limitations
Cowan et al. 2002	Case-control study of exposure to smoke from oil-well fires; DOD registry, Army only	873 with asthma vs 2,464 controls	Physician-assigned diagnosis of asthma 3-6 years after war	Self-reported exposure: OR 1.56 (95% CI 1.23-1.97) Cumulative modeled exposure: OR 1.24 (95% CI 1.00-1.55) for intermediate cumulative modeled exposure: OR 1.40 (95% CI 1.11-1.75) for high exposure; Number of days at > 65 µg/m^3: OR 1.22 (95% CI 0.99-1.51) for 1-5 days; 1.41 (95% CI 1.12-1.77) for 6-30 days	Sex, age, race, military rank, smoking history, self-reported exposure	Effect seen in former smokers and never-smokers, but not current smokers; key strength: modeled exposure rather than only self-reported exposure; limitations: self-selected population; no specified criteria for asthma diagnosis and no pulmonary function data; pre-exposure asthma status unknown
Lange et al. 2002	Cross-sectional study of exposure to smoke from oil-well fires; derived from cohort study	1,560 Iowa veterans	Asthma symptoms; bronchitis symptoms	For modeled exposure, adjusted ORs for quartiles of exposure, 0.77-1.26 with no dose-response relationship; for self-reported exposure, asthma ORs 1.77-2.83, bronchitis ORs 2.14-4.78	Sex, age, race, military rank, smoking history, military service, level of preparedness for war	Structured interviews conducted 5 years after war; key strengths: modeled exposure rather than only self-reported exposure, population-based sample; key limitation: symptom-based case definition of bronchitis and asthma

Study	Design	Population	Outcomes	Results	Adjustments	Comments or Limitations
Smith et al. 2002	DOD hospitalizations 1991-1999; exposure modeling for oil-well fire smoke	405,142 active-duty Gulf War veterans	ICD-9 CM codes Asthma Acute bronchitis Chronic bronchitis Emphysema Respiratory conditions due to chemical fumes and vapors Other respiratory diseases	Exposed vs non-exposed: OR 0.90 (95% CI 0.74-1.10) OR 1.09 (95% CI 0.62-1.90) OR 0.78 (95% CI 0.38-1.57) OR 1.36 (95% CI 0.62-2.98) OR 0.71 (95% CI 0.23-2.17) OR 1.45 (95% CI 0.86-2.46)	"Influential predictors" of $p<0.15$ included in analyses	Objective measure of disease not subject to recall bias; no issues with self-selection; however, only DOD hospitals, only active duty, no information on smoking or other exposures that may be related to respiratory symptoms, most adults with asthma or chronic bronchitis have never been hospitalized for that condition

Study of respiratory outcomes specifically associated with exposure to nerve agent

Study	Design	Population	Outcomes	Results	Adjustments	Comments or Limitations
Gray et al. 1999b	DOD hospitalizations 1991-1995, exposure to nerve agents at Khamisiyah based on 1997 DOD exposure models	Not exposed (n = 224,804), uncertain low-dose exposure (n = 75,717), exposed (n = 48,770)	Respiratory system disease (vs not exposed): Uncertain low dose <0.013 mg-min/m^3 0.013-0.097 mg-min/m^3 0.097-0.514 mg-min/m^3	OR 0.92 (95% CI 0.85-0.99) OR 0.90 (95% CI 0.77-1.04) OR 0.89 (95% CI 0.79-1.02) OR 1.26 (95% CI 1.05-1.51)	Sex, age group, prewar hospitalization, race, service type, marital status, pay grade, occupation	See Smith et al. 2002 also, probable substantial exposure misclassification as models were revised, lack of a clear dose-response pattern, little biologic plausibility given that no effect was seen for nervous system diseases

Note: DOD = Department of Defense; NS = not significant; PFT = pulmonary function test.

DISEASES OF THE DIGESTIVE SYSTEM
(ICD-10 K00-K93)

Primary Studies

There were no excess hospitalizations for digestive system disorders, as broadly defined by a range of ICD codes, according to the first postwar hospitalization study (1991-1993) of Gray and colleagues (1996). That study compared hospitalizations of almost 550,000 Gulf War veterans and almost 620,000 nondeployed veterans. The major limitations of the study are its focus on DOD hospitalizations (see earlier discussion) and its inability to capture any but the most severe digestive system disorders (most would be treated on an outpatient basis).

Another hospitalization study conducted by Gray (2000) covered the years 1991-1994 and examined DOD, VA, and California hospitals. The study examined hospitalizations at nonfederal hospitals in California to eliminate potential bias related to veterans seeking care outside DOD and VA facilities. Because of the unreliability of residence data in DOD and VA datasets, the authors could not directly compare rates of hospitalization in the three categories. Therefore, they compared proportional morbidity ratios (PMRs) of hospital-discharge diagnoses (14 diagnostic categories from ICD-9) in Gulf War-deployed and nondeployed veterans. PMRs of most disease categories were not increased in deployed veterans. However, digestive system diseases were increased in VA hospitals (PMR 1.12, 95% CI 1.05-1.18) and in California hospitals (PMR 1.11, 95% CI 0.97-1.24), but not in DOD hospitals. Dyspepsia was one of the conditions studied by Eisen and colleagues (2005), who conducted medical evaluations in phase III of the VA's nationally representative, population-based study. From 1999-2001, 1,061 deployed and 1,128 nondeployed veterans were evaluated. They had been randomly selected from 11,441 deployed and 9,476 nondeployed veterans, who had participated in the phase I questionnaire, which was used in 1995 (Kang et al. 2000). Researchers were blind to deployment status. The diagnosis of dyspepsia was made by in-person interviews according to a history or symptoms of dyspepsia (frequent heartburn and recurrent abdominal pain) and use of antacids, H2 blockers, or other medications to treat dyspepsia. One study limitation was that deployed veterans were significantly younger, less educated, less likely to be married, and of lower income, although the analysis adjusted for those factors. Another limitation was that, despite three recruitment waves, the participation rate in the 2005 study was low: only 53% of Gulf War veterans and 39% of nondeployed veterans participated. To determine nonparticipation bias, the authors obtained previously collected findings from participants and nonparticipants from the DOD Manpower Data Center and gathered sociodemographic and self-reported health findings from the 1995 VA study (Kang et al. 2000). Both deployed and nondeployed participants were more likely than nonparticipants to report heartburn or indigestion. That could limit the generalizability of findings, but the authors adjusted for the disparity in their analysis of population prevalence. The prevalence of dyspepsia was higher in deployed than in nondeployed veterans: 9.1% vs 6.0% (OR 1.87, 95% CI 1.16-2.99). See Table 5.14 for a summary of the primary studies reviewed above.

Secondary Studies

In a study by Sostek and colleagues (1996), veterans from the same unit, deployed and not deployed to the Gulf, were compared for gastrointestinal complaints. Fifty-seven deployed unit members were compared with 44 nondeployed, and a higher prevalence of chronic gastrointestinal symptoms was noted in the deployed group. Especially common were abdominal pain, loose or frequent stools, and excessive gas. Every nongastrointestinal symptom asked about was more common in the deployed group (Sostek et al. 1996).

Ishoy and colleagues (1999a) analyzed self-reported gastrointestinal symptoms in relation to Gulf War exposures among 686 deployed veterans and 257 controls who were not deployed to the gulf. The groups were matched for age, sex, and profession within the military. The study was a further analysis of Danish Gulf War veterans and matched controls (Ishoy et al. 1999b). The earlier study found that on a questionnaire, eight of 14 gastrointestinal symptoms were reported significantly more frequently by veterans than by controls. After adjustment for the interrelationship of variables, only two of the eight gastrointestinal symptoms remained significant: prevalence of recurrent diarrhea for one year and rumbling in the stomach more than two times per week. The later study used both symptoms as the combined main outcome measure and investigated its relationship to 24 environmental exposures. Fifteen of the 24 exposures were significantly associated ($p < 0.05$) with that combined measure. After multivariate adjustment, only two were significantly associated with the outcome, exposure to insecticides against cockroaches (OR 2.3, 95% CI 1.2-4.4) and the burning of waste or manure (OR 2.5, 95% CI 1.3-5.0). Two others were nearly significant ($p < 0.10$): teeth-brushing with water contaminated with chemicals or pesticides and bathing in or drinking contaminated water. When those four significant or nearly significant exposures were examined in relation to the main outcome, a dose-response trend was found: the prevalence of the main outcome was greater after three or four exposures (18.9%) than after one or two exposures (7.4 % and 12.8%, respectively). In separate analyses of population attributable risk, 85% of the main outcome could be attributed to either environmental exposures or neuropsychologic symptoms. The limitations of the study are the use of self-reported symptoms and exposures.

Summary and Conclusion

There were many reports of gastrointestinal disturbances in Gulf War-deployed veterans. The disturbances seem to be linked to contaminated water and burning of animal waste. The committee notes that several studies reported an increase in the rate of self-reported dyspepsia. There appears to be a higher prevalence of dyspepsia in deployed Gulf War veterans than in nondeployed veterans. See Table 5.14 for a summary of the primary studies reviewed above.

TABLE 5.14 Gastrointestinal Outcomes

Reference	Design	Population	Outcomes	Results	Adjustments	Comments or Limitations
Eisen et al. 2005	Population-based, cross-sectional, prevalence, medical evaluation	1,061 US deployed and 1,128 nondeployed	Dyspepsia	Deployed: 9.1%; nondeployed: 6.0%; OR 1.87 (95% CI 1.16-2.99)	Age, sex, race, years of education, smoking, duty type, service branch, rank	Low participation rates, especially among nondeployed
Gray et al. 2000	Retrospective cohort study (hospitalization records)	Gulf War-deployed (August 1990-July 1992, n = 652,979) and Gulf War-nondeployed (n = 652,922) stratified by California residence, service, and service branch of allnon deployed veterans (n = 2,912,737)	Digestive system diseases	VA hospitals: PMR 1.12 (95% CI 1.05-1.18); DOD hospitals: PMR 0.98 (95% CI 0.96-0.99); California hospitals: PMR 1.11 (95% CI 0.97-1.24)	Hospitalization records were matched on sex, age	Findings might be influenced by chance or by potential confounders, including health-registry participation
Gray et al. 1996	Retrospective cohort study (hospitalization records)	DOD hospitals: 547,076 Gulf War veterans; 618,335 era veterans who did not serve in Gulf War	Digestive system diseases	All ORs < 1.0	Hospitalization rates and rate ratios adjusted for age,sex; multiple logistic-regression models adjusted for all observed demographic differences between groups	Study data reflect only hospitalization experience of persons who remained on active duty through September 1993

NOTE: DOD = Department of Defense; PMR = proportional mortality ratio; VA = Department of Veterans Affairs.

DISEASES OF THE SKIN AND SUBCUTANEOUS TISSUE
(ICD-10 L00-L99)

Primary Studies

For dermatologic outcomes, the committee defined a primary study as one that used a dermatologic examination, whereas secondary studies were based on self-reports. In the most nationally representative US study, Eisen and colleagues (2005) divided dermatologic conditions into two categories. Group 1 consisted of many common conditions (for example, freckles, seborrheic keratoses, moles, cherry hemangiomas, skin tags, and surgical scars), and Group 2 consisted of more unusual diagnoses not included in Group 1. Diagnoses were determined by a board-certified dermatologist, who evaluated the Group 2 conditions through teledermatology, at least two digital photographs, and the results of a standardized history and physical examination. In adjusted analyses, two Group 2 skin conditions were diagnosed more frequently among deployed than among nondeployed veterans: atopic dermatitis (1.2% vs 0.3%, OR 8.1, 95% CI 2.4-27.7) and verruca vulgaris (warts) (1.6% vs 0.6%, OR 4.02, 95% CI 1.28-12.6).

UK researchers performed dermatologic evaluations on UK Gulf War veterans (111 disabled and 98 nondisabled) and 133 disabled non-Gulf War veterans; disability was defined as reduced physical functioning. They found no differences among groups in any dermatologic conditions other than seborrheic dermatitis (8.1% in deployed vs 2.3% in nondeployed), which was more common in Gulf War veterans, irrespective of disability status (Higgins et al. 2002). See Table 5.15 for a summary of the primary papers that the committee considered for dermatologic outcomes.

Secondary Studies

Two other studies found higher prevalences of dermatologic conditions. Kang et al. (2000), using a stratified random-sampling method, compared 693,826 Gulf War-deployed veterans with 800,680 non-Gulf War veterans from the DOD Defense Manpower Data Center. They found increases in self-reported eczema or psoriasis (7.7% vs 4.4%; 95% of rate difference CI 3.26-3.42) and other dermatitis (25.1% vs 12.0%; 95% CI of rate difference 13.04-13.28). Proctor et al. (1998) looked at 252 Gulf War-deployed veterans—186 from the Fort Devens cohort and 66 from the New Orleans cohort—and compared them with 48 veterans deployed to Germany during the Gulf War. They reported on increased prevalence (15.5% vs 11.7% vs 1.9%) of dermatologic conditions such as rashes, eczema, and skin allergies. However, both studies relied primarily on self-reports or questionnaires and are thus vulnerable to recall or reporting bias.

Summary and Conclusion

On the basis of the few studies of dermatologic conditions, unrelated skin conditions occur more frequently among Gulf War-deployed veterans, but the findings are not consistent among the studies. The committee notes that there is some evidence in Gulf War-deployed veterans of a higher prevalence of two distinct dermatologic conditions—atopic dermatitis and warts. See Table 5.15 for a summary of the primary papers that the committee considered for dermatologic outcomes.

TABLE 5.15 Dermatologic Outcomes

Reference	Design	Population	Outcomes	Results	Adjustments	Limitations/Comments
Eisen et al. 2005	Population-based, cross-sectional, prevalence, medical evaluation	1,061 US deployed and 1,128 nondeployed	Atopic dermatitis and verruca vulgaris (warts)	Atopic dermatitis: 1.2% vs 0.3%, OR 8.1 (95% CI 2.4-27.7); verruca vulgaris (warts): 1.6% vs 0.6%, OR 4.02 (95% CI 1.28-12.6)	Age, sex, race, years of education, smoking, duty type, service branch, rank	Low participation rates, especially among nondeployed
Higgins et al. 2002	Prospective case-comparison study	Disabled Gulf War veterans (n = 111), nondisabled Gulf War veterans (n = 98), and 133 disabled non-Gulf War veterans	Seborrheic dermatitis	8.1% in deployed vs 2.3% in nondeployed	Socioeconomic status, military rank, current service history, smoking, alcohol use	

DISEASES OF THE MUSCULOSKELETAL SYSTEM AND CONNECTIVE TISSUE (ICD-10 M00-M99)

Arthritis and Arthralgia

Arthritis is the most common form of joint disease. Several powerful risk factors are major trauma, repetitive joint use, and age. Arthritis is diagnosed according to a combination of clinical features and radiographic findings. Arthralgia, which is a self-reported symptom of arthritis, refers to painful joints. In the absence of other clinical features and radiographic findings, arthralgias are not necessarily diagnostic of arthritis.

Primary Studies

Arthralgias were one of 12 primary health outcome measures studied by Eisen and colleagues (2005). They conducted medical evaluations in phase III of VA's nationally representative, population-based study. From 1999-2001, 1,061 deployed and 1,128 nondeployed veterans were evaluated. They had been randomly selected from 11,441 deployed and 9,476 nondeployed veterans who had participated in the phase I questionnaire in 1995 (Kang et al. 2000). Researchers were blinded to deployment status. Arthralgias were defined as persistent and clinically significant bone or joint symptoms with or without joint effusion, and treatment with anti-inflammatory agents, narcotic pain medications, or nonnarcotic pain medications. There was no statistically significant difference in arthralgias between deployed and nondeployed veterans (OR 1.15, 95% CI 0.70-1.89).

One study limitation was that despite three recruitment waves, the participation rate in the 2005 study was low: only 53% of Gulf War veterans and 39% of nondeployed veterans participated. To determine nonparticipation bias, the authors obtained previously collected findings on participants and nonparticipants from the DOD Manpower Data Center and gathered sociodemographic and self-reported health findings from the 1995 VA study (Kang et al. 2000). Both deployed and nondeployed participants were more likely than nonparticipants to report arthritis of any kind. See Table 5.16 for a summary of the studies reviewed in this section.

Secondary Studies

Two other studies examined differences in prevalence of arthritis, but they relied on self-reporting. Kang et al. (2000), using a stratified random-sampling method, compared data from the DOD Defense Manpower Data Center on 693,826 Gulf War deployed veterans and 800,680 non-Gulf War veterans, and asked about arthritis as a self-reported condition. They found a significant difference in such reporting between Gulf War-deployed and non-Gulf War deployed veterans (22.5% vs 16.7%, rate difference of 5.87, 95% CI 5.74-6.00). Gray et al. (2002) looked at 3,831 Gulf War deployed veterans, 4,933 veterans deployed elsewhere, and 3,104 nondeployed Seabees. The authors found an increase in reporting of arthritis among Gulf War than among Seabees deployed elsewhere (5.87% vs 4.42%). The latter, in turn, were similar to other nondeployed Seabees (4.42% vs 4.38%). The OR for Gulf War veterans vs veterans deployed elsewhere was 1.44 (95% CI 1.17-1.76), and that for Gulf War-deployed vs non-deployed veterans was 1.63 (95% CI 1.29-2.08).

Summary and Conclusion

Among those examined, there was no statistically significant difference in arthralgias, a surrogate for arthritis, but data on self-reports indicate that arthritis was more common among those deployed to the gulf. The data suffer, however, from the problem of self reporting of a common condition that can be easily confused with other symptoms. There appears to be no statistically significant increase in the prevalence of arthralgias among veterans who underwent a medical examination.

TABLE 5.16 Arthralgia

Study	Design	Population	Outcomes	Results	Adjustments	Comments or Limitations
Eisen et al. 2005	Population-based, cross-sectional, prevalence, medical evaluation	1,061 deployed vs 1,128 nondeployed	Persistent and clinically significant bone or joint symptoms with or without joint effusion, and treatment with anti-inflammatory agents, narcotic pain medications, or nonnarcotic pain medications	Prevalence: 6.4% vs. 6.8%, OR 1.15 (95% CI 0.70-1.89)	Age, sex, race, years of education, cigarette-smoking, duty type, service branch, rank	Low participation rates, especially among nondeployed

FIBROMYALGIA

The hallmarks of fibromyalgia are widespread muscle and skeletal pain and tenderness at numerous soft tissue sites on the body upon palpation, according to classification criteria promulgated by the American College of Rheumatology (ACR) (Wolfe et al. 1990). The case definition requires both widespread pain (pain on both sides of the body, above and below the waist, and including axial skeletal pain) lasting for at least 3 months and pain (not just tenderness) in at least 11 of 18 tender point sites on palpation with an approximate force of 4 kg. The presence of a second clinical disorder does not exclude a diagnosis of fibromyalgia. Other symptoms of fibromyalgia include fatigue, sleep disturbance, morning stiffness, and cognitive impairment, but those are not sensitive and specific enough to use for classification (Wolfe et al. 1990). Early characterization of the condition as an inflammation of muscle (hence the label fibrositis) have not been borne out through research (Goldenberg 1999). There is no pathologic or laboratory test with which to confirm the diagnosis. And there are no widely accepted causative factors. Fibromyalgia's prevalence in the general population is about 3.4% in women, and 0.5% in men, so it is one of the more common rheumatologic disorders (Wolfe et al. 1995). Its prevalence increases with age (Wolfe et al. 1995). On the basis of longitudinal studies, the course is chronic but variable in intensity (Wolfe et al. 1997). It should be noted that the existence of fibromyalgia as a distinct disease entity is considered controversial by some expert commentators (Nimnuan et al. 2001; Pearce 2004).

Primary Studies

For consideration as a primary study, the basis of diagnosis of fibromyalgia has to include symptom reporting and physical examination, rather than only symptom-based criteria. Fibromyalgia was one of 12 primary health-outcome measures studied by Eisen and colleagues (2005), who conducted medical evaluations in phase III of VA's nationally representative, population-based study. From 1999- 2001, 1,061 Gulf War veterans and 1,128 non-Gulf War veterans were evaluated. They had been randomly selected from 11,441 deployed and 9,476 nondeployed veterans, who had participated in the phase I questionnaire in 1995 (Kang et al. 2000). Researchers were blinded to deployment status. The diagnosis of fibromyalgia was based on diffuse body pain and pain on physical examination, following the ACR criteria (Wolfe et al. 1990). Self-reported diagnoses of fibrositis or fibromyalgia did not vary between deployed and nondeployed veterans (0.6% and 0.8% respectively; OR 1.21, 95% CI 0.36-4.10, adjusted for age, sex, race, cigarette-smoking, duty type, service branch, and rank). However, fibromyalgia diagnosed on the basis of physical examination was present in 2.0% of deployed and 1.2% of nondeployed veterans (adjusted OR 2.32, 95% CI 1.02- 5.27). Strengths of the study include the population-based sampling strategy, blinding of evaluating physicians, and use of validated diagnostic criteria based on physical examination. Limitations include the potential for substantial selection bias due to modest participation rates of 53% of Gulf War veterans and 39% of non-Gulf War veterans and the deployed veterans being significantly younger, less educated, less likely to be married, and of lower income, although the analysis adjusted for most of those factors.

Smith and colleagues (Smith et al. 2000) performed a study of postwar hospitalizations (1991-1997) among 551,841 deployed and 1,478,704 nondeployed active duty personnel. The association of hospitalization for fibromyalgia (ICD-9 code 729) with deployment status was investigated with Cox proportional-hazards models. The study found higher risk of fibromyalgia hospitalization among the deployed (RR 1.23, 95% CI 1.05-1.43). However, survival curves showed that the higher rate of hospitalization for fibromyalgia among the deployed occurred only between the inception of the Comprehensive Clinical Evaluation Program (CCEP) on June 30, 1994, through the middle of 1995. During that roughly 1-year period, many CCEP participants were admitted to the hospital only for purposes of evaluation. In fact, CCEP participants had more than 26 times the risk of being hospitalized for fibromyalgia than did nonparticipants. For the 3-year period before the inception of the CCEP, hospitalization for fibromyalgia was unrelated to Gulf War status (RR 0.92, 95% CI 0.74-1.13). The Smith et al. study has the advantage of a large, population-based sample and good statistical power for the detection of an effect. Its major limitations are the inclusion of only active-duty personnel, changes in hospitalization rates for fibromyalgia in association with the practices of the CCEP, and the fact that few cases of fibromyalgia are severe enough to warrant hospitalization. The findings on fibromyalgia are summarized in Table 5.17.

Secondary Studies

The Iowa study (Iowa Persian Gulf Study Group 1997) surveyed 1,896 deployed and 1,799 non-deployed veterans who listed Iowa as their home state at the time of enlistment. Fibromyalgia was assessed according to the symptom criteria of Wolfe and colleagues (Wolfe et al. 1995). Those criteria include the presence of widespread pain for at least 3 months. No physical examinations were conducted. Symptoms of fibromyalgia were present in about 21% of deployed veterans and 11% of nondeployed veterans. The authors found a prevalence difference of 9.3% (95% CI 7.3-11.2) after adjustment for age, sex, race, branch of military, and rank. The main strength of the study was the population-based sample. The main study limitation was "diagnosis" of fibromyalgia based on reported symptoms without physical examination to assess pain on digital palpation.

Canada deployed more than 3,000 troops to the Gulf region. A survey of the entire cohort of Canadian Gulf War-deployed veterans found that they were more likely than nondeployed veterans—group-matched to cases on sex, age, and regular vs reserve status—to report symptoms of fibromyalgia (Goss Gilroy Inc. 1998). The criteria for fibromyalgia, adapted from several published studies (e.g., Wolfe et al. 1995), were having "overall" body pain lasting 3 months or longer and having pain at a level of 1 or higher (on a scale of 1-10) over the preceding 24 hours or self-report of fibromyalgia or fibrositis. Symptom-defined fibromyalgia was present in about 16% of deployed veterans and 10% of nondeployed veterans (adjusted OR 1.81, 95% CI 1.55-2.13). The main strength of the study was the population-based sample. The main study limitation was diagnosis of fibromyalgia based on symptoms or self-reported diagnosis without physical examination to assess pain on digital palpation.

Bourdette et al. (2001) studied 244 Oregon and Washington Gulf War veterans who had unexplained illness after clinical evaluation to exclude "explainable" illness. Fifty (20.8%) fulfilled the ACR criteria for fibromyalgia (although it is not stated, presumably these included clinical examination to see whether pain was present at at least 11 of 18 body points on digital palpation). Those 50 represent 2.5% of 2022 Gulf War veterans solicited for participation (of the 2,022 solicited, 1,760 were located, 1,119 responded, and 799 were deemed eligible for the clinical study). The study's main limitations are its lack of a nondeployed comparison group and lack of clarity about the nature of the clinical examination for fibromyalgia.

Summary and Conclusion

The diagnosis of fibromyalgia is based entirely on symptoms and physical examination; there are no pathologic or laboratory tests with which to confirm it. Among the available cross-sectional studies that include both Gulf War-deployed and non-deployed veterans, only Eisen and colleagues (2005) used the full ACR case definition of fibromyalgia, including criteria based on physical examination. Fibromyalgia was diagnosed in 2.0% of deployed and 1.2% of nondeployed veterans, for an adjusted OR of 2.32 (95% CI 1.02-5.27). A strength of this study is the population-based sampling strategy. An important limitation is the modest participation rates (53% of Gulf War veterans and 39% of non-Gulf War veterans) with the potential to introduce selection bias. After accounting for a 1-year period during which many CCEP participants were admitted to the hospital only for purposes of evaluation, the study by Smith and colleagues (2000) found no association between Gulf War deployment and hospitalization for fibromyalgia. That finding does not appear inconsistent with positive findings in the Eisen et al. study, in that few cases of fibromyalgia are severe enough to warrant hospitalization. Notably, the prevalence of a diagnosis of fibromyalgia in the Eisen et al. study is about 300 times the prevalence of hospitalization for fibromyalgia in the Smith et al. study. The Iowa study and the Canadian study both found significantly increased fibromyalgia symptoms among deployed veterans compared with nondeployed veterans. The findings of those two studies, although generally supportive of the findings of the Eisen et al. study, are of limited value owing to the lack of a physical examination to enable the use of the full criteria for diagnosis. The Bourdette et al. study, which had no nondeployed comparison group, estimated that at least 2.5% met the full ACR case definition of fibromyalgia. In conclusion, largely on the basis of the Eisen et al. study, which used the criteria of the ACR for diagnosis of fibromyalgia but could have been subject to unrecognized selection bias, there is a higher prevalence of fibromyalgia among deployed Gulf War veterans than among nondeployed veterans.

TABLE 5.17 Fibromyalgia

Study	Design	Population	Outcomes	Results	Adjustments	Comments or Limitations
Eisen et al. 2005	Population-based, cross-sectional, prevalence, medical evaluation	1,061 deployed, 1,128 nondeployed	Symptoms and physical examination using criteria of American College of Rheumatology (Wolfe et al. 1990)	Prevalence: 2.0% vs 1.2%, OR 2.32 (95% CI 1.02-5.27)	Age, sex, race, years of education, cigarette smoking, duty type, service branch, rank	Uses gold standard for diagnosis of fibromyalgia; low participation rates, especially among nondeployed
Smith et al. 2000	Postwar hospitalization study	551,841 deployed, 1,478,704 nondeployed	Hospitalization (1991-1997); Cox proportional-hazards models ICD-9 codes for fibromyalgia (729.1)	RR 1.23 (95% CI 1.05-1.43); however, survival curves indicate excess due to hospitalization only for purposes of evaluation during the CCEP; before CCEP: RR 0.92 (95% CI 0.74-1.13)	Sex, age, branch of service	No increase after accounting for CCEP effect; limited to active duty; most cases of fibromyalgia are not severe enough to warrant hospitalization

NOTE: CCEP = Comprehensive Clinical Evaluation Program.

BIRTH DEFECTS AND ADVERSE PREGNANCY OUTCOMES
(ICD-10 O00-Q99)

This section evaluates the findings on birth defects in the offspring of veterans, on adverse reproductive outcomes, on the risk of male infertility, and on sexual problems. As appropriate, the major results from each study are organized by whether the father or the mother served in the gulf and by outcome. Table 5.18 summarizes all the primary studies on birth defects and adverse reproductive outcomes reviewed by the committee.

Birth Defects

Birth defects occur in about 3% of live births. The numerous types of serious or disabling birth defects include structural defects, chromosomal abnormalities, and birth defect syndromes (California Birth Defects Monitoring Program 2006). Because of that diversity, epidemiologists attempting to calculate whether birth defects are increased in a particular group such as deployed veterans, encounter the problem of making multiple comparisons; that is, the greater the number or the more types of comparisons, the greater the likelihood that one or more of them will appear statistically significant when no true differences exists. Several statistical techniques are used to adjust for, or minimize, the problem of multiple comparisons but they are not foolproof.

Primary Studies

In the most comprehensive population-based study, Araneta and colleagues (2003) identified birth defects among infants of military personnel born from January 1, 1989 to December 31, 1993, from population-based birth defect registries in six states: Arizona, Hawaii, Iowa, and selected counties of Arkansas, California, and Georgia (metropolitan Atlanta). They compared the prevalence of 48 selected congenital anomalies diagnosed from birth to the age of 1 year between Gulf War veterans' and nondeployed veterans' infants conceived before the war; between Gulf War veterans' and nondeployed veterans' infants conceived during or after the war; and between Gulf War veterans' infants conceived before and after the war. The authors performed separate analyses on the basis of whether the mother or the father was engaged in military service. If both parents were in the military then the birth was categorized as an infant of a military mother. The study found higher prevalence of three cardiac defects (tricuspid valve insufficiency, aortic valve stenosis, and coarctation of the aorta), and one kidney defect (renal agenesis and hypoplasia) among infants conceived after the war to Gulf War veteran fathers. There also was a higher prevalence of hypospadias (malformation of the urethra and urethral groove), a genitourinary defect among sons conceived postwar to Gulf War veteran mothers compared to their nondeployed counterparts. Aortic valve stenosis, coarctation of aorta, and renal angenesis and hypoplasia were also elevated among infants conceived among the Gulf War veteran fathers postwar compared to those conceived prewar. There was only 1 birth defect recorded among 142 births conceived prewar to Gulf War veteran mothers, and this precludes comparisons with this group.

This study is particularly informative because it relies on medically confirmed outcomes diagnosed through the first year, rather than at birth, and uses information from population registries, as opposed to information from voluntary participation by study subjects. Because both nonmilitary and military hospitals participated in the registries in all states except California (nonmilitary only), births among reservists, National Guard, and former military personnel were eligible, as well as among those on active duty. The study also included comparisons of births to Gulf War veterans before and after deployment. One limitation is that the study relied on availability of unique personal identifiers in military and birth certificate data, which leads to the possibility that some military offspring might be missed among the cases, and that would make the observed prevalence more conservative than the actual. Another is the study's low power to assess individual defects that are rare. The authors also published the results of the pilot study of their method, which was performed in only Hawaii (Araneta et al. 2000); because the data are incorporated in the larger six-state study, we did not review them separately here.

Secondary Studies

Additional studies of birth defects are considered secondary either because they rely on self-reports (and thus introduce potential recall bias) or because they consider only groups of birth defects. Studying groups of birth defects, although useful in identifying patterns, makes it difficult to determine which specific defects may be increased. Doyle and colleagues (2004) evaluated the prevalence of self-reported birth defects among the offspring of all UK veterans (male and female) deployed to the gulf and among the offspring of nondeployed veterans who responded to a postal questionnaire. Response rates were higher among the Gulf War veterans (53% of men, 72% of women) than the comparison group (42% of men, 60% of women). They considered pregnancies conceived after deployment (after January 1, 1991, for nondeployed veterans) through November 8, 1997. Medical confirmation was requested for all fetal deaths at 16 weeks or more or of unknown gestation and for liveborn children in whom a congenital abnormality, serious childhood medical condition, or death was reported. Among infants conceived by fathers deployed to the gulf compared with infants of fathers not deployed, the OR for any malformation was 1.5 (95% CI 1.3-1.7). Elevated risks were observed specifically for malformations of the genital system, urinary system, musculoskeletal system, and cranial neural crest; for "other" malformations of the digestive system; for "other" non-chromosomal malformations; and for metabolic and single-gene defects. The risks of urinary system and musculoskeletal system defects remained increased when the cases were restricted to the 55% that had clinical confirmation. No statistically significant increased risk of birth defects in infants of mothers deployed to the gulf was found.

Three additional secondary studies assessed groups of birth defects. In a population-based survey in the United States, Kang et al. (2001) observed excess risks of self-reported "likely birth defects" and specifically "moderate to severe defects" among infants of Gulf War-deployed fathers and mothers compared with nondeployed fathers and mothers. Most defects were isolated anomalies, and no clear patterns were found. First pregnancies ending after June 30, 1991, were considered in this analysis. Another US study that included live births at 135 military hospitals from 1991-1993 did not find evidence of a statistically significant increased risk of "any birth defects" or "severe birth defects" in infants born to fathers or mothers deployed to the gulf (Cowan et al. 1997). However, because this study included only births at military hospitals, only parents on active duty at time of the birth were included, and it is likely that higher risk pregnancies were referred to civilian hospitals. The observed number of birth defects among children (liveborn and stillborn) born after deployment to National Guard personnel in two units in southeast Mississippi was not greater than expected on the basis of population-based registries (Penman et al. 1996).

A population-based study of male Canadian veterans (Goss Gilroy Inc. 1998) surveyed deployed and nondeployed veterans for self-reported birth defects. Overall, deployed veterans reported higher rates of birth defects (a combined category that includes births before, during, and after the Gulf War). Birth defects that occurred at similar rates include urogenital and kidney defects.

In a study of Kuwaiti nationals, a higher prevalence of congenital heart disease (CHD) was observed in babies born after invasion than before invasion (Abushaban et al. 2004). However, no data were available on the residence status of the parents during the war (inside or outside Kuwait), or of their proximity to potential environmental pollutants, for example, oil-well fires. It is also possible that there was underreporting of CHD before to the war. This study is considered secondary because the study population is not Gulf War veterans.

Goldenhar Syndrome

Anecdotal reports raised the possibility of increased prevalence of Goldenhar syndrome, a rare craniofacial abnormality, among children of Gulf War veterans. External features of the syndrome are ear abnormalities, such as microtia, anotia, and preauricular tags. Among infants conceived after the Gulf War (or December 31, 1990 for nondeployed veterans) through September 30, 2003 and born to active-duty personnel in military hospitals, five cases with Gulf War veteran fathers and two cases with NDV fathers were identified (Araneta et al. 1997). Given those small numbers, it is difficult to determine whether an excess risk is associated with service in the gulf. In a case-control study of hemifacial microsomia (of which Goldenhar syndrome is one type) in craniofacial clinics in 24 US cities, the OR of 2.4 for parental army service in general (95% CI 1.4-4.2) was statistically significant. The association with Gulf War army service in particular (OR 2.8, 95% CI 0.8-9.6) was increased but not statistically significant (Werler et al. 2005).

Summary and Conclusion

Primarily on the basis of the Araneta et al. and Doyle et al. studies, because of the availability of medical confirmation in those studies, there is some evidence of increased risk of birth defects among offspring of Gulf War veterans. However, with the possible exception of urinary tract abnormalities, the specific defects with increased prevalence in the two studies were not consistent. The reported association of Gulf War service with Goldenhar syndrome was inconclusive. Overall, the studies are difficult to interpret because specific birth defects are relatively rare, multiple comparisons were performed, and sample sizes were small when divided by timing of exposure (before or after conception) and whether the mother or the father was exposed.

Thus the committee concludes that there is no consistent pattern of higher prevalence of birth defects among offspring of male or female Gulf War veterans and that no single defect, except urinary tract abnormalities, has been found in more than one well-designed study.

Adverse Pregnancy Outcomes

Primary Study

The prevalence of spontaneous abortions, stillbirths, and ectopic pregnancies has been studied in deployed and nondeployed women. Araneta and colleagues (2004) recruited women admitted to military hospitals for pregnancy-related diagnoses (including livebirths, spontaneous and induced abortions, ectopic pregnancies, and pregnancy-related complications) from August 2, 1990-May 31, 1992. Among women who conceived after the war, the risk of spontaneous abortions and ectopic pregnancies was higher among deployed than among nondeployed women. The risk of those outcomes among so called "Gulf War exposed" conceptions was increased, but not statistically significantly. Self-reported outcomes were confirmed by hospital-discharge data. Because only military hospitals were included, only information on active-duty personnel was included.

Secondary Studies

Doyle and colleagues (2004) also studied the risk of self-reported miscarriages and stillbirths among Gulf War-deployed fathers and mothers. They observed no effect of Gulf War service on the risk of stillbirths or of miscarriages in pregnancies reported by female veterans. There was a 40% increase in the risk of miscarriages among pregnancies reported by male Gulf War veterans compared with their nondeployed counterparts (95% CI 30%-50%), and the effect was stronger for early miscarriages (OR 1.5, 95% CI 1.3-1.6). However, in a study of nuclear industry workers (the Nuclear Industry Family Study) there was evidence of underreporting of miscarriages among the nonexposed workers. The potential selection bias could explain the increased risks observed among the Gulf War veterans.

In the Kang et al. (2001) study described above, there was an excess prevalence of self-reported spontaneous abortions and stillbirths among pregnancies conceived by Gulf War veteran fathers and of spontaneous abortions among pregnancies conceived by Gulf War veteran mothers. As that study relied on self-reports and inconsisten participation rates, the results are difficult to interpret.

Gray and colleagues (1996) conducted a hospitalization study (1991-1993) in which they compared almost 550,000 Gulf War veterans with almost 620,000 nondeployed veterans. The study found increased hospitalizations, in 1991 only, for the broad category "genitourinary system diseases". More specifically, the increase was due to female veterans' being hospitalized for inflammatory diseases of the ovary, fallopian tube, pelvic cellular tissue, and peritoneum. Male Gulf War veterans were hospitalized for redundant prepuce and phimosis (ICD code 605). This code often accompanies hospitalizations for circumcision. The major limitation of this study is its focus on DOD hospitalizations.

Summary and Conclusion

Although the results from the Araneta et al. study, which had hospital-discharge data available, are suggestive of an increased risk of spontaneous abortions and ectopic pregnancies, the results may not be generalizable to deployed women who left the service or to pregnancy-related admissions to nonmilitary hospitals. Thus, it is difficult to conclude whether there is a higher prevalence of adverse pregnancy outcomes in Gulf War-deployed than in nondeployed veterans.

Male Fertility Problems and Infertility

Primary Studies

Two studies have addressed fertility problems among men who served in the Gulf War. A group of UK Gulf War veterans (drawn from the same population as the Doyle et al. study described above) who fathered or tried to father children after the war and before August 1997 reported excess risk of infertility (defined as consulting a doctor after trying unsuccessfully for more than 1 year) compared to their nondeployed counterparts. The risks of type 1 infertility (never achieving pregnancy) and type II infertility (never achieving a live birth) were also significantly higher. Furthermore, more Gulf War veterans than non-Gulf War veterans experienced time to conception for planned pregnancies of more than 1 year (Maconochie et al. 2004). Those results are difficult to interpret because of low response rates, possible recall bias, and lack of information on partners' fertility status.

Ishoy and colleagues evaluated serum concentrations of reproductive hormones in a study of 661 Danish Gulf War veterans and 215 nondeployed veterans. There were no statistically significant differences in concentrations of luteinizing hormone, follicle stimulating hormone, testosterone, or inhibin B between the two groups (Ishoy et al. 2001a). There was a higher prevalence of reporting of sexual problems, specifically decreased libido, in the deployed veterans (Ishoy et al. 2001b). Sexual problems were associated with self-reported combat-related exposures, such as being threatened with arms or witnessing killing or wounding of victims and colleagues.

Summary and Conclusion

For the most part, the findings on fertility and sexual problems relied on self-reports which entail a substantial opportunity for recall bias. There was no evidence of statistically significant differences in concentrations of male reproductive hormones between Gulf War veterans and nondeployed veterans.

Although it appears that there is no difference in the prevalence of male fertility problems or infertility between the deployed Gulf War veterans and their nondeployed counterparts, it is difficult to draw conclusions from the small number of available studies.

TABLE 5.18 Birth Defects and Adverse Reproductive Outcomes

Study	Design	Population	Outcomes	Results	Adjustments	Comments or Limitations
Birth defects						
Araneta et al. 2003	Prevalence, population-based, birth-defect registry (active surveillance all cases identified from birth to 1 year)	Infants of military personnel born 1/1/1989–12/31/1993 in Arizona, Iowa, Hawaii, and participating counties of Arkansas, California, Georgia to GWV mothers (n = 450), NDV mothers (n = 3,966), GWV fathers (n = 11,511), NDV fathers (n = 29,086)	48 birth defects identified by CDC as occurring frequently or of public health importance, excluding pulmonary artery anomalies and adding dextrocardia, chromosomal anomalies (other than trisomies 13, 18, and 21), and Goldenhar syndrome	Unadjusted RRs: Postwar conceptions, GWVs vs NDVs: father: tricuspid valve insufficiency, 10/4648 vs 9/11,164 (RR 2.7, 95% CI 1.1–6.6); aortic valve stenosis, 5/4,648 vs 2/11,164 (RR 6.0, 95% CI 1.2–31.0); coarctation of aorta, 5/4,648 vs 3/11,164 (RR 4.0, 95% CI 0.96–16.8); renal agenesis or hypoplasia, 5/4,648 vs 5/11,164 (RR 2.4, 95% CI 0.7–8.3) mother: hypospadias 4/154 vs 4/967 (RR 6.3, 95% CI 1.5–26.3) GWVs postwar vs prewar conceptions father: aortic valve stenosis 5/4,648 vs 0/6,863 (RR 16.3, 95% CI 0.9–294); coarctation of aorta, 5/4,648 vs 1/6,863 (RR 7.4, 95% CI 0.9–63.3); renal agenesis and hypoplasia, 5/4,648 vs 0/6,863 (RR 16.3, (95% CI 0.9–294); adjustment did not change results	State, maternal and paternal age, race, marital status, education, plurality, parity, prenatal visits, gestational weight gain, branch of service, military rank, prenatal alcohol exposure, intrauterine growth retardation, low birth weight, small for gestational age, pre-eclampsia	Limitations: California limited to diagnoses in nonmilitary hospitals; relies on availability of unique personal identifiers in military and birth certificate data limited power to assess individual defects, multiple comparisons, limited to live births Strengths: population-based, including reservists, National Guard, former military personnel; includes defects diagnosed through first year, medically confirmed as opposed to self-reports, comparisons with prewar experience

Study	Design	Population	Outcomes	Results	Adjustments	Comments or Limitations
Doyle et al. 2004	Prevalence	All UK GWVs and randomly selected cohort of NDVs responding to postal questionnaire; conceptions from postdeployment (for NDVs – conceived after 1/1/1991) through 11/8/1997 GWV fathers (n = 16,442) NDV fathers (n = 11,517) GWV mothers (n = 484) NDV mothers (n = 377) External comparison populations: (1) NIFS; (2) annual registered stillbirths abnormalities in England and Wales, 1991-1998	Fetal death: early and late miscarriage, stillbirth; congenital malformations excluding minor abnormalities among live births; self-report with clinical confirmation attempted for fetal deaths and livebirths with reported abnormalities	Adjusted ORs: GWVs vs NDVs father: all miscarriages 2,829/15,539 vs 1,525/10,988 (OR 1.4, 95% CI 1.3-1.5); any congenital malformation, 686/13,191 vs 342/9,758 (OR 1.5, 95% CI 1.3-1.7); other malformations of digestive system, 69/13,191 vs 31/9,758 (OR 1.6, 95% CI 1.0-2.5); genital system, 45/13,191 vs 19/9,758 (OR 1.8, 95% CI 1.0-3.0); urinary system[a], 103/13,191 vs 48/9,758 (OR 1.6, 95% CI 1.1-2.3); musculoskeletal system[a], 194/13,191 vs 78/9,758 (OR 1.8, 95% CI 1.4-2.4); other non-chromosomal malformations, 45/13,191 vs 19/9,758 (OR 1.7, 95% CI 1.0-3.0); cranial neural crest, 184/13,191 vs 101/9,758 (OR 1.3, 95% CI 1.0-1.7); metabolic and single gene defects, 22/13,191 vs 8/9,758 (OR 2.0, 95% CI 0.9-4.8); mothers: no significant associations	Stratum matched on branch of service, sex, age, serving status, rank; ORs adjusted by year of pregnancy end, paternal/maternal pregnancy order, maternal age, service, rank, previous fetal death, multiplicity	Response rates: GWVs: men 53%, women 72%; NDVs: men 42%, women 60% Limitations: poor response rates among men and response rates lower in NDVs, low numbers of miscarriages in NDVs population could mean participation and reporting bias; multiple comparisons Strengths: medical confirmation for some cases; fetal deaths as well as live births; external comparison groups to evaluate possible biases

199

Study	Design	Population	Outcomes	Results	Adjustments	Comments or Limitations
Goldenhar Syndrome						
Araneta et al. 1997	Prevalence	Infants conceived after GW (or 12/31/1990 for NDVs) through 9/30/2003, born to active-duty military personnel in military hospitals; GWVs (n = 34,069), NDVs (n = 41,345)	Goldenhar syndrome criteria: presence of microtia, anotia, or preauricular tag; presence of either hypoplasia of mandible or a physical feature peculiar to syndrome, such as equibulbar dermoid or coloboma of upper lid	Unadjusted RRs: GWV fathers (5 cases) vs NDV fathers (2 cases) (RR 3.03, 95% CI 0.63-20.57)	Sex, race, history of fetal loss, maternal and paternal age, military occupation	Limitations: only military hospitals, may miss high-risk pregnancies, included cases only among liveborn infants diagnosed at birth
Werler et al. 2005	Case-control	HFM cases ≤3 years old (born 1996-2002) from craniofacial clinics in 24 US cities (n = 232); controls matched by age and pediatrician (n = 832)	HFM, facial asymmetry, or Goldenhar syndrome and no evidence of Mendelian inherited or chromosomal anomaly	Adjusted ORs: cases vs controls; parental army service, 22/232 vs 45/832 (OR 2.4, 95% CI 1.4-4.2); parental GW army service, 4/232 vs 9/832 (OR 2.8, 95% CI 0.8-9.6)	Family income, race, BMI in early pregnancy, multiple gestation	Limitations: unmeasured lifestyle factors Strengths: included cases diagnosed up to of 3 years age
Adverse pregnancy outcomes						
Araneta et al. 2004	Prevalence	Deployed women admitted to military hospitals for pregnancy-related diagnoses (including live births, abortions, ectopic pregnancies, pregnancy-related complications) from 8/2/1990 to 5/31/1992 and who responded to mailed survey:	Self-reported stillbirths, spontaneous abortions, ectopic pregnancies, pregnancy-related complications (ICD-9-CM codes 640-676); confirmed by discharge	Adjusted RRs: mothers: GWV vs NDV postwar conceptions: spontaneous abortions, 68 vs 39 (RR 2.92, 95% CI 1.87-4.56); ectopic pregnancies, 32 vs 6 (RR 7.70, 95% CI 3.00-19.8); GWV vs NDV exposed conceptions:	Age, race, education, marital status, branch of service, military rank, parity, history of adverse outcome	Overall response rate: 50% Limitations: low response rate; no information on smoking, alcohol, caffeine, other known risk factors for fetal loss; possible limited generalizability due to

Study	Design	Population	Outcomes	Results	Adjustments	Comments or Limitations
		GW-exposed conceptions (n = 415), GW postwar conceptions (n = 298), NDVs (n = 427)	diagnostic data	spontaneous abortions, 48 vs 39 (RR 1.44, 95% CI 0.91-2.29); ectopic pregnancies, 10 vs 6 (RR 1.91, 95% CI 0.67-5.46)		restriction to military hospital admissions; recall bias Strengths: confirmation with discharge data, assessed GW-exposed and postwar conceptions

Male fertility problems and infertility

Study	Design	Population	Outcomes	Results	Adjustments	Comments or Limitations
Maconochie et al. 2004 (same cohort as Doyle et al. 2004)	Retrospective reproductive cohort	Male UK veterans fathering or trying to father pregnancies after GW and before 8/97 GWV (n = 10,465) NDV (n = 7,376)	Self-reported fertility problems: tried unsuccessfully for > 1 year and consulted doctor; type I infertility: never achieving pregnancy; type II infertility: never achieving live birth; semen quality; time to conception; attempted clinical confirmation from both partners' physicians	Adjusted ORs: fertility problems, 732/10,465 vs 370/7,376: (OR 1.38, 95% CI 1.20-1.60); type I 259/10,465 vs 122/7,376 (OR 1.41, 95% CI 1.05-1.89); type II 356/10,465 vs 166/7,376 (OR 1.50, 95% CI 1.18-1.89); time to conception > 1 year for planned pregnancies, 845/9,968 vs 528/7,408 (OR 1.18, 95% CI 1.04-1.34) (increase in risk stable with time since GW)	Maternal and paternal age at first infertility consult or post-GW conception, year of first consult or conception, pre-GW pregnancy history, military service and rank, smoking, alcohol, pregnancy order	Response rates: GWVs, 53%; NDVs, 42% Limitations: low response rates, possible recall bias, clinically evaluated only 40% Strengths: attempted clinical evaluation, information on nonresponders available
Ishoy et al. 2001a	Cross-sectional	Danish Gulf War Study, GWVs (n = 661) NDVs (n = 215)	Self-reports of sexual problems (e.g., including reduced libido); measured male reproductive hormones: serum concentrations of LH, FSH,	Male GWVs vs NDVs: self-reported sexual problems, 12.0% vs 3.7% (p<0.001); reproductive hormones, no significant difference; suspected oligospermia, FSH≥10 IU/L, inhibin B≤80 pg/ml,	Age; BMI available; stratified on deployment organization, duration of deployment	Participation rates: GWVs, 83.6%; NDVs, 57.8% Limitations: limited control for confounding, small numbers for study of fertility rates, congenital

Study	Design	Population	Outcomes	Results	Adjustments	Comments or Limitations
			testosterone, inhibin B	1.6% vs 1.6%; fertility rates, spontaneous abortion, congenital malformations: no differences		malformations Strengths: measurement of hormones objective and unbiased
Ishoy et al. 2001b (elaboration of findings in Ishoy et al. 2001a)	Cross-sectional	Danish Gulf War Study: GWVs (n = 661), NDVs (n = 215)	Self-reported sexual problems	Male GWVs vs NDVs: sexual problems (80% decreased libido), 79/661 vs 8/215 (OR 2.9, (95% CI 1.4-6.0) (among GWVs associated with "having seen killed or wounded victims"; "having been threatened with arms"; "having watched colleagues being seriously threatened or shot at"; water hygienic environment)	Age	Limitations: small study, self-reported soft outcomes and exposures
Fertility problems						
Ishoy et al. 1999b				Increased prevalence of self-reported sexual problems up to 6 years after war		

NOTE: BMI = body-mass index; CDC = Centers for Disease Control and Prevention; FSH = follicle-stimulating hormone; GW = Gulf War; GWV = Gulf War veteran; HFM = hemofacial microsomia; LH = luteinizing hormone; NDV = nondeployed veteran; NIFS = Nuclear Industry Family Study.
[a] Associations attenuated but still statistically significant with clinical confirmation

SYMPTOMS, SIGNS, AND ABNORMAL CLINICAL AND LABORATORY FINDINGS (ICD-10 R00-R99)

Unexplained Illness

Many of the symptom clusters that Gulf War veterans report are based on factor analysis of survey data. Those symptom clusters have been referred to as Gulf War Syndrome, chronic multisymptom illness, "unexplained" illnesses, etc. The term "unexplained" is not meant to imply that the illnesses are unique in being of unknown etiology, as that is true of many medical conditions, but rather that the illnesses do not fit into established medical diagnostic categories.

Several lines of inquiry, described and evaluated in this section, have tried to overcome reliance on established diagnoses by seeking to determine whether veterans have increased hospitalizations for any unexplained illness and are suffering from a potentially new syndrome. To uncover the existence of a new syndrome, most research has used a statistical technique known as factor analysis. The technique probes a cluster of symptoms to answer the question, Is the symptom cluster best studied and treated as a new and unique syndrome, or is it a variant of a known syndrome? The technique, its nomenclature, and its purposes are explained in Chapter 3.

Hospitalizations for Unexplained Illness

A hospitalization study (1991-1996) examined DOD's hospital discharge dataset to search for excess admissions for unexplained illnesses in deployed veterans (n = 552,111) and nondeployed veterans (n = 1,479,751) (Knoke and Gray 1998). The authors reasoned that their earlier study of hospitalizations (Gray et al. 1996) might have missed those for a new or poorly recognized syndrome. Hospital discharge coding might have inconsistently classified such hospitalizations by many diagnoses, thereby masking an effect if one were present. The study operationally defined unexplained illnesses as diagnoses in several catchall ICD-9 diagnostic categories that comprised nonspecific infections and other ill-defined conditions. It examined only first hospitalizations to avoid overcounting medical conditions that required repeated hospitalizations among the subset of patients who had at least one unexplained illness coded on the discharge summary. Up to eight discharge diagnoses were examined per hospitalization. The authors found that deployed active-duty military members were less likely to have been hospitalized for unexplained illnesses than nondeployed (RR 0.93, 95% CI 0.91-0.96) (Knoke and Gray 1998). That finding adjusted for a variety of covariates and removed the effect of participation in the CCEP after June 1, 1994. Participants in that voluntary program had been admitted to the hospital only for evaluation. The Knoke and Gray (1998) study has the advantage of large study groups and high statistical power for the detection of an effect. Its major limitations are its inclusion of only active duty personnel and its inability to detect illnesses not severe enough to warrant hospitalization.

In another hospitalization study Gray et al. (2000) extended the examination of hospitalizations (1991-1994) to cover not only active duty but also reserve and former military personnel who had been deployed to the Gulf War. The study investigated hospitalizations at DOD, VA, and nonfederal hospitals in California to eliminate potential bias related to veterans seeking care outside DOD and VA facilities. Because of the unreliability of state-of-residence data in DOD and VA datasets, the authors could not directly compare rates of hospitalization among the three sources. Therefore, they compared PMRs of hospital discharge diagnoses (14 diagnostic categories from ICD-9) in deployed vs nondeployed veterans. For VA hospitals, but not for DOD or California hospitals, the PMR was increased for ill-defined diseases (PMR 1.24, 95% CI 1.16-1.33).

Factor-Analysis Derived Syndromes

We focus now on the extensive literature using factor analysis and cluster analysis to determine whether veterans' symptoms might constitute a new syndrome or whether they are a variant of a known syndrome. Secondary studies are included in tables and text, a departure from most other sections of the report. Similar to the neurobehavioral section, secondary studies provide valuable supplementary information that helps to increase or decrease confidence in the conclusions drawn from the primary studies, thus secondary studies also are included in this section.

The largest and most nationally representative survey of US veterans, conducted by VA, found that nearly 30% of veterans meet a case definition of "multisymptom illness", compared with 16% of nondeployed veterans (Blanchard et al. 2006). Those figures indicate that unexplained illnesses are the most prevalent outcome following service in the Gulf War.

The committee identified five primary studies (Cherry et al. 2001b; Doebbeling et al. 2000; Forbes et al. 2004; Ismail et al. 1999; Kang et al. 2002) and four secondary studies. Those secondary studies have similar methodologic limitations that apply elsewhere in this volume, such as non-representative samples, selection bias, recall bias and small samples.

Primary Studies

University of Manchester (UK)

Through factor analysis, Cherry et al. (Cherry et al. 2001b) identified seven distinct factors in a large, population-based study of British Gulf War-era service members who answered 95 interval-scale symptom questions. Deployed veterans—two random samples of Gulf War veterans (main and validation cohorts)—were compared with a stratified sample of service members who had not been deployed. The seven factors, which accounted for 48% of the variance, could be found in all three groups separately and in combination: psychologic (24 symptoms), peripheral (10 symptoms), neurologic (13 symptoms), respiratory (11 symptoms), gastrointestinal (6 symptoms), concentration (10 symptoms) and appetite (5 symptoms). Deployed veterans' mean factor scores[7] were significantly higher for five factors: psychologic, peripheral, respiratory, gastrointestinal, and concentration. No difference was found in the neurologic factor, and appetite was significantly lower than in the nondeployed cohort. None of the factors was exclusive to Gulf War veterans, so the investigators concluded that their findings did not support a new syndrome (Cherry et al. 2001b). It was a large, diverse, population-based cohort study with high participation rates (86% overall) and had the added benefits of using a 21-point interval scale to record severity of symptoms and a mannequin to ease recording of peripheral dysesthesias and pain. The study included both retired and active-duty personnel and went to extraordinary lengths to find and account for all potential participants, thus minimizing the risk of selection bias. The investigators also divided the Gulf War veterans into two groups:

---

[7] Mean factor scores were computed by adding the sum of mean symptom scores (from 0-21) for each symptom that loaded onto the factor and dividing by the number of symptoms.

the main and validation cohorts. They found the same factors in both cohorts, suggesting consistency. Finally, they took the additional step of conducting cluster analysis (see the next section, Cherry et al. 2001b) to find out whether symptom complexes were similar if a different statistical technique was used. Nonetheless, the analyses, as in other studies, relied on self-reported symptom data and thus are subject to usual concerns about recall bias.

Department of Veteran Affairs

The nationally representative VA study searched for potentially new syndromes through factor analysis by Kang and colleagues (Kang et al. 2002). Data were from a study drawn from 15,000 deployed and 15,000 nondeployed active-duty, reserve, National Guard, and retired service members from all four branches. The authors inquired, through questionnaires, about 47 symptoms on a three-point ordinal scale. Factor analysis of the deployed and nondeployed cohorts yielded the same six factors.[8] The six factors were fatigue and depression[9]; neurologic[10]; musculoskeletal and rheumatologic[11]; gastrointestinal[12]; pulmonary[13]; and upper respiratory[14]. Several symptoms loaded onto more than one factor. In the deployed group, the fatigue or depression factor had an eigenvalue of 12.82 and accounted for 79% of the variance, and the neurologic factor had an eigenvalue of 1.27 and accounted for 8% of the variance. In the nondeployed group the corresponding numbers were 10.39 and 71% and 1.39 and 10%, respectively. Rather than examining differences between factors that loaded or did not load between the deployed and nondeployed groups, the authors examined which symptoms within each factor loaded and did not load. In the neurologic factor, four symptoms—loss of balance or dizziness, speech difficulty, blurred vision, and tremors or shaking—loaded for the deployed but not for the nondeployed group. A group of 277 deployed veterans (2.4%) and a group of 43 nondeployed veterans (0.45%) met a case definition that subsumed the four symptoms. The authors interpreted their findings as suggesting a possible unique neurologic syndrome related to Gulf War deployment that would require objective supporting clinical evidence. The study was a large, diverse, population-based study that had a relatively high response rate of 70%. The use of low eigenvalues to create a six-factor solution was unconventional and may have led to overinterpretation of the data. In the authors' words, "as with the U.K. study (Ismail et al. 1999), we observed that the Gulf War and non-Gulf War veterans displayed virtually identical factor solution in the 5-factor analysis. Only the 6-factor solution model produced different results for the Gulf War and non-Gulf War veterans." Notably the six-factor solution necessitated the use of very low eigenvalues. In addition, the authors' examination of symptoms that loaded onto the neurologic factor for deployed veterans was unconventional in the factor analysis literature but was a unique approach to extracting potentially important symptom differences.

---

[8] It should be noted that in reaching the six-factor solution, Kang et al. went far below the conventional eigenvalue cutoff value of 1.0 (the so-called Kaiser-Guttman rule) and included four factors with eigenvalues below 1.0 for the deployed group and three for the nondeployed group.

[9] Awakening tired and worn out, concentration and memory problems, excessive fatigue, fatigue more than 24 hours after exertion; feeling anxious, irritable, or upset; feeling depressed or "blue"; sleep difficulty; sleepiness during daytime.

[10] Blurred vision, concentration or memory problems, irregular heartbeat, loss of balance or dizziness, speech difficulty, sudden loss of strength, tremors or shaking, excessive fatigue, fatigue more than 24 hours after exertion.

[11] Back pain or spasms, generalized muscle aches, joint aches, numbness in hands or feet, swelling in joints, swelling in extremities.

[12] Constipation, diarrhea, nausea; reflux, heartburn, or indigestion; stomach or abdominal pain, vomiting.

[13] Coughing, irregular heartbeat, shortness of breath, tightness in chest, wheezing.

[14] Coughing, runny nose, sore throat, swollen glands, trouble swallowing.

A later study by VA researchers (Blanchard et al. 2006) sought to determine the prevalence of unexplained illnesses in the VA study. This study was not a factor analysis itself but applied CDC's definition of chronic multisymptom illness, which had been derived in part by factor analysis (Fukuda et al. 1998), to determine its prevalence. The Fukuda et al. study also provided prevalence figures, but because of its nonrepresentative sample, the committee considered it a secondary study. The VA investigators first assessed veterans' responses to the 47 symptom questions. Participants were asked about symptoms in face-to-face interviews 10 years after the war. Several other publications resulting from VA's study are discussed elsewhere in this volume (for example, Eisen et al. 2005). Participants who reported one or more symptoms from each of three clusters (fatigability, mood and cognition, and musculoskeletal) were considered to meet the case definition.[15] Cases were classified as severe if at least one symptom in each cluster was rated as severe. The investigators found that overall 29% of deployed participants met the criteria for chronic multisymptom illness as opposed to 16% of nondeployed participants. Severe chronic multisymptom illness was found in 7% of deployed veterans and 1.6% of nondeployed veterans. Among deployed veterans a higher score on the Combat Exposure Scale was associated with chronic multisymptom illness. Among nondeployed veterans, female sex, less than a college education, and a higher score on the Combat Exposure Scale were associated with chronic multisymptom illness. Deployed veterans who met the case definition had lower mean scores on the SF-36 for physical and mental health, more nonroutine clinic visits and a higher mean number of prescriptions, and they were more likely to be using psychotropic medications. Chronic multisymptom illness in both deployed and nondeployed veterans was more likely to be associated with fibromyalgia syndrome, CFS, symptomatic arthralgias, dyspepsia, the metabolic syndrome, PTSD, anxiety disorders, major depression, nicotine dependence, and more than one psychiatric diagnosis during the year preceding the examination but, with the exception of CFS, was no more likely to be associated with deployment than with nondeployment. The report was based on the same cohort as Fukuda et al. and had the same problems of low participation rates and self-reporting of symptoms. Nonetheless, it provides evidence that the cluster of symptoms previously studied by Fukuda et al. also existed in this group and 10 years after the Gulf War was twice as common in participants who had been deployed as in those who had not been.

The Iowa Cohort

The Iowa study (Iowa Persian Gulf Study Group 1997) was the first major population-based study to group symptoms into categories suggestive of existing syndromes or disorders, such as fibromyalgia or depression. Its finding of a considerably higher prevalence among Gulf War veterans of symptom groups suggestive of fibromyalgia, depression, and cognitive dysfunction motivated the first applications of factor analysis to group and classify veterans' symptoms. Several years later, the same team of Iowa investigators performed a factor analysis on the Iowa cohort (Doebbeling et al. 2000). They studied the frequency and severity of 137 self-reported symptoms among 1,896 Gulf War veterans and 1,799 veterans who had not been deployed. They randomly divided the Gulf War veterans into two groups that they called the derivative sample and the validation sample. They identified three symptom factors in deployed veterans in the derivative sample that accounted for 35% of the variance: somatic distress (joint

---

[15] Persistent fatigue for at least 24 hours after exertion was the single symptom in Cluster A (fatigability). Feeling depressed; feeling irritable; difficulty in thinking or concentrating; feeling worried; tense, or anxious; problems in finding words; and problems in going to sleep were the symptoms included in Cluster B (mood and cognition). Joint pain and muscle pain were the two symptoms included in Cluster C (musculoskeletal).

stiffness, myalgia, polyarthralgia, numbness or tingling, headaches, and nausea), psychologic distress (feeling nervous, worrying, feeling distant or cut off, depression, and anxiety), and panic (anxiety attacks; a racing, skipping or pounding heart; attacks of chest pain or pressure; and attacks of sweating). They confirmed them in a separate factor analysis in the validation sample. They also conducted factor analysis in the nondeployed group and found the same three factors, which accounted for 29% of the variance. Thus, the study did not support the existence of a new syndrome. The authors noted the difficulty of attributing to a single condition the increased reporting of nearly every symptom in every bodily system. The strengths of this study included the size and diversity of its study population and the inclusion of nondeployed veterans. It also had a substantially higher participation rate (90.7%) than earlier studies.

Guy's, King's, St. Thomas's Schools of Medicine (UK)

Ismail and colleagues (Ismail et al. 1999) applied factor analysis to a large representative sample of UK veterans. They were able to identify three fundamental factors, which they classified as related to mood and cognition (headaches, irritability or outbursts of anger, sleeping difficulties, feeling jumpy or easily startled, unrefreshing sleep, fatigue, feeling distant or cut off from others, forgetfulness, loss of concentration, avoiding doing things or situations, and distressing dreams), the respiratory system (unable to breathe deeply enough, faster breathing than normal, feeling short of breath at rest, and wheezing), and the peripheral nervous system (tingling in fingers and arms, tingling in legs and arms, and numbness or tingling in fingers or toes). The pattern of symptom reporting by Gulf War veterans differed little from the patterns reported by Bosnia and Gulf War-era comparison groups, although the Gulf War cohort reported a higher frequency of symptoms and greater symptom severity. The UK authors interpreted their results as arguing against the existence of a unique Gulf War syndrome. Strengths of the study were its two comparison groups and its ability to compare how well its three-factor solution fit its Bosnian and nondeployed ("era") cohorts. As with the study by Haley et al. (1997b, see below), however, a lower-than-ideal response rate of 65% may have introduced selection bias.

Australian Cohort

In a population-based study of all Australian Gulf War veterans, Forbes and colleagues (Forbes et al. 2004) applied factor analysis to findings from a 62-item symptom questionnaire. Symptom reporting was ordinal: "none", "mild", "moderate", and "severe". Three factors were found that accounted for 47.1% of the variance: psychophysiologic distress (23 symptoms), cognitive distress (20 symptoms), and arthroneuromuscular distress (six symptoms). They were broadly similar to factors in previous analyses and were the same as factors found among nondeployed Australian veterans. However, although the prevalence was similar among deployed and nondeployed veterans, factor scores were higher among the deployed than among the nondeployed. That indicates greater severity of symptoms. The authors concluded that there was no evidence of a unique pattern of self-reported symptoms in deployed veterans. One limitation of this study is that most members of the Australian cohort were from the Navy, so its generalizability to services and personnel from other countries, particularly the United States, may be limited. Nonetheless, its inclusion of all Australian Gulf War veterans and a stratified random sample of nondeployed Gulf War-era Defence Force personnel eliminated the potential for selection bias that other studies had more difficulty in controlling. It is also valuable in setting a baseline of unexplained illness as an effect of deployment itself without the overlay of direct combat and environmental exposures more commonly encountered on land.

Secondary Studies

Air National Guard

Fukuda and colleagues (1998) used factor analysis and other methods to assess the health status of Gulf War Air Force veterans in response to a request from DOD, VA, and the state of Pennsylvania. Their focus was to assess the prevalence and causes of an unexplained illness in members of one Air National Guard unit. By studying that unit and three comparison Air Force populations, the investigators aimed to organize symptoms into a case definition and to carry out clinical evaluations on participants from the index Air National Guard unit. They administered a 35-item symptom inventory that included symptom severity (mild, moderate, or severe) and duration (less than 6 months or 6 months or longer) and divided the 3,255 participants who had answered all symptom questions into two subsamples of 1,631 and 1,624. They conducted an exploratory factor analysis of the first subsample that yielded 10 factors with eigenvalues greater than 1.0; three of the factors accounted for 39.1% of the total variance. When the three were examined in a confirmatory factor analysis in the second subsample, two could be confirmed. The first, called mood-cognition-fatigue, consisted of the symptoms: feeling depressed, feeling anxious, feeling moody, difficulty in remembering or concentrating, trouble in finding words, difficulty in sleeping, and fatigue. The second, called musculoskeletal, consisted of the symptoms: joint stiffness, joint pain, and muscle pain. They used those 10 symptoms from the two confirmed factors to develop a preliminary case definition having a combined factor score in the top 25th percentile. That was compared with an alternative clinical case definition of having one or more symptoms in each of two of three symptom categories: fatigue, mood-cognition, and musculoskeletal. Forty-five percent of deployed veterans met the factor-score-based case definition, whereas only 15% of nondeployed veterans met it. The same percentages met the clinical case definition, which because of its greater clinical simplicity than the factor-score-based case definition, was then used to create a case definition of chronic multisymptom illness. The new definition, which was used in later studies, was having one or more chronic symptoms (present for 6 months or longer) from at least 2 of the 3 categories: fatigue, mood-cognition[16], and musculoskeletal[17]. A case was classified as severe if each symptom reported that was used to meet the case definition was rated as severe.

Of the participants surveyed, those deployed to the Gulf War experienced a higher prevalence of chronic symptoms (33 of 35 symptoms with more than 6-month duration were reported to be more prevalent) than nondeployed veterans. According to the case definition of chronic multisymptom disease, 39% of Gulf War-deployed veterans and 14% of nondeployed veterans had mild-to-moderate cases, and 6% and 0.7%, respectively, had severe cases. On the basis of a total of 158 clinical examinations performed in one unit, there were no abnormal physical or laboratory findings that differentiated those who met the case definition from those who did not meet the case definition. Cases, however, reported significantly lower functioning and well-being.

Because such a large fraction (14%) of nondeployed veterans met the mild-to-moderate case definition, the investigators concluded that the case definition could not specifically characterize Gulf War veterans who had unexplained illnesses (Fukuda et al. 1998). The study, however, had several limitations, the most important of which was its coverage of only current

---

[16] Symptoms of feeling depressed, difficulty in remembering or concentrating, feeling moody, feeling anxious, trouble finding words or difficulty in sleeping.

[17] Symptoms of joint pain, joint stiffness, or muscle pain.

Air Force personnel several years after the Gulf War (Air National Guard, Air Force Reserve, and-active duty personnel), which limits its generalizability to other branches of service and to those who left the service possibly because of illness. The use of self-reported symptoms introduced the possibility of reporting bias, and the low participation rates in two of the four units (62% and 35%) introduced the possibility of selection bias. Nonetheless, symptom reporting and prevalence were similar among the four units. A particular strength of this study was its use of a symptom inventory rather than asking veterans about specific diagnoses, such as CFS, MCS, depression, and various neurologic abnormalities. Its use of a more intensive examination of Gulf War veterans from the index unit—including an additional clinical questionnaire; interviewer-administered modules on major depression, somatization disorder, and panic disorder; a screening physical examination with blinded examiners; and a variety of laboratory tests—provided important additional data even though participation rates were low (62%).

Seabee Cohort

Knoke et al. (Knoke et al. 2000) applied factor analysis to active-duty Seabees in response to the factor analysis conducted by Haley et al. (1997b). The study population was drawn from US Navy construction battalion personnel (Seabees) who were on active duty in 1990 and remained on active duty in 1994, when the study was conducted. The instrument contained 98 symptom questions. Among the 524 Gulf War veterans and 935 nondeployed Seabees, Knoke and colleagues performed three factor analyses: the first on the deployed Seabees, the second on the nondeployed Seabees, and the third on both. Each factor analysis identified five factors that accounted for 80%, 89%, and 93% of the total variance respectively. The factors were insecurity or minor depression (27 symptoms), somatization (13 symptoms), depression (10 symptoms), obsessive-compulsive (seven symptoms), and malaise (seven symptoms). Scores among the three analyses were similar for insecurity or minor depression; higher in Gulf War veterans for somatization, depression and obsessive-compulsive; and higher in nondeployed Seabees for malaise.[18] Somatization, depression, and obsessive-compulsive affected an excess of about 20% of Gulf War veterans. The findings were similar to those of Doebbeling (2000), Fukuda (1998) and Ismail (1999) and consistent with findings in a civilian population with CFS (Nisenbaum et al. 1998). They concluded that, unlike the results of the Haley et al. (1997b) study of Seabee reservists from one unit, there was no evidence of a unique spectrum of neurologic injury. While the Knoke et al. study used a larger population than Haley et al., this study still used personnel from a single service, so its generalizability is limited. Because participants were active-duty personnel by design, the results cannot be generalized to retired or reserve personnel who might have been more symptomatic. Nonetheless, the authors' careful examination of the methods and findings of Haley et al. in a very similar population makes this study quite useful.

Haley et al. Seabee Cohort and Validation Study

Haley and collaborators (Haley et al. 1997b) studied a battalion of naval reservists called to active duty for the Gulf War (n = 249). More than half the battalion had left the military by the time of the study. Of those participating, 70% reported having had a serious health problem since returning from the Gulf War and about 30% reported having no serious health problems. The

---

[18] Factor scores used to compare the groups were computed from the regression coefficients of the Gulf War veteran factor analysis, standardized for both groups by subtracting the median and dividing by the semi-interquartile range of the score for the Gulf War veteran group.

study was the first to examine groupings of symptoms in Gulf War veterans using factor analysis. Through standardized symptom questionnaires and a two-stage exploratory factor analysis, the investigators defined what they considered to be either six syndromes or six variants of a single syndrome, which they labeled impaired cognition, confusion-ataxia, arthromyoneuropathy, phobia-apraxia, fever-adenopathy, and weakness-incontinence. One-fourth of the veterans in this uncontrolled study (n = 63) were classified as having one of the six syndromes. The first three syndromes had the strongest factor clustering of symptoms. The study was limited by its lack of a comparison group; the authors were unable to comment on the uniqueness of the factors in relation to other groups of veterans. The findings were based entirely on symptoms self-reported in a mail survey; no in-person interview, physical examination, or laboratory data assisted in the characterization of the factors. In addition, the low participation rate (58%) could have introduced selection bias in that people more or less symptomatic may have participated preferentially. The authors comment that the only evidence they found of differential participation was that nonparticipants were less likely to report a serious illness and to be unemployed. If that can be generalized to all nonparticipants, it would mean that the prevalence of symptoms among participants may have been systematically overestimated. Finally, the study population came from a single unit, so findings cannot necessarily be generalized to all Gulf War veterans.

Haley and colleagues (2001) attempted to replicate their factor analysis findings in a validation cohort, which was separate from their original cohort of Seabees. The validation cohort (n = 335) consisted of veterans living in North Texas who had registered with a VA clinic in Dallas or who were recruited by advertising. In comparison with the Seabee cohort, participants in the validation cohort were more likely to have served in the Army and in general to be more representative of those who served in the gulf with regard to racial and ethnic background, age, and wartime military status. In this study, the authors used more sophisticated questionnaires than in the earlier Seabee cohort in an effort to replicate the earlier findings. They undertook a series of analyses to test whether the latent syndrome structure they found in the earlier cohort could be replicated in the larger and more representative cohort. They allowed, by design, only four symptom scales per syndrome factor to load onto five models and compared the five models with their earlier findings by using structure equation models. The five models had either 12 or 16 measured variables, which loaded onto three first-order factors and zero or one higher-order factor. In two models, the four additional variables (or symptom factors) were allowed to load onto the primary or higher-order factors. The three primary syndrome factors were impaired cognition, confusion-ataxia, and central pain (termed arthromyoneuropathy in the original study); and the four additional variables or secondary symptom factors were chronic watery diarrhea, chronic fatigue involving excessive muscle weakness, chronic fever and night sweats, and middle and terminal insomnia. The higher-order factor was the presence of an underlying single Gulf War syndrome that could explain all variance and covariance among the three first-order factors. Overall, 29% of participants had one or more of the three first-order factors, defined by dichotomizing the syndrome factor scale at 1.5, as in the original study. They found that the apparent three-factor solution, originally demonstrated in the Seabee cohort, was also present in this new cohort (Model 1); that the three syndrome factors probably represented a higher-order syndrome, such as a single Gulf War syndrome (Model 2); and that some additional symptoms (the four secondary symptom factors) appeared in all three syndrome variants. They suggested that the confusion-ataxia syndrome may represent a more severe form of a single Gulf

War syndrome of which impaired cognition and central pain variants (the other two syndrome factors) were less severe forms.

The small sample may have limited exploration of less common symptoms, and the nonrandom sample may have limited the generalizability of some of the results, such as syndrome prevalence, to some degree; but the detailed questionnaires, the substantially more refined symptom measures, and the external validation of the findings through comparison with the Seabee cohort were strengths of the second study. Note that this study, by design, had no comparison group. The authors were seeking to validate the presence of a symptom complex in deployed veterans, rather than to examine its prevalence in deployed vs nondeployed forces. The authors concluded by recommending study of a national randomly selected sample of deployed and nondeployed Gulf War-era military populations with their methods of symptom measurement and syndrome definition.

Portland Area Veterans

Investigators studied clusters of unexplained symptoms in a population-based study of Portland area veterans by creating a new case definition of unexplained illness (Storzbach et al. 2000). Cases were identified through questionnaires as meeting a threshold number and combination of symptoms (cognitive and psychologic, and musculoskeletal) and on the duration of fatigue. Veterans whose symptom clusters remained unexplained at clinical examination (after exclusion of established diagnoses) were defined as constituting cases. Controls were those who at the time of clinical examination had no history of case-defining symptoms during or after their service in the Gulf War. In an analysis of the 241 cases and 113 controls, investigators found small but statistically significant deficits in cases on some neurobehavioral tests of memory, attention, and response speed. Cases also were statistically significantly more likely to report increased distress and psychiatric symptoms (Storzbach et al. 2000). Finally, more than half the veterans with unexplained musculoskeletal pain met symptom-based criteria for fibromyalgia, and a large proportion met symptom-based criteria for CFS (Bourdette et al. 2001). The study also undertook a factor analysis, which initially loaded 48 of the 69 symptoms and accounted for 21.6% of the variance. The researchers then re-examined the 48 symptoms in a second factor analysis. Three were retained for rotation and further analysis. These three factors loaded 35 symptoms; cognitive and psychologic, mixed somatic, and musculoskeletal accounted for 34.2% of the common variance. Rather than using those factors as a working case definition to explore symptom differences between deployed and nondeployed veterans, the authors used their three-factor solution to validate their a priori case definition composed of 35 symptoms encompassing musculoskeletal pain, cognitive and psychologic changes, gastrointestinal complaints, skin or mucous membrane lesions, or unexplained fatigue.

There were two major findings when the researchers compared the three-factor solution to the a priori case definition of Gulf War unexplained illnesses. First, the factor analysis did not include any symptoms related to the gastrointestinal system, the skin, or mucous membranes. Second, three symptoms that had loaded onto the musculoskeletal factor—numbness in fingers or toes, clumsiness, and dizziness—were not included in the case definition. They then used that information to assess the accuracy of their case definition. The three-factor solution identified 103 (91%) of the 113 controls and 189 (78%) of the 241 cases in their clinically evaluated subsample. They also tested their three-factor solution against a modification of the clinical case definition-approach used by Fukuda and colleagues (1998)[19] and found that the three-factor solution predicted 103 (91%) of 113 controls and 108 (94%) of 115 cases in their study. They concluded that their factor analysis confirmed the finding of a cognitive and psychologic factor found by Haley et al. (1997b), Fukuda et al. (1998), and Ismail et al. (1999), and the finding of a musculoskeletal factor reinforced the finding of a musculoskeletal factor by Haley et al. and Fukuda et al. This study, like others, had a relatively low response rate (64%), which introduced the possibility of selection bias. As a result of its case-control design, the study population by definition comprised only Gulf War veterans; this eliminated the possibility of examining differences between deployed and nondeployed veterans. Nonetheless, its careful clinical and psychologic examinations of a sample of the responders is a strength, and its use of factor analysis to validate its a priori case definition rather than to create one de novo is unique among the factor analysis studies.

Shapiro and colleagues (Shapiro et al. 2002) performed factor analysis to determine whether there was a unique syndrome among veterans in the Oregon cohort who witnessed the Khamisiyah demolition and might thereby have been exposed to nerve agents. They divided their population-based sample into three groups: witnesses to Khamisiyah demolitions, nonwitnesses who had been deployed to the Gulf War, and nondeployed veterans. Their analysis of 25 symptoms among three groups of 1,779 veterans identified three factors that accounted for 46.7-52.2% of the overall variance, depending on the exposure group. Three factors were common to all three groups: cognitive or psychologic (in the exposed group, this included unusual irritability or anger; mood swings; changes in memory; persistent fatigue, tiredness, or weakness; difficulty in concentrating; and depression), dysesthesia (in the exposed group, tingling, burning sensation of pins and needles, and numbness or lack of feeling), and vestibular dysfunction (in the exposed group, loss of balance or coordination and dizzy spells). There were slight differences in symptoms that loaded in the nonexposed and nondeployed groups. There were no differences in distributions of factor scores between the three groups in terms of the cognitive or psychologic and vestibular dysfunction subgroups, but in log-linear analysis and logistic regression analysis the dysesthesia factor was significantly associated with having witnessed the demolition.

Shapiro and colleagues provide two important cautions in the use of factor analysis. The first is that—because different investigators use different lists of symptoms, different samples, and different factor extraction and rotation techniques and factor-loading cutoffs—results from different studies are not directly comparable. The second is related to the use of factor analysis with dichotomous variables. As noted above, factor analysis traditionally has used interval-level data or at least ordinal data. However, most Gulf War syndrome investigators have used dichotomous data to assess symptom presence or absence. To examine the potential pitfalls inherent in that approach, Shapiro and her colleagues created a dataset of 19 dichotomous variables typically used in Gulf War syndrome research, randomly generated values for them whose row and column totals corresponded to the frequencies observed in the study, and applied factor analysis. They repeated that 500 times and found that their random datasets could result in five factors that explained 30% of variance and loading more than 95% of the time at the traditional 0.4 cutoff. That potentially has profound implications for the interpretation of data from factor analysis. As Shapiro et al. note, "in the absence of more robust decision rules for

---

[19] In an initial clinical case definition, Fukuda et al. used any symptom reported for 6 months or longer by at least 25% of Gulf War veterans and at least 2.5 times more frequent among deployed veterans than among nondeployed veterans. For their analysis, Bourdette et al. modified the Fukuda et al. case definition to any symptom reported by 50% of Gulf War veterans reported for at least 1 month in the preceding 3 months.

these kinds of data, the resulting factor may be a rich mixture of randomness, which could lead investigators down uninformative paths."

Department of Veterans Affairs Gulf War Health Registry

Hallman and colleagues (2003) examined patterns of reported symptoms among a sample of persons who participated in the VA Gulf War Health Registry. The study population consisted of a state-based random sample of 2,011 veteran registry members residing in Delaware, Illinois, New Jersey, New York, North Carolina, Ohio, and Pennsylvania who were not participating in other studies. Questionnaires included 48 symptoms, which were rated on a three-point ordinal scale, and were returned by 1,161 veterans (58% of the sample). The investigators divided the participants into two groups and conducted five factor analyses in each group to ensure consistency. They identified four factors that accounted for 50.2% of the variance. The factors were mood-memory-fatigue (depression, anxiety, sudden mood changes, problems concentrating and remembering, unexplained weakness, sleep problems, and unexplained fatigue), musculoskeletal (pain or numbness in joints or muscles), gastrointestinal (abdominal pain and gas, diarrhea, nausea, and vomiting) and throat-breathing (difficulty in swallowing, swollen glands, nose or sinus problems, coughing, difficulty in breathing, and difficulty in tasting). Like Cherry et al. (2001b), they also conducted a cluster analysis (see below) to examine consistency between the two different statistical methods. The principal limitation of the study is the lack of a nondeployed control group, which limits its ability to identify factors that may have been peculiar to exposure to the Gulf War. However, by starting with presumably the most symptomatic subset of Gulf War veterans (those who had left the service and registered with the Gulf War Health Registry), it also had power to identify clusters peculiar to symptomatic Gulf War veterans. However, the four factors it identified were largely similar to factors identified both by other Gulf War investigators and in civilian populations (Gillespie et al. 1999; Nisenbaum et al. 1998).

Cluster Analysis

A somewhat related technique, cluster analysis (see Chapter 3), has been used in three cohorts to determine how groups of patients with particular symptoms may be related to one another (Cherry et al. 2001b; Everitt et al. 2002; Hallman et al. 2003).

University of Manchester Cohort (UK)

Cherry and colleagues (Cherry et al. 2001b) sequentially partitioned members of the three cohorts, using scores from the 95 symptoms reported. Convergence was reached within 200 iterations. Participants divided into six clusters, which the authors then compared with their seven-factor solution and the standardized mean factor scores. Cluster 1 was composed primarily of well people and had a smaller proportion of Gulf War veterans (36.4%) than nondeployed veterans (48.5%). Clusters 2 and 3 had similar prevalences in both groups. The final three clusters accounted for 23.8% of Gulf War veterans but only 9.8% of nondeployed veterans and included clusters with high scores on respiratory and gastrointestinal illnesses (cluster 4), on psychologic ill health (cluster 5), and both overall and especially on neurologic symptoms (cluster 6). Thus, there was an excess of 14% of Gulf War veterans in the three least healthy clusters.

Guy's, King's, St. Thomas's Schools of Medicine

Everitt and colleagues (2002) randomly sampled 500 participants from among the three cohorts (Gulf War veterans, Bosnian veterans and nondeployed Gulf War-era controls). They regrouped the original 50 ordinal-scale symptoms into 10 categories, retaining the same four-point severity score. They also used a technique, known as the gap statistic, that can be used to suggest the number of clusters that best describe the data (Tibshirani 2001). They identified five clusters by using conventional cluster analysis. Cluster 1 had low scores for all symptoms, cluster 2 had the highest scores for musculoskeletal symptoms and high scores for neuropsychologic, cluster 3 had high scores for neuropsychologic and higher scores for the remaining nine symptom groups, cluster 4 had high scores only for musculoskeletal symptoms, and cluster 5 had high scores in all 10 symptom groups, especially musculoskeletal and neuropsychologic. Gulf War veterans were 3-4 times more likely to fall into cluster 2 and 11 times more likely to fall into cluster 5, although cluster 5 contained only 26 people from all three cohorts combined. With the gap statistic, two clusters were identified, one with low scores in each symptom group and another with higher mean scores for the musculoskeletal and neuropsychologic groups. Some 72% percent of Gulf War veterans, 87% of Bosnian veterans, and 94% of era-deployed veterans were classified in cluster 1. The authors interpreted their findings to mean that there was no convincing evidence of a new unique Gulf War syndrome. The study's strengths included a random sample and standardized symptom measurements. However, the authors caution that the finding of nonspecific symptoms may have been distorted by reporting bias.

Department of Veterans Affairs Gulf War Health Registry

Hallman and colleagues (2003) conducted cluster analysis in their examination of 1,161 veterans participating in the VA Gulf War Health Registry. They used the mean factor scores from their factor analysis to group respondents on the basis of severity of symptoms. Examining the two randomly divided subsamples a total of five times each but using cluster analysis, they identified two stable clusters. Cluster 1, making up 60.4% of the sample, consisted of veterans who reported no or mild symptoms in each of the four factors. Cluster 2, the remaining 39.6% of participants, consisted of veterans with moderate-to-severe factor scores in the mood-memory-fatigue and musculoskeletal factors and mild-to-moderate scores in the gastrointestinal and throat-breathing factors. People classified in cluster 2 reported twice as many symptoms, reported more severe problems, were in poorer health, and had a greater reduction in mean activity as people in cluster 1 (37.2% vs 17.8%).

Summary and Conclusion

Factor analysis has been performed on all the major cohorts covered in Chapter 4, and cluster analysis has been performed on three cohorts, one of which was studied by both methods. The findings, despite methodologic differences, are quite similar. There seem to be similar groups of symptoms that fall roughly into factors that describe neurocognitive symptoms, musculoskeletal symptoms, and peripheral nervous system symptoms. Less commonly reported are factors that involve gastrointestinal and respiratory symptoms. Well-conducted factor analysis starts with representative samples with high participation rates. Several studies fall short on those two criteria, for instance, by including members of only one branch of the service (e.g., Fukuda et al. 1998; Haley et al. 1997b; Knoke et al. 2000), small samples (e.g., Haley et al. 1997b), or largely symptomatic groups of veterans (Hallman et al. 2003). Another problem is the lack of a comparison group in some of the studies, which limits investigators' ability to compare the presence of factors in deployed and nondeployed groups (e.g., Bourdette et al. 2001; Haley et al. 1997b). Although results of the studies are valuable and add rich detail to the epidemiologic literature surrounding Gulf War veterans, other studies are more representative and hence more generalizable (Ismail et al. 1999; Cherry et al. 2001b; Doebbeling et al. 2000; Kang et al. 2002). In those studies, the findings were quite similar, broadly describing neurologic, psychologic, cognitive, fatigue, and musculoskeletal symptoms. One exception is the Kang et al. study's last three factors—gastrointestinal, pulmonary, and upper respiratory—which may have been the result of using eigenvalues less than 1.0 in constructing the models. The three studies that compared the factors that emerged in the most representative deployed and nondeployed groups (Cherry et al. 2001b; Doebbeling et al. 2000; Ismail et al. 1999) found factors that were remarkably similar between the deployed and nondeployed groups and did not suggest a unique complex of symptoms that existed in the deployed group but not in the nondeployed group. However, in each of those studies, as in many of the less generalizable studies, symptoms were more severe in the deployed than in the nondeployed groups.

The three studies that used cluster analysis had broadly similar findings. The two studies that included nondeployed comparison groups (Cherry et al. 2001b; Everitt et al. 2002) failed to identify a unique cluster of symptoms that might represent a unique Gulf War syndrome, but each did identify a highly symptomatic cluster of patients that had a statistically significantly higher proportion of Gulf War veterans than non-Gulf War veterans, ranging from 14% in Cherry et al. (Cherry et al. 2001b) to 22.2% in Everitt et al. (Everitt et al. 2002) (two-cluster solution comparing Gulf War with Gulf War-era veterans). The cluster analyses confirm the finding from the most representative factor-analysis studies: although there is not a unique symptom complex among Gulf War veterans, they are clearly more symptomatic than their nondeployed counterparts.

In the end, studies using those designs have their limitations as noted by Shapiro (2002), and should be viewed in the context of their inherent limitations. They demonstrate that deployed veterans report more symptoms and more severe symptoms than their nondeployed counterparts. However, there is no symptom complex peculiar to deployed Gulf War veterans. The primary and secondary studies are summarized in Table 5.19.

TABLE 5.19 Factor Analyses of Gulf War Veteran Cohorts (Primary and Secondary Studies)

Reference	Population	Type of Data	Method	Rotation	Factor Loading Cutoff and Eigenvalue Cutoff	No. Factors Isolated	Percentage of Variance Explained	Factors Identified	Unique Factors in Deployed Veterans?
Cherry et al. 2001b (primary)	Active and retired n = 11,914	Interval (21 points)	Principal components	Orthogonal	> 0.40, Not stated	7	48	Psychologic, peripheral, neurologic, respiratory, gastrointestinal, concentration, appetite	All present; mean factor scores higher in GWVs for psychologic, peripheral, respiratory, gastrointestinal, concentration; lower for appetite
Kang et al. 2002 (primary)	Active and retired n = 19,383	Ordinal	Iterated principal factors	Orthogonal	> 0.30, extracted factors had to have at least two symptoms > 0.40; ≥ 0.57 for GWVs, 0.53 for nondeployed	6		Fatigue or depression, neurologic, musculoskeletal/rheumatologic, gastrointestinal, pulmonary, upper respiratory	Factors similar but 4 neurologic symptoms loaded on neurologic factor for deployed but not for non-deployed
Doebbeling et al. 2000 (primary)	Active and reserve n = 3,695	Ordinal and dichotomous	Unknown	Varimax and promax	≥ 0.35 Not stated	3	35 in both samples of deployed, 30 in non-deployed	Somatic distress; Psychological distress; Panic	Correlation between derivative and validation samples, same factors in non-deployed. Prevalence not stated.
Ismail et al. 1999 (primary)	Active n = 3,214	Ordinal and dichotomous	Principal factors	Orthogonal	> 0.40 > 1.0	3	~20	Mood-cognition; Respiratory system; Peripheral nervous system	No but 3-factor solution fit less well in Bosnian cohort than

Reference	Population	Type of Data	Method	Rotation	Factor Loading Cutoff and Eigenvalue Cutoff	No. Factors Isolated	Percentage of Variance Explained	Factors Identified	Unique Factors in Deployed Veterans?
									in GW deployed and less well in non-deployed than in Bosnian cohort. Prevalence not mentioned
Forbes et al. 2004 (primary)	Active and retired n = 2,781	Ordinal	Unknown	Orthagonal and oblique	> 0.40 > 1.0	3	47.1	Psycho-physiological distress; cognitive distress; arthro-neuromuscular distress	No Prevalence similar but severity higher in GWV
Fukuda et al. 1998 (secondary)	Active and reserve (Air Force) n = 3,255	Ordinal	Principal components	Oblique (Promax, in subsample 1 and Procrustes in subsample 2)	> 0.40 > 1.0	3	39.1	Fatigue; mood-cognition; musculoskeletal pain	NA 45% of deployed met factor score-based case definition of chronic multisymptom illness vs 15% of non-deployed
Knoke et al. 2000 (secondary)	Active (Navy) n = 1,459	Ordinal and dichotomous	Principal factors	Orthagonal	> 0.40 > 1.0	5	80-93	Insecurity; somatization; depression; obsessive-compulsive; malaise	Somatization, depression, obsessive-compulsive. About 3 times as common

Reference	Population	Type of Data	Method	Rotation	Factor Loading Cutoff and Eigenvalue Cutoff	No. Factors Isolated	Percentage of Variance Explained	Factors Identified	Unique Factors in Deployed Veterans?
Haley et al. 1997b (secondary)	Active and retired (Navy) n = 249	Interval	Principal axes	Orthagonal	> 0.40 1.0	6	71	Impaired cognition; Confusion-ataxia; Arthromyoneuropathy; Phobia-apraxia; Fever-adenopathy; Weakness-incontinence	NA
Haley et al. 2001 (secondary)	Active, reserve, retired n = 335	Continuous	Principal factors (in developmental sample)	Orthogonal and oblique	> 0.40 1.0 (in developmental sample)	Forced into five models with 3 syndrome factors	29 of any of the three syndrome factors	Impaired cognition; Confusion-ataxia; Central pain	Compared fit of factor analysis with that found in earlier Seabee study (Haley 1997b) using structural estimating equations. Some models also fitted higher order factor "Gulf War syndrome" and loaded four additional symptoms (chronic fatigue involving excessive muscle weakness, chronic fever and night sweats, middle and terminal insomnia, chronic watery diarrhea) onto

Reference	Population	Type of Data	Method	Rotation	Factor Loading Cutoff and Eigenvalue Cutoff	No. Factors Isolated	Percentage of Variance Explained	Factors Identified	Unique Factors in Deployed Veterans?
Bourdette et al. 2001 (secondary)	Active and reserve n = 443	Dichotomous	Principal components	Orthagonal	≥ 0.30 > 1.0	3	34.2	Cognitive/psychologic; Mixed somatic; Musculoskeletal	higher-order factor NA
Shapiro et al. 2002 (secondary)	Active and retired n = 1779	Dichotomous	Principal components	Orthagonal	> 0.60 > 1.0	3	46.7 among exposed, 49.8 among non-exposed deployed, 52.2 among non-deployed	Cognitive-psychologic; Dysesthesia; Vestibular dysfunction	Higher odds of having witnessed demolition among those reporting dysesthesia
Hallman et al. 2003 (secondary)	Retired, participants in VA Gulf War Health Registry n = 1,161	Ordinal	Principal axis	Oblique	Not stated	4	50.2	Mood-memory-fatigue; musculoskeletal; gastrointestinal; throat-breathing	NA

NOTE: GW = Gulf War; GWV = Gulf War Veteran; VA = Department of Veterans Affairs.

INJURY AND EXTERNAL CAUSES OF MORBIDITY AND MORTALITY
(ICD-10 S00-Y98)

Primary Studies

The first large mortality study of nearly all Gulf War-deployed veterans (n = 695,516) identified no excess postwar mortality, from all causes combined, compared with nondeployed veterans (n = 746,291). One particular cause of mortality, from motor-vehicle accidents, was somewhat higher (RR 1.31, 95% CI 1.14-1.49), but the risk was lower than the expected rate based on overall US mortality (SMR 0.82, 95% CI 0.75-0.89) after adjustment for age, sex, race, and year of death (Kang and Bullman 1996). The study examined mortality patterns from 1991 through 1993 by using two databases: the VA Beneficiary Identification and Records Locator Subsystem (BIRLS) and deaths reported to the Social Security Administration.[20] It compared deployed veterans with a cohort of similar size of veterans who did not serve in the Gulf War. Adjustments were made for age, race, marital status, branch of service, and type of unit. It also found no increase in suicide or homicide among Gulf War veterans. The second publication by the authors found that by 1994, the excess mortality risk from motor-vehicle accidents had disappeared. That finding is consistent with the mortality pattern after the Vietnam War (CDC 1987; Thomas et al. 1991; Watanabe and Kang 1995). The study found no overall differences in mortality between deployed and nondeployed veterans; additionally the mortality risk in both the deployed and nondeployed was less than half of what was expected in their civilian counterparts (Kang and Bullman 2001).

A study of all UK veterans of the Gulf War (n = 53,462) in relation to contemporaneous controls found no increase in mortality other than a small and nonsignificant increase in accidental death (RR 1.18, 95% CI 0.98-1.42) (Macfarlane et al. 2000). In that study, controls (n = 53,450) were matched by sex, age, branch of service, and level of fitness in an attempt to control for the healthy-warrior effect. Accidental-death increases were due primarily to motor-vehicle accidents (RR 1.25, 95% CI 0.91-1.72), air and space accidents (RR 1.77, 95% CI 0.86-3.81), and accidents caused by submersion, suffocation, or ingestion of foreign bodies (RR 3.25, 95% CI 1.00-13.69). There were no increases in suicide or homicide. The study covered the years 1991-1999 and did not address changes over time in excess external cause (or motor-vehicle) mortality.

As part of continuing mortality surveillance, the UK Defence Analytical Services Agency periodically publishes its cumulative mortality figures for deployed veterans vs Gulf War-era controls. From 1991 to June 30, 2005, there was no increase in mortality other than a small and nonsignificant increase in transportation accidents, which include land, water, and air accidents (SMR 1.21, 95% CI 0.96-1.51). "Other external causes of accidental injury"—a category including falls, drowning, and poisoning—also showed a small and nonsignificant increase (SMR 1.07, 95%CI 0.74-1.54) (Defence Analytical Services Agency 2005). Compared with earlier surveillance, those data show that the differences between deployed veterans and Gulf War-era controls in deaths from external causes disappeared about 10 years after the war. There also were a small and statistically nonsignificant increase in intentional self-harm (SMR 1.08,

---

[20] The degree of completeness of these record systems was assessed with a validation study that used state vital-statistics data. Ascertainment was estimated at 89% of all deaths in the Gulf War cohort and comparison group.

95% CI 0.85-1.39) and a small and statistically nonsignificant reduction in risk of death from assault (SMR 0.46, 95% CI 0.15-1.38).

Finally, one hospitalization study addresses the question of transportation-related injuries. A study of armed services personnel on active duty during the Gulf War was conducted after the war (1991-1994) at DOD, VA, and nonfederal hospitals in California (Gray et al. 2000). The purpose of including the latter type of hospitalization was to eliminate potential bias related to veterans' seeking care outside DOD and VA facilities. The authors found increased rates of hospitalization for the category "injury and poisoning" in DOD hospitals (PMR 1.03, 95% CI 1.01-1.05) and California hospitals (PMR 1.11, 95% CI 1.04-1.18), but not in VA hospitals. Table 5.20 summarizes the results of the primary mortality studies.

Secondary Studies

A mortality study of US active-duty military personnel focused exclusively on the Gulf War period of Operations Desert Storm and Desert Shield (1990-1991). It compared noncombat mortality among troops stationed in the Gulf War and troops on active duty elsewhere. There was no excess noncombat mortality in deployed veterans except for unintentional injury due to vehicle accidents and other causes (Writer et al. 1996). A similar study of noncombat injuries in navy and marine personnel during the Gulf War found that most patient visits were for injuries and poisonings[21] (Shaw et al. 1991).

A post-Gulf War population-based study of Iowa veterans (1995-1996) found that self-reported traumatic injuries were more likely in deployed than in nondeployed veterans (Zwerling et al. 2000). However, in a large UK study, self-reported "accidental injuries" were lower in Gulf War veterans than in nondeployed veterans, but the category was very broad (Simmons et al. 2004); the authors interpreted this unexpected result as due to inclusion of both major and minor injuries and to the possibility that Gulf War veterans are more inclined to report illnesses than injuries because of the belief that illnesses are associated with deployment.

Gackstetter et al. (2002) performed a nested case-control study of the large Gulf War-deployed and nondeployed population assembled by Kang and colleagues (1996). Deployed veterans who died in motor-vehicle accidents through 1995 (n = 1,343) were more likely to be male, younger, less educated, and never married than nondeployed controls (10 controls/case). They were also more likely to be enlisted, have combat occupations, and be in the National Guard or reserves and not in the Air Force. One of the datasets used by Gackstetter et al. (2002) examined prior morbidity patterns to determine underlying physical and mental health among 980 deployed veterans and nondeployed veterans who died in motor-vehicle accidents. After adjustment for demographic factors and military characteristics, the authors found that prior treatment for mental-health problems, particularly drug or alcohol abuse was strongly associated with such deaths, particularly among nondeployed veterans. The only predictor of motor-vehicle deaths among deployed veterans was prior motor-vehicle injury.

Summary and Conclusion

The committee found that various studies have looked at mortality in Gulf War veterans but have numerous limitations. The principal limitation is the short duration of their followup

---

[21] As noted earlier, poisoning is one of the standard codes used on death certificates under "other external causes of accidental injury". The category includes falls, drowning, and poisoning.

observation period. More time must elapse before investigators will be able to assess increased mortality that would result from illnesses with long latency, such as cancer, or with a gradually deteriorating course, such as cardiovascular disease. Another potential limitation in comparing deployed and nondeployed personnel is the healthy-warrior effect. That might result in selection bias, insofar as chronically ill or less fit members of the armed forces might be less likely than more fit members to have been deployed. Thus, there might have been nonrandom assignment of those selected for deployment and not selected for deployment. That is demonstrated by the excess number of deaths from HIV infection among nondeployed veterans reported by Kang and Bullman (2001).

Some studies provide evidence of a modest increase in transportation-related injuries among deployed compared with nondeployed Gulf War veterans in the decade immediately after deployment. That increase in mortality appears to have been restricted to the first several years after the war.

TABLE 5.20 Mortality and Injury Studies

Study	Population	Outcomes	Results	Adjustment	Comments or Limitations
Kang and Bullman 1996; Kang and Bullman 2001	695,516 Gulf War veterans vs 746,291 non-Gulf War veterans	Mortality 1991-1997; Cox proportional hazards models	Increased deaths from motor-vehicle accidents in Kang and Bullman 1996 (RR 1.31, 95% CI 1.14-1.49) RRs became nonsignificant in Kang and Bullman 2001 (RR 1.17, 95% CI 0.98-1.4) in 1994-1995; Increased HIV deaths in non-Gulf War veterans; no difference in potential nerve gas exposure; no homicide or suicide increase	Sex, age, race, marital status, branch of service, type of unit	Short duration of followup; healthy warrior effect may obscure difference
Macfarlane et al. 2000	53,462 Gulf War veterans vs 53,450 Gulf War-era cohort, UK	Mortality 1991-1999	Higher mortality in Gulf War veterans from external causes (RR 1.18, 95% CI 0.98-1.42); no increase in homicide or suicide	Matching by sex, age, branch, fitness for service	
Defence Analytical Services Agency 2005	53,409 Gulf War veterans vs 53,143 Gulf War-era cohort, UK	Mortality 1991-June 2005	No increase in mortality except small and nonsignificant increase in "transport accidents" (SMR 1.21, 95% CI 0.96-1.51); "other external causes of accidental injury" (SMR 1.07, 95% CI 0.74-1.54); higher deaths from external causes disappeared about 10 years after Gulf War	Matching by sex, age, branch	
Gray et al. 2000	652,979 Gulf War veterans vs 652,922 randomly selected nondeployed veterans	Morbidity 1991 to 1994	Increased rates of hospitalizations for the "injury and poisoning" in DOD hospitals (PMR 1.03, 95% CI 1.01-1.05) and California (PMR 1.11, 95% CI 1.04-1.18; decreased rates for VA hospitals (PMR 0.89, 95% CI 0.83-0.96)	DOD hospitals adjusted for age, sex, and race; California and VA hospitals adjusted for age and sex	

ALL-CAUSE HOSPITALIZATION STUDIES

This section concentrates on all-cause hospitalization to determine whether there is an excess risk of hospitalization among Gulf War veterans. Hospitalizations for specific causes, although noted here, are discussed in more detail throughout the report. The primary studies are summarized in Table 5.21.

Primary Studies

Studies of differences in rates of hospitalization between deployed and nondeployed populations can indicate excess morbidity associated with Gulf War service. Although they are less able to detect subtle differences than studies that measure morbidity directly or examine outpatient morbidity, they are less crude than studies of differential mortality. Overall hospitalization (that is, for all causes) and cause-specific hospitalization were the subject of several large studies, mostly of active-duty personnel discharged from DOD hospitals.

The first study (1991-1993) compared the hospitalizations of almost 550,000 Gulf War veterans and almost 620,000 nondeployed veterans and found no consistent differences over time in all-cause hospitalizations after the war (Gray et al. 1996). There were increased rates of hospitalization of Gulf War veterans in some diagnostic categories in some years (for example, neoplasms in 1991 and diseases of the blood in 1992), but the rates were not consistently increased, except rates of hospitalization for mental illness in 1992 and 1993. The study also found increased hospitalization, in 1991 only, for the broad category "genitourinary system diseases". The authors found, more specifically, that the increase was due to female veterans being hospitalized for inflammatory diseases of the ovary, fallopian tube, pelvic cellular tissue, and peritoneum. Those increases could be explained by deferral of care, postwar pregnancies, and some psychiatric disorders (alcohol dependence, nondependent drug abuse, and adjustment reactions). The study also examined reasons for separation from the armed services in 1991-1993. Contrary to expectations, the study found that deployed veterans were less likely than nondeployed veterans to have separated for reasons of medical disqualification, dependence or hardship, entry into officer programs, retirement, or behavior or performance failure.

A second hospitalization study extended the study period (1991-1996) and re-examined the dataset to search for excess hospital admissions for unexplained illnesses (Knoke and Gray 1998). The authors reasoned that the first study might have missed hospitalizations for a new or poorly recognized syndrome.

In the third study, Gray et al. (2000) examined hospitalizations (1991-1994) of active-duty, reserve, and former military personnel who had been deployed to the Gulf War. The study examined hospitalizations at DOD, VA, and nonfederal California hospitals to eliminate potential bias related to veterans' seeking care outside DOD and VA facilities. Because of the unreliability of state-of-residence data in DOD and VA datasets, the authors could not directly compare rates of hospitalization among the three sources. Rather, they compared PMRs of hospitalization-discharge diagnoses (14 diagnostic categories from ICD-9) in Gulf War-deployed and nondeployed veterans. PMRs of most disease categories were not increased. However, four categories were increased in VA patients (but not in active-duty military or California veterans): respiratory (PMR 1.19, 95% CI 1.10-1.29), digestive (PMR 1.12, 95% CI 1.05-1.18), skin (PMR 1.14, 95% CI 1.00-1.27), and ill-defined diseases (PMR 1.24, 95% CI 1.16-1.33). Among respiratory diseases, the authors reported increases in asthma, but no data were shown. The study is antecedent to a more detailed study of respiratory hospitalizations in relation to exposure to smoke from oil-well fires (Smith et al. 2002). The authors also found increased rates of hospitalization for the category "injury and poisoning" in DOD (PMR 1.03, 95% CI 1.01-1.05) and California hospitals (PMR 1.11, 95% CI 1.04-1.18). Each finding is discussed in the relevant section of this chapter.

Other hospitalization studies are reviewed in this chapter and address specific causes of hospitalizations, for example, in relation to exposure to oil-well fires and respiratory outcomes (Smith et al. 2002) or exposure to nerve agents and specific hospitalizations (Gray et al. 1999b; Smith et al. 2003). For the most part, the studies did not find increased hospitalization in relation to the exposures. A study that did find an increase in hospitalizations for acute psychiatric disorders (Dlugosz et al. 1999) is discussed in the psychiatric section. Table 5.22 summarizes the results of the all-cause hospitalization studies.

Summary and Conclusion

The all-cause hospitalization studies provide some reassurance that excess hospitalizations did not occur among veterans of the Gulf War who remained on active duty through 1994. The studies, however, have several limitations, including that they were largely of active-duty personnel and cannot be generalized to the entire cohort of Gulf War veterans, inasmuch as it been noted that Gulf War veterans who left the military reported worse health outcomes than those who remained (Ismail et al. 2000). As is the case for mortality studies, it is too soon to capture hospitalizations from illnesses that might have longer latency, such as some cancers. In addition, hospitalization data might be incomplete on people separated from the military and admitted to nonmilitary (VA and civilian) hospitals. The studies did not measure the use of outpatient treatment and thus detected only illnesses that required hospitalization.

225

TABLE 5.21 All-Cause Hospitalization Studies

Study	Population	Outcomes	Results	Adjustment	Comments or Limitations
Gray et al. 1996	579,931 US Gulf War veterans who were on regular active duty vs about 700,000 randomly selected controls; of these, 1,165,411 had complete data for 1991 (with losses in each later year)	Hospitalization records: DOD only, 1991-1993	Increased rates of hospitalizations for respiratory disease (1991); genitourinary disease (1991), neoplasm (1991), blood (1992), mental disorders (1992-1993); decreased rates for infection (1991-1992), endocrine or metabolic (1991-1993), circulatory (1992), digestive (1991-1993), musculoskeletal (1991-1992), ill-defined conditions (1991-1992)	Gulf War service, sex, age, race, marital status, branch and length of service, occupation, rank, salary	Active duty only, no assessment of outpatient treatment, respiratory findings removed after adjustment for VA screening-program attendance
Knoke and Gray 1998	552,111 deployed vs 1,479,751 nondeployed service members in service during Gulf War and remaining there through 1996	Hospitalization records: DOD only, 1991-1996, ICD 799.9 (unexplained illness)	No excess in hospitalizations in this period when effect of CCEP was eliminated	Race, rank, salary, military branch, occupation, prewar hospitalization, sex	Active duty only, no assessment of outpatient treatment, respiratory findings removed after adjustment for VA screening-program attendance
Gray et al. 2000	652,979 veterans vs 652,922 random controls; active duty plus National Guard	Hospitalization records: DOD plus VA plus California hospitals, 1991-1994	Increased PMRs: DOD, injury and poisoning; VA, respiratory, digestive, ill-defined conditions; California, injury and poisonings	Stratified by age, sex, and ethnicity	Could not identify multiple admissions for same subject across databases
Gray et al. 1999b	124,487 Army Gulf War veterans (active duty and National Guard) with possible low exposure to chemical munitions vs 224,804 other Army Gulf War veterans deployed at same time	Postwar hospitalizations, 1991-1995	Group with low modeled exposure to nerve agent had highest rate of hospitalizations for all causes and for neoplasms	Prewar hospitalization, reserve, sex, age group, marital status, race, pay grade, occupation	Modeling of exposures, short time of followup

Study	Population	Outcomes	Results	Adjustment	Comments or Limitations
Smith et al. 2003	431,762 regular and reserve personnel in Army and Air Force in gulf during Khamisiyah demolitions	Postwar morbidity, 1991-2000	Cardiac arrythmias for nerve-agent exposure (RR 1.23, CI 1.04-1.44) No increase for other agents	Sex, age, status, prewar hospitalization, pay grade, race, branch, days deployed, marital status, occupation	

NOTE: CCEP = Comprehensive Clinical Evaluation Program; DOD = Department of Defense; VA = Department of Veterans Affairs.

MULTIPLE CHEMICAL SENSITIVITY

Multiple chemical sensitivity[22] is a controversial condition that can be loosely defined by a person's inability to tolerate multiple chemically unrelated compounds. Although it has been described by physicians since the 1950s, major medical associations have questioned the existence of MCS (American Academy of Allergy 1999; American College of Physicians 1989; American Medical Association 1992). In contrast, a recent evaluation of the biomedical literature commissioned at the request of the UK Health and Safety Executive, found "suggestive" evidence that MCS exists (Graveling et al. 1999). Still, there are no pathologic or laboratory tests and it is often a diagnosis reached by exclusion when no other cause for the symptoms can be identified. There is no validated questionnaire for this symptom complex.

Researchers have developed a set of criteria for the diagnosis of MCS (for example, the Cullen criteria[23]). There are an array of symptoms such as fatigue, cognitive impairment, and headaches) that might be elicited by relatively low concentrations of chemicals with diverse structures and mechanisms of action. For example, symptomatic individuals often report that their symptoms are caused and later triggered by exposure to pesticides, fuels, combustion products, perfumes and other chemical agents (Caress et al. 2002; Kipen and Fiedler 2002). People, including Gulf War veterans, who have MCS symptoms report functional impairment and disability (Black et al. 1999; Fiedler et al. 1996; Jason et al. 2000). About 2-6% of the US population reports having MCS according to various definitions used in population-based studies (Caress and Steinemann 2003; Caress et al. 2002; Kreutzer et al. 1999).

Primary Studies

There is no validated model for MCS and no universally adopted definition of it; therefore, for the purposes of this section, a primary study is one that includes a comparison group, an appropriate questionnaire—that is, a study that uses generally recognized criteria for MCS. Primary studies are summarized in Table 5.22.

Using previously collected symptom reporting from their population-based UK cohort, Reid and colleagues (2001) estimated the prevalence of MCS and CFS and their relationships to Gulf War exposures. (The findings on CFS are presented earlier in this chapter.) One control group was veterans deployed to Bosnia, and the other was a group of Gulf War-era veterans who were deployed elsewhere. In contrast with some studies, the nondeployed control groups were recruited from among the subset of nondeployed service members who were fit for combat duty; this avoided selection bias from the healthy-warrior effect. A case of MCS was defined by using symptom criteria of Simon and colleagues (1993). The prevalence of MCS in deployed veterans was 1.3% (95% CI 1.0-1.7)—a higher figure than that in the two comparison groups, the Bosnia

---

[22] Multiple chemical sensitivity is not listed as a condition in the ICD-10.

[23] Cullen's definition is widely used. It includes four elements: (1) the syndrome is acquired after a documented environmental exposure that might have caused objective evidence of health effects; (2) the symptoms are referable to multiple organ systems and vary predictably in response to environmental stimuli; (3) the symptoms occur in relation to measurable concentrations of chemicals, but the concentrations are below those known to harm health; and (4) no objective evidence of organ damage can be found (Cullen 1987).

and Gulf War-era cohorts. The OR of MCS in Gulf War-deployed vs Bosnia-deployed was 4.5 (95% CI 1.7-11.8), and in Gulf War-deployed vs Gulf War era-deployed was 7.2 (95% CI 2.8-18.2). MCS was associated with the majority of the exposures, but self-reported pesticide exposure was among the strongest. Limitations of the study were self-reported symptoms and exposures. The study on which the data are based (Unwin et al. 1999) estimated the prevalence of MCS at 0.8% in Gulf War veterans, a prevalence similar to that in Bosnia veterans (0.4%) and Gulf War-era veterans (0.3%). The differences between the three cohorts were not significant. The Unwin study did not construct a case of MCS from symptom criteria, like that of Reid et al., but rather asked respondents to self-report their medical disorders, one of which was "multiple chemical sensitivity".

As part of the large population-based Iowa study, Black and colleagues (2000) sought to determine symptom prevalence of and risk factors for MCS. The case criteria were developed by expert consensus. A total of 3,695 veterans were surveyed with structured telephone interviews in 1995-1996. The details of the study are described in Chapter 4. The prevalence of MCS was 5.4% among deployed and 2.6% among nondeployed. Determined through multivariate analysis, the independent risk factors for developing MCS were deployment to the Gulf War, numerous sociodemographic factors (such as, age, male sex, marital status, and education), psychiatric history, and current psychiatric conditions. The sample size was a strength of the study, as was the use of an expert consensus working case definition as described in the study.

As part of the same study, Black and colleagues (1999) surveyed veterans to determine the effect of MCS on disability. Of the total sample of 169 subjects who met their case definition of MCS, the authors found high levels of disability in comparison to those (n = 3,526) who did not meet their case criteria. MCS cases reported more than 12 days in bed due to disability (OR 3.2, 95% CI 1.7-6.3), receipt of VA disability status (OR 3.5, 95% CI 2.1-5.9), receipt of VA disability compensation (OR 3.9, 95% CI 1.9-7.8), receipt of medical disability status (OR 7.3, 95% CI 1.0-50.9), and unemployment (OR 9.8, 95% CI 4.8-20.1). Adjustments were made for age, sex, branch of military, rank, and whether a veteran had regular military or reserve status. A study limitation is that it did not have external validation of disability status.

Canada deployed more than 3,000 sea, land, and air forces to the gulf region. A large proportion of them participated in a naval blockade and they were responsible for one-fourth of enemy interceptions in the gulf. A survey of the entire cohort found that deployed veterans were 4 times as likely as nondeployed veterans (OR 4.01, 95% CI 2.43-6.62) to report symptoms of MCS (Goss Gilroy Inc. 1998).

Secondary Studies

Studying an Army cohort at Fort Devens (n = 180) in 1995, Proctor and colleagues (2001) conducted in-person interviews to determine the prevalence of presumptive MCS, chemical sensitivity, and CFS. The comparison population (n = 46) was an air ambulance company deployed to Germany during the Gulf War. During an environmental interview, if subjects reported health symptoms triggered by chemical odors, they were questioned to determine whether they met MCS case criteria adapted from Cullen (1987). The subjects were not excluded for having self-reported asthma, as well as concurrent psychiatric diagnosis (by psychiatric diagnostic interviews via the SCID). Among deployed veterans, there was a nonsignificant increased prevalence of presumptive MCS of 2.9% vs 0% among nondeployed. An important limitation of the study is the small sample; a strength was that it adopted the Cullen criteria for MCS.

In 1999, Gray and colleagues (2002) surveyed all Seabees (n = 18,945) who had been on active duty during the time of the Gulf War regardless of whether they remained on active duty, were in the reserve, or had separated from the service. There were 11,868 respondents, who were divided into three groups: 3,831 Seabees deployed to the Gulf War, 4,933 Seabees deployed elsewhere, and 3,104 Seabees not deployed. MCS was included in a checklist of about 20 physician-diagnosed conditions about which veterans were asked. In comparison with nondeployed Seabees, the Gulf War-deployed Seabees reported being more than 4 times as likely to have been given an MCS diagnosis (OR 4.47, 95% CI 2.30-8.69). The odds were similarly increased when Gulf War-deployed veterans were compared with those deployed elsewhere (OR 4.08, 95% CI 2.29-7.24). The strength of this study was the large and homogeneous population.

Summary and Conclusion

Overall, the rates of MCS are similar in deployed and civilian populations. MCS or MCS-like symptoms have, as noted earlier, neither a validated questionnaire nor a standard definition. Diagnosis is often by exclusion. Several large or population-based studies of Gulf War veterans found, by questionnaire, that the prevalence of MCS-like symptoms ranged from 2-6%. Most studies found that the prevalence in Gulf War veterans was about 2-4 times higher than that in nondeployed veterans. However, none of the primary studies used the same definition, so it is difficult to compare them. Furthermore, none performed medical evaluations. Although direct physician interaction with patients is always preferable, it should be recognized that there are no physical findings with MCS.

TABLE 5.22 Multiple Chemical Sensitivity (MCS)

Study	Design	Population	Outcomes	Results	Adjustments	Comments
Reid et al. 2001	Population-based, cross-sectional, prevalence	3,531 Gulf war-deployed vs 2,050 Bosnia-deployed vs 2,614 Gulf War-era deployed	Symptoms meeting criteria of Simon et al. 1993 by questionnaire	Gulf vs Bosnia: OR 4.5 (95% CI 1.7-11.8); Gulf vs Gulf War-era: OR 7.2 (95% CI 2.8-18.2)	Sex, age, marital status, education, rank, employment status on followup	Self-reported symptoms and functioning, unvalidated case definition
Unwin et al. 1999	Population-based, cross-sectional, prevalence	4,248 Gulf war-deployed vs 4,250 Bosnia-deployed vs 4,246 Gulf War-era deployed	Self-reported medical condition on questionnaire	Gulf vs Bosnia: OR 1.9 (95% CI 0.8-4.4); Gulf vs Gulf War-era: OR 2.2 (95% CI 1.0-4.9)	Age, smoking, alcohol consumption, marital status, educational attainment, officer or other rank, employment status, civilian or military status	
Black et al. 2000	Population-based, cross-sectional, prevalence	1,896 deployed vs 1,799 nondeployed	Structured telephone survey, case criteria for MCS by expert consensus	Independent risk factors for MCS by multivariate analysis, significant at 5%: deployment to Gulf War (OR 1.94); age, male sex, rank, branch of service, previous psychiatric treatment (OR 2.31); current mental illness	Age, sex, branch of military, rank	Strengths are large sample, use of an expert consensus, working case definition

Study	Design	Population	Outcomes	Results	Adjustments	Comments
Black et al. 1999	Population-based, cross-sectional, prevalence	169 with MCS vs 3,526 without MCS	Structured telephone survey, case criteria for MCS by expert consensus, SF-36	>12 days in bed to disability (OR 3.2 95%, CI 1.7-6.3); VA disability status (OR 3.5, 95% CI 2.1-5.9); VA disability compensation (OR 3.9, 95% CI 1.9-7.8); medical disability (OR 7.3, 95% CI 1.0-50.9); unemployment (OR 9.8, 95% CI 4.8-20.1); greater health services use	Age, sex, branch of military, rank, regular military or reserve status	No external validation of disability status
Goss Gilroy Inc. 1998	Population-based, cross-sectional, prevalence	3,113 deployed vs 3,439 nondeployed	Algorithm requiring physical illness with routine exposure to several substances, at least two substances making person ill, avoidance of at least one activity, positive response to two different sets of eight systemic symptoms, cognitive dysfunction	OR 4.01 (95% CI 2.43-6.62)	Rank, income, age	Self-reported symptoms, unvalidated case definition

NOTE: VA = Department of Veterans Affairs.

REFERENCES

Abul AT, Nair PC, Behbehanei NA, Sharma PN. 2001. Hospital admissions and death rates from asthma in Kuwait during pre- and post-Gulf War periods. *Annals of Allergy, Asthma, and Immunology* 86(4):465-468.

Abushaban L, Al-Hay A, Uthaman B, Salama A, Selvan J. 2004. Impact of the Gulf war on congenital heart diseases in Kuwait. *International Journal of Cardiology* 93(2-3):157-162.

ACS (American Cancer Society). 2006. *Statistics for 2006*. [Online]. Available: http://www.cancer.org/docroot/stt/stt_0.asp [accessed March 21, 2006].

Al-Khalaf B. 1998. Pilot study: The onset of asthma among the Kuwaiti population during the burning of oil wells after the Gulf War. *Environment International* 24(1-2):221-225.

The ALS Association. 2006. *The ALS Association*. [Online]. Available: http://www.alsa.org/; [accessed January 17, 2006].

Amato AA, McVey A, Cha C, Matthews EC, Jackson CE, Kleingunther R, Worley L, Cornman E, Kagan-Hallet K. 1997. Evaluation of neuromuscular symptoms in veterans of the Persian Gulf War. *Neurology* 48(1):4-12.

American Academy of Allergy, Asthma, and Immunology (AAAAI). 1999. Idiopathic environmental intolerances. *Journal of Allergy and Clinical Immunology* 103(1 Pt 1):36-40.

American College of Physicians. 1989. Clinical ecology. *Annals of Internal Medicine* 111(2):168-178.

American Medical Association. 1992. Clinical ecology *Journal of the American Medical Association* 268(24):3465-3467.

Anger WK. 2003. Neurobehavioural tests and systems to assess neurotoxic exposures in the workplace and community. *Occupational and Environmental Medicine* 60(7):531-538, 474.

Anger WK, Storzbach D, Binder LM, et al. 1999. Neurobehavioral deficits in Persian Gulf veterans: Evidence from a population-based study. *Journal of the International Neuropsychological Society.* 5(3):203-212.

Annegers JF, Appel SH, Perkins P, Lee J. 1991. Amyotrophic lateral sclerosis mortality rates in Harris County, Texas. *Advances in Neurology* 56:239-243.

Araneta MR, Moore CA, Olney RS, Edmonds LD, Karcher JA, McDonough C, Hiliopoulos KM, Schlangen KM, Gray GC. 1997. Goldenhar syndrome among infants born in military hospitals to Gulf War veterans. *Teratology* 56(4):244-251.

Araneta MR, Destiche DA, Schlangen KM, Merz RD, Forrester MB, Gray GC. 2000. Birth defects prevalence among infants of Persian Gulf War veterans born in Hawaii, 1989-1993. *Teratology* 62(4):195-204.

Araneta MR, Schlangen KM, Edmonds LD, Destiche DA, Merz RD, Hobbs CA, Flood TJ, Harris JA, Krishnamurti D, Gray GC. 2003. Prevalence of birth defects among infants of Gulf War veterans in Arkansas, Arizona, California, Georgia, Hawaii, and Iowa, 1989-1993. *Birth Defects Research* 67(4):246-260.

Araneta MR, Kamens DR, Zau AC, Gastanaga VM, Schlangen KM, Hiliopoulos KM, Gray GC. 2004. Conception and pregnancy during the Persian Gulf War: The risk to women veterans. *Annals of Epidemiology* 14(2):109-116.

Armon C. 2003. An evidence-based medicine approach to the evaluation of the role of exogenous risk factors in sporadic amyotrophic lateral sclerosis. *Neuroepidemiology* 22(4):217-228.

Armon C. 2004a. Amyotrophic Lateral Sclerosis. In: Nelson LM, Tanner CM, Van Den Eeden SK, McGuire VM, Editors. *Neuroepidemiology: From Principles to Practice*. New York: Oxford University Press. Pp. 162-187.

Armon C. 2004b. Occurrence of amyotrophic lateral sclerosis among Gulf War veterans. *Neurology* 62(6):1027.

Axelrod BN, Milner IB. 1997. Neuropsychological findings in a sample of Operation Desert Storm veterans. *Journal of Neuropsychiatry and Clinical Neurosciences* 9(1):23-28.

Barrett DH, Doebbeling CC, Schwartz DA, Voelker MD, Falter KH, Woolson RF, Doebbeling BN. 2002. Posttraumatic stress disorder and self-reported physical health status among US Military personnel serving during the Gulf War period: A population-based study. *Psychosomatics* 43(3):195-205.

Binder LM, Storzbach D, Anger WK, Campbell KA, Rohlman DS. 1999. Subjective cognitive complaints, affective distress, and objective cognitive performance in Persian Gulf war veterans. *Archives of Clinical Neuropsychology* 14(6):531-536.

Black DW, Carney CP, Forman-Hoffman VL, Letuchy E, Peloso P, Woolson RF, Doebbeling BN. 2004a. Depression in veterans of the first gulf war and comparable military controls. *Annals of Clinical Psychiatry* 16(2):53-61.

Black DW, Carney CP, Peloso PM, Woolson RF, Schwartz DA, Voelker MD, Barrett DH, Doebbeling BN. 2004b. Gulf War veterans with anxiety: Prevalence, comorbidity, and risk factors. *Epidemiology* 15(2):135-142.

Black DW, Doebbeling BN, Voelker MD, Clarke WR, Woolson RF, Barrett DH, Schwartz DA. 1999. Quality of life and health-services utilization in a population-based sample of military personnel reporting multiple chemical sensitivities. *Journal of Occupational and Environmental Medicine* 41(10):928-933.

Black DW, Doebbeling BN, Voelker MD, Clarke WR, Woolson RF, Barrett DH, Schwartz DA. 2000. Multiple chemical sensitivity syndrome: Symptom prevalence and risk factors in a military population. *Archives of Internal Medicine* 160(8):1169-1176.

Blanchard EB, Jones-Alexander J, Buckley TC, Forneris CA. 1996. Psychometric properties of the PTSD Checklist (PCL). *Behaviour Research and Therapy* 34(8):669-673.

Blanchard MS, Eisen SA, Alpern R, Karlinsky J, Toomey R, Reda DJ, Murphy FM, Jackson LW, Kang HK. 2006. Chronic Multisymptom Illness Complex in Gulf War I Veterans 10 Years Later. *American Journal of Epidemiology* 163(1):66-75.

Bombardier CH, Buchwald D. 1996. Chronic fatigue, chronic fatigue syndrome, and fibromyalgia. Disability and health-care use. *Medical Care* 34(9):924-930.

Bourdette DN, McCauley LA, Barkhuizen A, Johnston W, Wynn M, Joos SK, Storzbach D, Shuell T, Sticker D. 2001. Symptom factor analysis, clinical findings, and functional status in a population-based case control study of Gulf War unexplained illness. *Journal of Occupational and Environmental Medicine* 43(12):1026-1040.

Brailey K, Vasterling JJ, Sutker PB. 1998. Psychological Aftermath of Participation in the Persian Gulf War. Lundberg A, Editor. *The Environment and Mental Health: A Guide for Clinicians*. London: Lawrence Erlbaum Associates. Pp. 83-101.

Brown RG, Scott LC, Bench CJ, Dolan RJ. 1994. Cognitive function in depression: Its relationship to the presence and severity of intellectual decline. *Psychological Medicine* 24(4):829-847.

Buchwald D, Garrity D. 1994. Comparison of patients with chronic fatigue syndrome, fibromyalgia, and multiple chemical sensitivities. *Archives of Internal Medicine* 154(18):2049-2053.

Buchwald D, Pearlman T, Umali J, Schmaling K, Katon W. 1996. Functional status in patients with chronic fatigue syndrome, other fatiguing illnesses, and healthy individuals. *American Journal of Medicine* 101(4):364-370.

Bullman TA, Mahan CM, Kang HK, Page WF. 2005. Mortality in US Army Gulf War Veterans Exposed to 1991 Khamisiyah Chemical Munitions Destruction. *American Journal of Public Health* 95(8):1382-1388.

Bunegin L, Mitzel HC, Miller CS, Gelineau JF, Tolstykh GP. 2001. Cognitive performance and cerebrohemodynamics associated with the Persian Gulf Syndrome. *Toxicology and Industrial Health* 17(4):128-137.

California Birth Defects Monitoring Program. 2006. *Discoveries and Data Center*. [Online]. Available: http://www.cbdmp.org/bd_intro.htm [accessed April 10, 2006].

Caress SM, Steinemann AC. 2003. A review of a two-phase population study of multiple chemical sensitivities. *Environmental Health Perspectives* 111(12):1490-1497.

Caress SM, Steinemann AC, Waddick C. 2002. Symptomatology and etiology of multiple chemical sensitivities in the southeastern United States. *Archives of Environmental Health* 57(5):429-436.

CDC (Centers for Disease Control). 1987. Postservice mortality among Vietnam veterans. *Morbidity and Mortality Weekly Report* 36(5):61-64.

CDC. 1995. Unexplained illness among Persian Gulf War veterans in an Air National Guard Unit: Preliminary report--August 1990-March 1995. *Journal of the American Medical Association* 274(1):16-17.

Chalder T, Berelowitz G, Pawlikowska T, Watts L, Wessely S, Wright D, Wallace EP. 1993. Development of a fatigue scale. *Journal of Psychosomatic Research* 37(2):147-153.

Cherry N, Creed F, Silman A, Dunn G, Baxter D, Smedley J, Taylor S, Macfarlane GJ. 2001a. Health and exposures of United Kingdom Gulf war veterans. Part II: The relation of health to exposure. *Occupational and Environmental Medicine* 58(5):299-306.

Cherry N, Creed F, Silman A, Dunn G, Baxter D, Smedley J, Taylor S, Macfarlane GJ. 2001b. Health and exposures of United Kingdom Gulf war veterans. Part I: The pattern and extent of ill health. *Occupational and Environmental Medicine* 58(5):291-298.

Chio A, Benzi G, Dossena M, Mutani R, Mora G. 2005. Severely increased risk of amyotrophic lateral sclerosis among Italian professional football players. *Brain* 128(Pt 3):472-476.

Coffman CJ, Horner RD, Grambow SC, Lindquist J. 2005. Estimating the occurrence of amyotrophic lateral sclerosis among Gulf War (1990-1991) veterans using capture-recapture methods. *Neuroepidemiology* 24(3):141-150.

Cohen J. 1992. A power primer. *Psychological Bulletin* 112:155-159.

Coombe MD, Drysdale SF. 1993. Assessment of the effects of atmospheric oil pollution in post war Kuwait. *Journal of the Royal Army Medical Corps* 139(3):95-97.

Cowan DN, DeFraites RF, Gray GC, Goldenbaum MB, Wishik SM. 1997. The risk of birth defects among children of Persian Gulf War veterans. *New England Journal of Medicine* 336(23):1650-1656.

Cowan DN, Lange JL, Heller J, Kirkpatrick J, DeBakey S. 2002. A case-control study of asthma among U.S. Army Gulf War veterans and modeled exposure to oil well fire smoke. *Military Medicine* 167(9):777-782.

Cullen MR. 1987. The worker with multiple chemical sensitivities: An overview. *Occupational Medicine* 2(4):655-661.

David AS, Farrin L, Hull L, Unwin C, Wessely S, Wykes T. 2002. Cognitive functioning and disturbances of mood in UK veterans of the Persian Gulf War: A comparative study. *Psychological Medicine* 32(8):1357-1370.

Davis LE, Eisen SA, Murphy FM, Alpern R, Parks BJ, Blanchard M, Reda DJ, King MK, Mithen FA, Kang HK. 2004. Clinical and laboratory assessment of distal peripheral nerves in Gulf War veterans and spouses. *Neurology* 63(6):1070-1077.

DeBakey S, Paxton M, Weaver R, Lange J, Cowan D, Kang H, Hooper T, Gackstetter G. 2002. #52 Risk of motor vehicle fatality associated with prior morbidity among gulf war era veterans. *Annals of Epidemiology* 12(7):509.

Defence Analytical Services Agency. 2005. 1990/1991 Gulf Conflict—UK Gulf Veterans Mortality Data: Causes of Death. *National Statistics*.

Department of Health and Human Services. 1999. *Mental Health: A Report of the Surgeon General*. [Online]. Available: http://www.surgeongeneral.gov/library/mentalhealth/home.html [accessed April 10, 2006].

Dlugosz LJ, Hocter WJ, Kaiser KS, Knoke JD, Heller JM, Hamid NA, Reed RJ, Kendler KS, Gray GC. 1999. Risk factors for mental disorder hospitalization after the Persian Gulf War: U.S. Armed Forces, June 1, 1991-September 30, 1993. *Journal of Clinical Epidemiology* 52(12):1267-1278.

Dobie DJ, Kivlahan DR, Maynard C, Bush KR, McFall M, Epler AJ, Bradley KA. 2002. Screening for post-traumatic stress disorder in female Veteran's Affairs patients: Validation of the PTSD checklist. *General Hospital Psychiatry* 24(6):367-374.

Doebbeling BN, Clarke WR, Watson D, Torner JC, Woolson RF, Voelker MD, Barrett DH, Schwartz DA. 2000. Is there a Persian Gulf War syndrome? Evidence from a large population-based survey of veterans and nondeployed controls. *American Journal of Medicine* 108(9):695-704.

Doyle P, Maconochie N, Davies G, Maconochie I, Pelerin M, Prior S, Lewis S. 2004. Miscarriage, stillbirth and congenital malformation in the offspring of UK veterans of the first Gulf war. *International Journal of Epidemiology* 33(1):74-86.

Draxler RR, McQueen JT, Stunder BJB. 1994. An evaluation of air pollutant exposures due to the 1991 Kuwait oil fires using a Lagrangian model. *Atmospheric Environment* 28(13):2197-2210.

Eisen SA, Kang HK, Murphy FM, Blanchard MS, Reda DJ, Henderson WG, Toomey R, Jackson LW, Alpern R, Parks BJ, Klimas N, Hall C, Pak HS, Hunter J, Karlinsky J, Battistone MJ, Lyons MJ. 2005. Gulf War veterans' health: Medical evaluation of a US cohort. *Annals of Internal Medicine* 142(11):881-890.

Epstein KR. 1995. The chronically fatigued patient. *Medical Clinics of North America* 79(2):315-327.

Everitt B, Ismail K, David AS, Wessely S. 2002. Searching for a Gulf War syndrome using cluster analysis. *Psychological Medicine* 32(8):1371-1378.

Fiedler N, Kipen H, Natelson B, Ottenweller J. 1996. Chemical sensitivities and the Gulf War: Department of Veterans Affairs Research Center in basic and clinical science studies of environmental hazards. *Regulatory Toxicology and Pharmacology* 24(1 Pt 2):S129-S138.

Forbes AB, McKenzie DP, Mackinnon AJ, Kelsall HL, McFarlane AC, Ikin JF, Glass DC, Sim MR. 2004. The health of Australian veterans of the 1991 Gulf War: Factor analysis of self-reported symptoms. *Occupational and Environmental Medicine* 61(12):1014-1020.

Forbes D, Creamer M, Biddle D. 2001. The validity of the PTSD checklist as a measure of symptomatic change in combat-related PTSD. *Behaviour Research and Therapy* 39(8):977-986.

Franse LV, Valk GD, Dekker JH, Heine RJ, van Eijk JT. 2000. 'Numbness of the feet' is a poor indicator for polyneuropathy in Type 2 diabetic patients. *Diabetic Medicine* 17(2):105-110.

Fukuda K, Straus SE, Hickie I, Sharpe MC, Dobbins JG, Komaroff A. 1994. The chronic fatigue syndrome: A comprehensive approach to its definition and study. International Chronic Fatigue Syndrome Study Group. *Annals of Internal Medicine* 121(12):953-959.

Fukuda K, Nisenbaum R, Stewart G, Thompson WW, Robin L, Washko RM, Noah DL, Barrett DH, Randall B, Herwaldt BL, Mawle AC, Reeves WC. 1998. Chronic multisymptom illness affecting Air Force veterans of the Gulf War. *Journal of the American Medical Association* 280(11):981-988.

Gackstetter G, DeBakey S, Cowan D, Paxton M, Weaver R, Lange J, Kang H, Bullman T, Lincoln A, Hooper T. 2002. Fatal motor vehicle crashes among veterans of the gulf war era: A nested case-control study. *Annals of Epidemiology* 12(7):509.

Gillespie N, Kirk KM, Heath AC, Martin NG, Hickie I. 1999. Somatic distress as a distinct psychological dimension. *Social Psychiatry and Psychiatric Epidemiology* 34(9):451-458.

Goldenberg DL. 1999. Fibromyalgia syndrome a decade later: What have we learned? *Archives of Internal Medicine* 159(8):777-785.

Goldstein G, Beers SR, Morrow LA, Shemansky WJ, Steinhauer SR. 1996. A preliminary neuropsychological study of Persian Gulf veterans. *Journal of the International Neuropsychological Society* 2(4):368-371.

Goshorn RK. 1998. Chronic fatigue syndrome: A review for clinicians. *Seminars in Neurology* 18(2):237-242.

Goss Gilroy Inc. 1998. *Health Study of Canadian Forces Personnel Involved in the 1991 Conflict in the Persian Gulf.* Ottawa, Canada: Goss Gilroy Inc. Department of National Defence.

Graveling RA, Pilkington A, George JP, Butler MP, Tannahill SN. 1999. A review of multiple chemical sensitivity. *Occupational and Environmental Medicine* 56(2):73-85.

Gray GC, Coate BD, Anderson CM, Kang HK, Berg SW, Wignall FS, Knoke JD, Barrett-Connor E. 1996. The postwar hospitalization experience of U.S. veterans of the Persian Gulf War. *New England Journal of Medicine* 335(20):1505-1513.

Gray GC, Kaiser KS, Hawksworth AW, Hall FW, Barrett-Connor E. 1999a. Increased postwar symptoms and psychological morbidity among U.S. Navy Gulf War veterans. *American Journal of Tropical Medicine and Hygiene* 60(5):758-766.

Gray GC, Smith TC, Knoke JD, Heller JM. 1999b. The postwar hospitalization experience of Gulf War Veterans possibly exposed to chemical munitions destruction at Khamisiyah, Iraq. *American Journal of Epidemiology* 150(5):532-540.

Gray GC, Smith TC, Kang HK, Knoke JD. 2000. Are Gulf War veterans suffering war-related illnesses? Federal and civilian hospitalizations examined, June 1991 to December 1994. *American Journal of Epidemiology* 151(1):63-71.

Gray GC, Reed RJ, Kaiser KS, Smith TC, Gastanaga VM. 2002. Self-reported symptoms and medical conditions among 11,868 Gulf War-era veterans: The Seabee Health Study. *American Journal of Epidemiology* 155(11):1033-1044.

Haley RW. 2003. Excess incidence of ALS in young Gulf War veterans. *Neurology* 61(6):750-756.

Haley RW, Hom J, Roland PS, Bryan WW, Van Ness PC, Bonte FJ, Devous MD Sr, Mathews D, Fleckenstein JL, Wians FH Jr, Wolfe GI, Kurt TL. 1997a. Evaluation of neurologic function in Gulf War veterans. A blinded case-control study. *Journal of the American Medical Association* 277(3):223-230.

Haley RW, Kurt TL, Hom J. 1997b. Is there a Gulf War Syndrome? Searching for syndromes by factor analysis of symptoms. *Journal of the American Medical Association* 277(3):215-222.

Haley RW, Billecke S, La Du BN. 1999. Association of low PON1 type Q (type A) arylesterase activity with neurologic symptom complexes in Gulf War veterans. *Toxicology and Applied Pharmacology* 157(3):227-233.

Haley RW, Luk GD, Petty F. 2001. Use of structural equation modeling to test the construct validity of a case definition of Gulf War syndrome: Invariance over developmental and validation samples, service branches and publicity. *Psychiatry Research* 102(2):175-200.

Haley RW, Fleckenstein JL, Marshall WW, McDonald GG, Kramer GL, Petty F. 2000a. Effect of basal ganglia injury on central dopamine activity in Gulf War syndrome: Correlation of proton magnetic resonance spectroscopy and plasma homovanillic acid levels. *Archives of Neurology* 57(9):1280-1285.

Haley RW, Marshall WW, McDonald GG, Daugherty MA, Petty F, Fleckenstein JL. 2000b. Brain abnormalities in Gulf War syndrome: Evaluation with 1H MR spectroscopy. *Radiology* 215(3):807-817.

Haley RW, Vongpatanasin W, Wolfe GI, Bryan WW, Armitage R, Hoffmann RF, Petty F, Callahan TS, Charuvastra E, Shell WE, Marshall WW, Victor RG. 2004. Blunted circadian variation in autonomic regulation of sinus node function in veterans with Gulf War syndrome. *American Journal of Medicine* 117(7):469-478.

Hallman WK, Kipen HM, Diefenbach M, Boyd K, Kang H, Leventhal H, Wartenberg D. 2003. Symptom patterns among Gulf War registry veterans. *American Journal of Public Health* 93(4):624-630.

Hardt J, Buchwald D, Wilks D, Sharpe M, Nix WA, Egle UT. 2001. Health-related quality of life in patients with chronic fatigue syndrome: An international study. *Journal of Psychosomatic Research* 51(2):431-434.

Higgins EM, Ismail K, Kant K, Harman K, Mellerio J, Du Vivier AW, Wessely S. 2002. Skin disease in Gulf war veterans. *QJM* 95(10):671-676.

Holmes DT, Tariot PN, Cox C. 1998. Preliminary evidence of psychological distress among reservists in the Persian Gulf War. *Journal of Nervous and Mental Disease* 186(3):166-173.

Holmes GP, Kaplan JE, Gantz NM, Komaroff AL, Schonberger LB, Straus SE, Jones JF, Dubois RE, Cunningham-Rundles C, Pahwa S, et al. 1988. Chronic fatigue syndrome: A working case definition. *Annals of Internal Medicine* 108(3):387-389.

Hom J, Haley RW, Kurt TL. 1997. Neuropsychological correlates of Gulf War syndrome. *Archives of Clinical Neuropsychology* 12(6):531-544.

Horner RD, Kamins KG, Feussner JR, Grambow SC, Hoff-Lindquist J, Harati Y, Mitsumoto H, Pascuzzi R, Spencer PS, Tim R, Howard D, Smith TC, Ryan MA, Coffman CJ, Kasarskis EJ. 2003. Occurrence of amyotrophic lateral sclerosis among Gulf War veterans. *Neurology* 61(6):742-749.

Hotopf M, Mackness MI, Nikolaou V, Collier DA, Curtis C, David A, Durrington P, Hull L, Ismail K, Peakman M, Unwin C, Wessely S, Mackness B. 2003. Paraoxonase in Persian Gulf War veterans. *Journal of Occupational and Environmental Medicine* 45(7):668-675.

Hotopf M, David A, Hull L, Nikalaou V, Unwin C, Wessely S. 2004. Risk factors for continued illness among Gulf War veterans: A cohort study. *Psychological Medicine* 34(4):747-754.

Hull L, Farrin L, Unwin C, Everitt B, Wykes T, David AS. 2003. Anger, psychopathology and cognitive inhibition: A study of UK servicemen. *Personality and Individual Differences* 35(5):1211.

Ikin JF, Sim MR, Creamer MC, Forbes AB, McKenzie DP, Kelsall HL, Glass DC, McFarlane AC, Abramson MJ, Ittak P, Dwyer T, Blizzard L, Delaney KR, Horsley KWA, Harrex WK, Schwarz H. 2004. War-related psychological stressors and risk of psychological disorders in Australian veterans of the 1991 Gulf War. *British Journal of Psychiatry* 185:116-126.

Iowa Persian Gulf Study Group. 1997. Self-reported illness and health status among Gulf War veterans: A population-based study. *Journal of the American Medical Association* 277(3):238-245.

Ishoy T, Suadicani P, Guldager B, Appleyard M, Gyntelberg F. 1999a. Risk factors for gastrointestinal symptoms. The Danish Gulf War Study. *Danish Medical Bulletin* 46(5):420-423.

Ishoy T, Suadicani P, Guldager B, Appleyard M, Hein HO, Gyntelberg F. 1999b. State of health after deployment in the Persian Gulf. The Danish Gulf War Study. *Danish Medical Bulletin* 46(5):416-419.

Ishoy T, Andersson AM, Suadicani P, Guldager B, Appleyard M, Gyntelberg F, Skakkebaek NE, Danish Gulf War Study. 2001a. Major reproductive health characteristics in male Gulf War Veterans. The Danish Gulf War Study. *Danish Medical Bulletin* 48(1):29-32.

Ishoy T, Suadicani P, Andersson A-M, Guldager B, Appleyard M, Skakkebaek N, Gyntelberg F. 2001b. Prevalence of male sexual problems in the Danish Gulf War Study. *Scandinavian Journal of Sexology* 4(1):43-55.

Ismail K, Blatchley N, Hotopf M, Hull L, Palmer I, Unwin C, David A, Wessely S. 2000. Occupational risk factors for ill health in Gulf veterans of the United Kingdom. *Journal of Epidemiology and Community Health* 54(11):834-838.

Ismail K, Everitt B, Blatchley N, Hull L, Unwin C, David A, Wessely S. 1999. Is there a Gulf War syndrome? *Lancet* 353(9148):179-182.

Jason LA, Taylor RR, Kennedy CL. 2000. Chronic fatigue syndrome, fibromyalgia, and multiple chemical sensitivities in a community-based sample of persons with chronic fatigue syndrome-like symptoms. *Psychosomatic Medicine* 62(5):655-663.

Joseph TK, Foster L, Pasquina PF. 2004. Decreased prevalence of peripheral nerve pathology by electrodiagnostic testing in Gulf War veterans. *Military Medicine* 169(11):868-871.

Kang HK, Bullman TA. 1996. Mortality among US veterans of the Persian Gulf War. *New England Journal of Medicine* 335(20):1498-1504.

Kang HK, Bullman TA. 2001. Mortality among US veterans of the Persian Gulf War: 7-year follow-up. *American Journal of Epidemiology* 154(5):399-405.

Kang HK, Mahan CM, Lee KY, Magee CA, Murphy FM. 2000. Illnesses among United States veterans of the Gulf War: A population-based survey of 30,000 veterans. *Journal of Occupational and Environmental Medicine* 42(5):491-501.

Kang H, Magee C, Mahan C, Lee K, Murphy F, Jackson L, Matanoski G. 2001. Pregnancy outcomes among US Gulf War veterans: A population-based survey of 30,000 veterans. *Annals of Epidemiology* 11(7):504-511.

Kang HK, Mahan CM, Lee KY, Murphy FM, Simmens SJ, Young HA, Levine PH. 2002. Evidence for a deployment-related Gulf War syndrome by factor analysis. *Archives of Environmental Health* 57(1):61-68.

Kang HK, Natelson BH, Mahan CM, Lee KY, Murphy FM. 2003. Post-traumatic stress disorder and chronic fatigue syndrome-like illness among Gulf War veterans: A population-based survey of 30,000 veterans. *American Journal of Epidemiology* 157(2):141-148.

Karlinsky JB, Blanchard M, Alpern R, Eisen SA, Kang H, Murphy FM, Reda DJ. 2004. Late prevalence of respiratory symptoms and pulmonary function abnormalities in Gulf War I Veterans. *Archives of Internal Medicine* 164(22):2488-2491.

Kelsall HL, Sim MR, Forbes AB, Glass DC, McKenzie DP, Ikin JF, Abramson MJ, Blizzard L, Ittak P. 2004a. Symptoms and medical conditions in Australian veterans of the 1991 Gulf War: Relation to immunisations and other Gulf War exposures. *Occupational and Environmental Medicine* 61(12):1006-1013.

Kelsall HL, Sim MR, Forbes AB, McKenzie DP, Glass DC, Ikin JF, Ittak P, Abramson MJ. 2004b. Respiratory health status of Australian veterans of the 1991 Gulf War and the effects of exposure to oil fire smoke and dust storms. *Thorax* 59(10):897-903.

Kessler RC, Sonnega A, Bromet E, Hughes M, Nelson CB. 1995. Posttraumatic stress disorder in the National Comorbidity Survey. *Archives of General Psychiatry* 52(12):1048-1060.

Kessler RC, Chiu WT, Demler O, Merikangas KR, Walters EE. 2005. Prevalence, severity, and comorbidity of 12-month DSM-IV disorders in the National Comorbidity Survey Replication. *Archives of General Psychiatry* 62(6):617-627.

Kipen HM, Fiedler N. 2002. Environmental factors in medically unexplained symptoms and related syndromes: The evidence and the challenge. *Environmental Health Perspectives* 110 (S4):597-599.

Knoke JD, Gray GC. 1998. Hospitalizations for unexplained illnesses among US veterans of the Persian Gulf War. *Emerging Infectious Diseases* 4(2):211-219.

Knoke JD, Gray GC, Garland FC. 1998. Testicular cancer and Persian Gulf War service. *Epidemiology* 9(6):648-653.

Knoke JD, Smith TC, Gray GC, Kaiser KS, Hawksworth AW. 2000. Factor analysis of self-reported symptoms: Does it identify a Gulf War syndrome? *American Journal of Epidemiology* 152(4):379-388.

Komaroff AL, Fagioli LR, Doolittle TH, Gandek B, Gleit MA, Guerriero RT, Kornish RJ 2nd, Ware NC, Ware JE Jr, Bates DW. 1996. Health status in patients with chronic fatigue syndrome and in general population and disease comparison groups. *American Journal of Medicine* 101(3):281-90.

Kreutzer R, Neutra RR, Lashuay N. 1999. Prevalence of people reporting sensitivities to chemicals in a population-based survey. *American Journal of Epidemiology* 150(1):1-12.

Kroenke K, Koslowe P, Roy M. 1998. Symptoms in 18,495 Persian Gulf War veterans. Latency of onset and lack of association with self-reported exposures. *Journal of Occupational and Environmental Medicine* 40(6):520-528.

Lange G, Tiersky LA, Scharer JB, Policastro T, Fiedler N, Morgan TE, Natelson BH. 2001. Cognitive functioning in Gulf War Illness. *Journal of Clinical and Experimental Neuropsychology* 23(2):240-249.

Lange JL, Schwartz DA, Doebbeling BN, Heller JM, Thorne PS. 2002. Exposures to the Kuwait oil fires and their association with asthma and bronchitis among gulf war veterans. *Environmental Health Perspectives* 110(11):1141-1146.

Levine PH, Young HA, Simmens SJ, Rentz D, Kofie VE, Mahan CM, Kang HK. 2005. Is testicular cancer related to Gulf War deployment? Evidence from a pilot population-based study of Gulf War era veterans and cancer registries. *Military Medicine* 170(2):149-153.

Lezak M, Loring D, Howieson D. 2004. *Neuropsychological Assessment*. New York: Oxford University Press.

Lindem K, Heeren T, White RF, Proctor SP, Krengel M, Vasterling J, Sutker PB, Wolfe J, Keane TM. 2003a. Neuropsychological performance in Gulf War era veterans: Traumatic stress symptomatology and exposure to chemical-biological warfare agents. *Journal of Psychopathology and Behavioral Assessment* 25(2):105-119.

Lindem K, Proctor SP, Heeren T, Krengel M, Vasterling J, Sutker PB, Wolfe J, Keane TM, White RF. 2003b. Neuropsychological performance in Gulf War era veterans: Neuropsychological symptom reporting. *Journal of Psychopathology and Behavioral Assessment* 25(2):121-127.

Lindem K, White RF, Heeren T, Proctor SP, Krengel M, Vasterling J, Wolfe J, Sutker PB, Kirkley S, Keane TM. 2003c. Neuropsychological performance in Gulf War era veterans: Motivational factors and effort. *Journal of Psychopathology and Behavioral Assessment* 25(2):129-138.

Lucchini R, Albini E, Benedetti L, Alessio L. 2005. Neurobehavioral science in hazard identification and risk assessment of neurotoxic agents—what are the requirements for further development? *International Archives of Occupational and Environmental Health* 78(6):427-437.

Macfarlane GJ, Thomas E, Cherry N. 2000. Mortality among UK Gulf War veterans. *Lancet* 356(9223):17-21.

Macfarlane GJ, Biggs AM, Maconochie N, Hotopf M, Doyle P, Lunt M. 2003. Incidence of cancer among UK Gulf war veterans: Cohort study. *British Medical Journal* 327(7428):1373-1375.

Mackness B, Mackness MI, Arrol S, Turkie W, Durrington PN. 1997. Effect of the molecular polymorphisms of human paraoxonase (PON1) on the rate of hydrolysis of paraoxon. *British Journal of Pharmacology* 122(2):265-268.

Maconochie N, Doyle P, Carson C. 2004. Infertility among male UK veterans of the 1990-1 Gulf war: Reproductive cohort study. *British Medical Journal* 329(7459):196-201.

Magruder KM, Frueh BC, Knapp RG, Davis L, Hamner MB, Martin RH, Gold PB, Arana GW. 2005. Prevalence of posttraumatic stress disorder in Veterans Affairs primary care clinics. *General Hospital Psychiatry* 27(3):169-179.

Matarazzo JD. 1972. *Wechsler's Measurement and Appraisal of Adult Intelligence.* 5th ed. Baltimore, Md: Williams and Wilkins

McCauley LA, Joos SK, Lasarev MR, Storzbach D, Bourdette DN. 1999. Gulf War unexplained illnesses: Persistence and unexplained nature of self-reported symptoms. *Environmental Research* 81(3):215-223.

McCauley LA, Lasarev M, Sticker D, Rischitelli DG, Spencer PS. 2002. Illness experience of Gulf War veterans possibly exposed to chemical warfare agents. *American Journal of Preventive Medicine* 23(3):200-206.

McGuire V, Longstreth WT Jr, Koepsell TD, van Belle G. 1996a. Incidence of amyotrophic lateral sclerosis in three counties in western Washington state. *Neurology* 47(2):571-573.

McKenzie DP, Ikin JF, McFarlane AC, Creamer M, Forbes AB, Kelsall HL, Glass DC, Ittak P, Sim MR. 2004. Psychological health of Australian veterans of the 1991 Gulf War: An assessment using the SF-12, GHQ-12 and PCL-S. *Psychological Medicine* 34(8):1419-1430.

McQueen JT, Draxler RR. 1994. Evaluation of model back trajectories of the Kuwait oil fires smoke plume using digital satellite data. *Atmospheric Environment* 28(13):2159-2174.

Merck Manuals Online Medical Library. *Polyneuropathy: Peripheral Nerve Disorders.* [Online]. Available: http://www.merck.com/mmhe/sec06/ch095/ch095h.html [accessed July 11, 2006].

Nelson LM, McGuire V, Longstreth WT Jr, Matkin C. 2000. Population-based case-control study of amyotrophic lateral sclerosis in western Washington State. I. Cigarette smoking and alcohol consumption. *American Journal of Epidemiology* 151(2):156-163.

Nicolson GL, Nasralla MY, Haier J, Pomfret J. 2002. High frequency of systemic mycoplasmal infections in Gulf War veterans and civilians with Amyotrophic Lateral Sclerosis (ALS). *Journal of Clinical Neuroscience* 9(5):525-529.

Nimnuan C, Rabe-Hesketh S, Wessely S, Hotopf M. 2001. How many functional somatic syndromes? *Journal of Psychosomatic Research* 51(4):549-557.

NINDS (National Institute of Neurological Disorders and Stroke). 2006. *National Institute of Neurological Disorders and Stroke*. [Online]. Available: http://www.ninds.nih.gov/ [accessed January 17, 2006].

Nisenbaum R, Reyes M, Mawle AC, Reeves WC. 1998. Factor analysis of unexplained severe fatigue and interrelated symptoms: Overlap with criteria for chronic fatigue syndrome. *American Journal of Epidemiology* 148(1):72-77.

Nisenbaum R, Ismail K, Wessely S, Unwin C, Hull L, Reeves WC. 2004. Dichotomous factor analysis of symptoms reported by UK and US veterans of the 1991 Gulf War. *Population Health Metrics* 2(1):8.

O'Toole BI, Marshall RP, Grayson DA, Schureck RJ, Dobson M, Ffrench M, Pulvertaft B, Meldrum L, Bolton J, Vennard J. 1996. The Australian Vietnam Veterans Health Study: III. psychological health of Australian Vietnam veterans and its relationship to combat. *International Journal of Epidemiology* 25(2):331-340.

Pearce JM. 2004. Myofascial pain, fibromyalgia or fibrositis? *European Neurology* 52(2):67-72.

Penman AD, Tarver RS, Currier MM. 1996. No evidence of increase in birth defects and health problems among children born to Persian Gulf War Veterans in Mississippi. *Military Medicine* 161(1):1-6.

Petruccelli BP, Goldenbaum M, Scott B, Lachiver R, Kanjarpane D, Elliott E, Francis M, McDiarmid MA, Deeter D. 1999. Health effects of the 1991 Kuwait oil fires: A survey of US army troops. *Journal of Occupational and Environmental Medicine* 41(6):433-439.

Pizarro J, Silver RC, Prause J. 2006. Physical and mental health costs of traumatic war experiences among Civil War veterans. *Archives of General Psychiatry* 63(2):193-200.

Proctor SP, Heeren T, White RF, Wolfe J, Borgos MS, Davis JD, Pepper L, Clapp R, Sutker PB, Vasterling JJ, Ozonoff D. 1998. Health status of Persian Gulf War veterans: Self-reported symptoms, environmental exposures and the effect of stress. *International Journal of Epidemiology* 27(6):1000-1010.

Proctor SP, Heaton KJ, White RF, Wolfe J. 2001. Chemical sensitivity and chronic fatigue in Gulf War veterans: A brief report. *Journal of Occupational and Environmental Medicine* 43(3):259-264.

Proctor SP, White RF, Heeren T, et al. 2003. Neuropsychological Functioning in Danish Gulf War Veterans. *Journal of Psychopathology and Behavioral Assessment* 25(2):85-93.

Reid S, Hotopf M, Hull L, Ismail K, Unwin C, Wessely S. 2001. Multiple chemical sensitivity and chronic fatigue syndrome in British Gulf War veterans. *American Journal of Epidemiology* 153(6):604-609.

Ries, LAG, Eisner, MP, Kosary, CL, Hankey, BF, Miller, BA, Clegg, L, Mariotto, A, Feuer, EJ, and Edwards, BK, eds. 2005. *SEER Cancer Statistics Review, 1975-2002*. [Online]. Available: http://seer.cancer.gov/csr/1975_2002/ [accessed 2006].

Rivera-Zayas J, Arroyo M, Mejias E. 2001. Evaluation of Persian Gulf veterans with symptoms of peripheral neuropathy. *Military Medicine* 166(5):449-451.

Roland PS, Haley RW, Yellin W, Owens K, Shoup AG. 2000. Vestibular dysfunction in Gulf War syndrome. *Otolaryngology—Head and Neck Surgery* 122(3):319-329.

Rose MR, Sharief MK, Priddin J, Nikolaou V, Hull L, Unwin C, Ajmal-Ali R, Sherwood RA, Spellman A, David A, Wessely S. 2004. Evaluation of neuromuscular symptoms in UK Gulf War veterans: A controlled study. *Neurology* 63(9):1681-1687.

Rostker, B. 2000. *US Demolition Operations at Khamisiyah*. [Online]. Available: http://www.gulflink.osd.mil/khamisiyah_ii/ [accessed August 6, 2004].

Rowland LP. 2000. Hereditary and Acquired Motor Neuron Diseases. In: Rowland LP, Editor. *Merritt's Neurology*. 10th ed. Philadelphia, PA: Lippincott Williams and Wilkins. Pp. 708-714.

Roy-Byrne P, Arguelles L, Vitek ME, Goldberg J, Keane TM, True WR, Pitman RK. 2004. Persistence and change of PTSD symptomatology—a longitudinal co-twin control analysis of the Vietnam Era Twin Registry. *Social Psychiatry and Psychiatric Epidemiology* 39(9):681-685.

Shapiro SE, Lasarev MR, McCauley L. 2002. Factor analysis of Gulf War illness: What does it add to our understanding of possible health effects of deployment? *American Journal of Epidemiology* 156(6):578-585.

Sharief MK, Priddin J, Delamont RS, Unwin C, Rose MR, David A, Wessely S. 2002. Neurophysiologic analysis of neuromuscular symptoms in UK Gulf War veterans: A controlled study. *Neurology* 59(10):1518-1525.

Shaw E, Hermansen L, Pugh W, White M. 1991. *Disease and Non-Battle Injuries Among Navy and Marine Corps Personnel During Operation Desert Shield/Desert Storm*. US: US Naval Health Research Center.

Siddique N, Sufit R, Siddique T. 1999. Degenerative Motor, Sensory, and Autonomic Disorders. In: Goetz CG, Pappert EJ, Editors. *Textbook of Clinical Neurology*. 1st ed. Philadelphia, PA: W.B. Saunders Company. Pp. 695-717.

Sillanpaa MC, Agar LM, Milner IB, Podany EC, Axelrod BN, Brown GG. 1997. Gulf War veterans: A neuropsychological examination. *Journal of Clinical and Experimental Neuropsychology* 19(2):211-219.

Simmons R, Maconochie N, Doyle P. 2004. Self-reported ill health in male UK Gulf War veterans: A retrospective cohort study. *BMC Public Health* 4(1):27.

Simon GE, Daniell W, Stockbridge H, Claypoole K, Rosenstock L. 1993. Immunologic, psychological, and neuropsychological factors in multiple chemical sensitivity: A controlled study. *Annals of Internal Medicine* 119(2):97-103.

Smith TC, Gray GC, Knoke JD. 2000. Is systemic lupus erythematosus, amyotrophic lateral sclerosis, or fibromyalgia associated with Persian Gulf War service? An examination of Department of Defense hospitalization data. *American Journal of Epidemiology* 151(11):1053-1059.

Smith TC, Heller JM, Hooper TI, Gackstetter GD, Gray GC. 2002. Are Gulf War veterans experiencing illness due to exposure to smoke from Kuwaiti oil well fires? Examination of Department of Defense hospitalization data. *American Journal of Epidemiology* 155(10):908-917.

Smith TC, Gray GC, Weir JC, Heller JM, Ryan MA. 2003. Gulf War veterans and Iraqi nerve agents at Khamisiyah: Postwar hospitalization data revisited. *American Journal of Epidemiology* 158(5):457-467.

Sostek MB, Jackson S, Linevsky JK, Schimmel EM, Fincke BG. 1996. High prevalence of chronic gastrointestinal symptoms in a National Guard Unit of Persian Gulf veterans. *American Journal of Gastroenterology* 91(12):2494-2497.

Steele L. 2000. Prevalence and patterns of Gulf War illness in Kansas veterans: Association of symptoms with characteristics of person, place, and time of military service. *American Journal of Epidemiology* 152(10):992-1002.

Sterne JA, Davey Smith G. 2001. Sifting the evidence-what's wrong with significance tests? *British Medical Journal* 322(7280):226-231.

Storzbach D, Campbell KA, Binder LM, McCauley L, Anger WK, Rohlman DS, Kovera CA. 2000. Psychological differences between veterans with and without Gulf War unexplained symptoms. *Psychosomatic Medicine* 62(5):726-735.

Storzbach D, Rohlman DS, Anger WK, Binder LM, Campbell KA. 2001. Neurobehavioral deficits in Persian Gulf veterans: Additional evidence from a population-based study. *Environmental Research* 85(1):1-13.

Straus SE. 1991. History of chronic fatigue syndrome. *Reviews of Infectious Diseases* 13(S1):S2-S7.

Stretch RH, Bliese PD, Marlowe DH, Wright KM, Knudson KH, Hoover CH. 1996a. Psychological health of Gulf War-era military personnel. *Military Medicine* 161(5):257-261.

Stretch RH, Marlowe DH, Wright KM, Bliese PD, Knudson KH, Hoover CH. 1996b. Post-traumatic stress disorder symptoms among Gulf War veterans. *Military Medicine* 161(7):407-410.

Sutker PB, Davis JM, Uddo M, Ditta SR. 1995. War zone stress, personal resources, and PTSD in Persian Gulf War returnees. *Journal of Abnormal Psychology* 104(3):444-452.

Tasman A, Lieberman J, Kay J. 2003. *Psychiatry.* 2nd ed. London, UK: Wiley

Thomas TL, Kang HK, Dalager NA. 1991. Mortality among women Vietnam veterans, 1973-1987. *American Journal of Epidemiology* 134(9):973-980.

Tibshirani R WGHT. 2001. Estimating the number of clusters in a data set via the gap statistic. *Journal of the Royal Statistical Society: Series B (Statistical Methodology)* 63(2):411-423.

Unwin C, Blatchley N, Coker W, Ferry S, Hotopf M, Hull L, Ismail K, Palmer I, David A, Wessely S. 1999. Health of UK servicemen who served in Persian Gulf War. *Lancet* 353(9148):169-178.

Valenti M, Pontieri FE, Conti F, Altobelli E, Manzoni T, Frati L. 2005. Amyotrophic lateral sclerosis and sports: A case-control study. *European Journal of Neurology* 12(3):223-225.

Vasterling JJ, Brailey K, Tomlin H, Rice J, Sutker PB. 2003. Olfactory functioning in Gulf War-era veterans: Relationships to war-zone duty, self-reported hazards exposures, and psychological distress. *Journal of the International Neuropsychological Society* 9(3):407-418.

Watanabe KK, Kang HK. 1995. Military service in Vietnam and the risk of death from trauma and selected cancers. *Annals of Epidemiology* 5(5):407-412.

Weathers F, Ford J. 1996. Psychometric properties of the PTSD Checklist (PCL-C, PCL-S; PCL-M; PCL-PR). In: Stamm BH, Editor. *Measurement of Stress Trauma and Adaptation*. Lutherville, Maryland: Sidron Press.

Werler MM, Sheehan JE, Mitchell AA. 2005. Gulf war veterans and hemifacial microsomia. *Birth Defects Research. Part A, Clinical and Molecular Teratology* 73(1):50-52.

Wessely S. 1998. The epidemiology of chronic fatigue syndrome. *Epidemiologia e Psichiatria Sociale* 7(1):10-24.

Wessely S. 2005. Risk, psychiatry and the military. *British Journal of Psychiatry* 186:459-466.

White RF, Proctor SP, Heeren T, Wolfe J, Krengel M, Vasterling J, Lindem K, Heaton KJ, Sutker P, Ozonoff DM. 2001. Neuropsychological function in Gulf War veterans: Relationships to self-reported toxicant exposures. *American Journal of Industrial Medicine* 40(1):42-54.

Wikipedia. *Polyneuropathy*. [Online]. Available: http://en.wikipedia.org/wiki/Polyneuropathy; [accessed July 11, 2006].

Winkenwerder W. 2002. *US Demolition Operations at Khamisiyah*. Washington, DC: Department of Defense.

Wolfe F, Smythe HA, Yunus MB, Bennett RM, Bombardier C, Goldenberg DL, Tugwell P, Campbell SM, Abeles M, Clark P, et al. 1990. The American College of Rheumatology 1990 Criteria for the Classification of Fibromyalgia. Report of the Multicenter Criteria Committee. *Arthritis and Rheumatism* 33(2):160-172.

Wolfe F, Ross K, Anderson J, Russell IJ, Hebert L. 1995. The prevalence and characteristics of fibromyalgia in the general population. *Arthritis and Rheumatism* 38(1):19-28.

Wolfe F, Anderson J, Harkness D, Bennett RM, Caro XJ, Goldenberg DL, Russell IJ, Yunus MB. 1997. Health status and disease severity in fibromyalgia: Results of a six-center longitudinal study. *Arthritis and Rheumatism* 40(9):1571-1579.

Wolfe J, Erickson DJ, Sharkansky EJ, King DW, King LA. 1999a. Course and predictors of posttraumatic stress disorder among Gulf War veterans: A prospective analysis. *Journal of Consulting and Clinical Psychology* 67(4):520-528.

Wolfe J, Proctor SP, Erickson DJ, Heeren T, Friedman MJ, Huang MT, Sutker PB, Vasterling JJ, White RF. 1999b. Relationship of psychiatric status to Gulf War veterans' health problems. *Psychosomatic Medicine* 61(4):532-540.

Writer J, DeFraites R, Brundage J. 1996. Comparative Mortality among US Military Personnel in the Persian Gulf and Worldwide during Operations Desert Shield and Desert Storm. *Journal of the American Medical Association* 275(2):118-121.

Zwerling C, Torner JC, Clarke WR, Voelker MD, Doebbeling BN, Barrett DH, Merchant JA, Woolson RF, Schwartz DA. 2000. Self-reported postwar injuries among Gulf War veterans. *Public Health Reports* 115(4):346-349.

6

CONCLUSIONS AND RECOMMENDATIONS

The committee was established to review, evaluate, and summarize the peer-reviewed scientific and medical literature addressing the health status of Persian Gulf War veterans. This chapter summarizes what the literature collectively tells us about the veterans' symptoms and illnesses.

QUALITY OF THE STUDIES

Overall the studies of Gulf War veterans' health are of varied quality. Although they have provided valuable information, many of them have limitations that hinder accurate assessment of the veterans' health status. There is a detailed discussion of the limitations of the studies of Gulf War veterans in Chapter 4. The issues under discussion include the lack of representativeness of the entire Gulf War population in some studies, low participation rates in most studies, studies that might be too narrow in their assessment of health status, instruments that might have been too insensitive to detect abnormalities in deployed veterans, and the timing of the investigation relative to the latency for some health outcomes (for example, cancer). In addition, many of the US studies are cross-sectional, and this limits the opportunity to learn about symptom duration and chronicity, latency of onset, and prognosis. Those limitations make it difficult to interpret the results of the findings particularly when several well-conducted studies produce inconsistent results. Furthermore, most of the studies rely on self-reports rather than objective measures of symptoms and exposures.

OVERVIEW OF HEALTH OUTCOMES

In looking at health outcomes in Gulf War-deployed veterans, numerous researchers have attempted to determine whether a set of symptoms reported by veterans could be defined as a unique syndrome or illness. Investigators have attempted, by using factor or cluster analysis, to find a unique outcome, but none has been identified. Veterans of the Gulf War, from the US, the UK, Canada, Australia, and Denmark report higher rates of nearly all symptoms or sets of symptoms than their nondeployed counterparts; that finding was reported consistently in every study reviewed by this committee. Some of the symptoms have been associated with neurobehavioral decrements on neurocognitive tests.

Not surprisingly, given the global excess reporting of symptoms among Gulf War-deployed veterans, the rates of individual symptoms as well as the rates of chronic multisymptom illnesses were higher among deployed veterans than nondeployed in many studies. Multisymptom-based medical conditions reported to occur more frequently among deployed Gulf War veterans include fibromyalgia, chronic fatigue syndrome (CFS), and multiple chemical sensitivity (MCS). However, the case definitions for those conditions are based on symptom reports, and there are no objective diagnostic criteria that can be used to validate the findings, so it is not clear whether the literature supports a true excess of those conditions or whether the associations are spurious and result from the increased reporting of symptoms across the board. The literature also demonstrates that deployment places veterans at increased risk for symptoms that meet diagnostic criteria for a number of psychiatric illnesses, particularly posttraumatic stress disorder (PTSD), anxiety, depression, and substance abuse. In addition, comorbidities have been reported, for example, veterans reporting symptoms of both PTSD and depression. The committee felt confident that several studies validated the increased risk of psychiatric disorders.

Some studies indicate that Gulf War veterans are at increased risk for amyotrophic lateral sclerosis (ALS). With regard to birth defects, there is weaker evidence that Gulf War veterans' offspring might be at risk for some birth defects; the findings are inconsistent. Finally, long-term exacerbation of asthma appeared to be associated with oil-well fire smoke, but there were no objective measures of pulmonary function in the studies.

The health outcomes presented above are discussed in some detail in the following pages. They are grouped according to whether the findings were based primarily on self-reporting of symptoms or on objective measures and diagnostic medical tests.

Outcomes Based Primarily on Symptoms and Self-Reports

The largest and most nationally representative survey of US veterans found that nearly 29% of deployed veterans met a case definition of "multisymptom illness", compared with 16% of nondeployed veterans (Blanchard et al. 2006). Those figures indicate that unexplained illnesses are the most prevalent health outcome of service in the Gulf War. Several researchers, using factor or cluster analyses, have tried to determine whether or not the symptoms that have been reported by Gulf War veterans cluster in such a way as to make up a unique syndrome, such as "Gulf War Illness".

Numerous studies (Cherry et al. 2001; Doebbeling et al. 2000; Everitt et al. 2002; Forbes et al. 2004; Kang et al. 2002) have used statistical techniques, such as factor and cluster analyses to search for such symptom clusters or syndromes. Those studies have demonstrated that deployed veterans report more symptoms and more severe symptoms than their nondeployed counterparts, but they did not find a unique symptom complex (or syndrome) in deployed Gulf War veterans. What those studies have found is a global increase in symptoms reported by Gulf War-deployed veterans compared to their counterparts—global in that the increased symptom rates occur in every category of health outcome.

Among the many symptoms reported by Gulf War veterans are deficits in neurocognitive ability. Obviously such reports are of concern because of the potential for those deficits to have adverse effects on the lives of the veterans. Primary studies found nonsignificant trends of poorer neurobehavioral performance when Gulf War veterans were compared to nondeployed veterans or veterans deployed to Germany. However, when PTSD (White et al. 2001) or depressed mood (David et al. 2002) was treated as a confounder in the statistical analyses those trends disappeared, but that adjustment might be inappropriate because of the possibility of overcontrolling a variable that might lie on the causal pathway.

One study concluded that Gulf War veterans who reported symptoms associated with the Gulf conflict performed more poorly on neurobehavioral tests than veterans who did not report symptoms (Storzbach et al. 2000); another study found substantial neurobehavioral deficits in deployed veterans but had intentionally recruited veterans who experienced a high prevalence of post-Gulf War illness (Hom et al. 1997). That study failed to adjust for key confounders and for the large number of statistical comparisons in their study, raising doubt about the validity of their findings.

In conclusion, primary studies of deployed Gulf War veterans vs non-Gulf War deployed do not demonstrate differences in cognitive and motor measures as determined through neurobehavioral testing. However, returning Gulf War veterans with at least one symptom commonly reported by Gulf War veterans (such as, fatigue, memory loss, confusion, inability to concentrate, mood swings, somnolence, GI distress, muscle and joint pain, or skin or mucous membrane complaints) demonstrated poorer performance on cognitive tests when compared to returning veterans who did not report such symptoms.

Several studies focused on multisymptom-based medical conditions: fibromyalgia, CFS, and MCS. Those conditions have several features in common: they do not fit a precise diagnostic category; case definitions are symptom-based; there are no objective criteria for validating the case definitions; and the symptoms among those syndromes overlap to some extent. Gulf War-deployed veterans report higher rates of symptoms that are consistent with case definitions of MCS, CFS, and fibromyalgia.

Several large or population-based studies of Gulf War veterans found, by questionnaire, that prevalence of MCS-like symptoms ranged from 2% to 6% (Black et al. 1999; Black et al. 1999; 2000; Black et al. 2000; Goss Gilroy Inc. 1998; Goss Gilroy Inc. 1998; Goss Gilroy Inc. 1998; Reid et al. 2001; Unwin et al. 1999). Most studies found that the prevalence in Gulf War veterans was about 2-4 times higher than that in nondeployed veterans. However, no two of the primary studies used the same definition of MCS, so it is difficult to compare them, and none performed medical evaluations to exclude other explanations as would be required by the case definition of MCS.

The prevalence of CFS among Gulf War veterans is highly variable from study to study; most studies used the Centers for Disease Control and Prevention case definition. One primary study (Eisen et al. 2005) demonstrated a higher prevalence of CFS in deployed than in nondeployed veterans (odds ratio [OR] 40.6, 95% confidence interval [CI] 10.2-161.15).

Secondary studies also showed a higher prevalence of CFS and CFS-like illnesses among veterans deployed to the Persian Gulf than among their counterparts who were not deployed or who were deployed elsewhere.

The diagnosis of fibromyalgia is based on symptoms and a very limited physical examination that consists of determining whether pain is elicited by pressing on several points on the body; there are no laboratory tests with which to confirm the diagnosis. Only one of the available cross-sectional studies, Eisen and colleagues (2005), included both Gulf War-deployed and nondeployed veterans and used the full American College of Rheumatology (ACR) case definition of fibromyalgia, including the physical-examination criteria (other studies used a case definition based on symptoms alone). That study found a statistically significant difference in the prevalence of fibromyalgia between deployed and nondeployed veterans (2.0% vs 1.2%; adjusted OR[1], 2.32; 95% CI, 1.02- 5.27). The study by Smith and colleagues (2000) found no association between Gulf War deployment and hospitalization for fibromyalgia, but the committee did not find this to be inconsistent with the positive findings in the Eisen et al. study because very few cases of fibromyalgia are severe enough to warrant hospitalization (notably,

---

[1] Adjusted for differences in age, sex, race, years of education, smoking, type of duty, service branch, and rank.

the prevalence of a diagnosis of fibromyalgia in the Eisen et al. study is about 300 times the prevalence of hospitalization for fibromyalgia in the Smith et al. study). The Iowa study (Iowa Persian Gulf Study Group 1997) and the Canadian study (Goss Gilroy Inc. 1998) both found significantly more fibromyalgia symptoms among deployed veterans than among nondeployed. The findings of those two studies, although generally supportive of the findings of the Eisen et al. study, are of limited value because the lack of a physical examination prohibits the use of the full criteria for diagnosis. The Bourdette study (2001), which did not have a nondeployed-veteran comparison group, estimated a minimum prevalence of 2.47% in the deployed veterans and used the full ACR case definition of fibromyalgia. In conclusion, largely on the basis of the Eisen et al. (2005) study, which used the criteria of the American College of Rheumatology for diagnosis of fibromyalgia but which could have been subject to unrecognized selection bias, there is a higher prevalence of fibromyalgia among deployed Gulf War veterans than among nondeployed veterans.

Other symptoms that are self-reported more often by deployed veterans are gastrointestinal symptoms, particularly dyspepsia; dermatologic conditions, particularly atopic dermatitis and warts; and joint pains.

There were many reports of gastrointestinal symptoms in Gulf War-deployed veterans. Those symptoms seem to be linked to reports of exposures to contaminated water and burning of animal waste in the war theater. The committee notes that several studies (e.g., Eisen et al. 2005) reported a higher rate of self-reported dyspepsia in Gulf War-deployed veterans than in nondeployed veterans. In the context of nearly all symptoms being reported more frequently in Gulf War veterans, it is difficult to interpret these findings.

For dermatologic conditions, a few studies included an examination of the skin and thus were more reliable than self-reports (e.g., Eisen et al. 2005); those have reported that a few unrelated skin conditions occurred more frequently among Gulf War-deployed veterans; however, the findings are not consistent. There is some evidence of a higher prevalence of two distinct dermatologic conditions, atopic dermatitis and verruca vulgaris (warts), in Gulf War-deployed veterans.

Arthralgias (joint pains) were more frequently reported among Gulf War veterans. Likewise, self-reports of arthritis were more common among those deployed to the gulf. Again, in the context of global reporting increases, such data are difficult to interpret. Moreover, studies that included a physical examination did not find evidence of a statistically significant increase in arthritis (Eisen et al. 2005).

Finally, Gulf War veterans consistently have been found to suffer from a variety of psychologic conditions. Two well-designed studies using validated interview-based assessments reported that several psychiatric disorders, most notably PTSD and depression, are 2-3 times more likely in Gulf War-deployed than in nondeployed veterans (Black et al. 2004; Wolfe et al. 1999). Moreover, comorbidities were reported among a number of veterans, with co-occurrence of PTSD, depression, anxiety, or substance abuse. Most of the other studies administered well-validated symptom questionnaires and their findings were remarkably similar: an overall two to three-fold increase in the prevalence of psychiatric disorders. When traumatic war exposures were assessed with symptoms, studies characteristically showed higher rates, particularly of PTSD, in veterans who had more traumatic war experiences than in those with lower levels of traumatic exposure. In other words, studies found a dose-response relationship between the degree of traumatic war exposure and PTSD. The finding of such a dose-response relationship provides increased confidence in the association with deployment.

Outcomes with Objective Measures or Diagnostic Medical Tests

In reviewing the studies of mortality, the committee found numerous limitations. The principal one is the short duration of followup observation. More time must elapse before investigators will be able to assess increased mortality that would result from illnesses with long latency, such as cancer, or that would have a gradually deteriorating course, such as cardiovascular disease. Another potential limitation of comparing deployed with nondeployed personnel is the healthy-warrior effect, which might result in selection bias, insofar as chronically ill or less fit members of the armed forces might be less likely to have been deployed than more fit members.

A number of studies examined rates of injuries in Gulf War veterans (Kang and Bullman 1996; Kang and Bullman 2001; Macfarlane et al. 2000). Those studies provide evidence of a modest increase in transportation-related injuries and mortality among deployed than among nondeployed Gulf War veterans in the decade immediately after deployment. However, studies with longer followup indicate that the increased injury rate was likely to have been restricted to the first several years after the war (Kang and Bullman 2001).

With regard to all causes of hospitalization, studies provide some evidence that excess hospitalizations did not occur among veterans of the Gulf War who remained on active duty through 1994 (Gray et al. 1996). Those studies have certain limitations, however, as they were largely of active-duty personnel and cannot be generalized to the entire cohort of Gulf War veterans; it has been noted that Gulf War veterans who left the military reported worse health outcomes than those who remained (Ismail et al. 2000). It also might be too soon to capture hospitalizations from illnesses that might have longer latency, such as some cancers. In addition, hospitalization data on people separated from the military and admitted to nonmilitary (Department of Veterans Affairs [VA] and civilian) hospitals or on those who used outpatient facilities might be incomplete.

Veterans are understandably concerned about increases in cancer, and the studies reviewed did not demonstrate consistent evidence of increased overall cancer in the Gulf War veterans compared with nondeployed veterans (Kang and Bullman 2001; Macfarlane et al. 2003). However, many veterans are young for cancer diagnoses, and, for most cancers, the time since the Gulf War is probably too short to expect the onset of cancer. Incidence of and mortality from cancer in general, and brain and testicular cancer in particular, have been assessed in cohort studies. An association of brain-cancer mortality with possible nerve-agent exposure (as modeled by the Department of Defense [DOD] exposure model of 2000) was observed in one study (Bullman et al. 2005), but, as discussed in more detail in Chapter 2, there were many uncertainties in the exposure model. Results for testicular cancer were mixed: one study concluded that there was no evidence of an excess risk (Knoke et al. 1998), and another, small, registry-based study (Levine et al. 2005) suggested that there might be an increased risk.

Another concern for veterans has been whether amyotrophic lateral sclerosis (ALS) is increased in Gulf War veterans. Two primary studies and one secondary study found that deployed veterans appear to be at increased risk for ALS. The primary study by Horner et al. (2003), which had the possibility of underascertainment of cases in the nondeployed population, was confirmed by a secondary analysis by Coffman et al. (2005) that documented a nearly 2-fold increase in risk. A secondary study by Haley (2003) used general population estimates as the comparison group and found a slightly higher relative risk.

Peripheral neuropathy has also been studied in Gulf War veterans. One large, well-designed study conducted by VA (Davis et al. 2004), which used a thorough and objective evaluation and a stringent case definition, did not find statistically significant evidence of excess peripheral neuropathy. Several other secondary studies supported no excess risk. Some studies (e.g., Cherry et al. 2001) do report higher rates of peripheral neuropathy, but they use self-reports, which the committee did not accept as a reliable measure of peripheral neuropathy. Thus, there does not appear to be an increase in the prevalence of peripheral neuropathy in deployed vs nondeployed veterans, as defined by history, physical examination, and electrophysiologic studies.

With regard to cardiovascular disease, primary studies found no statistically significant differences between deployed and nondeployed veterans in rates of hypertension (Fukuda et al. 1998). One study did report a small but significant increase in hospitalizations due to cardiovascular disease among a subset of deployed veterans who were possibly exposed to nerve agents from the Khamisiyah plume compared with Gulf War-deployed veterans who were not in the possible exposure area (Smith et al. 2003). The increased hospitalizations were due entirely to an increase in cardiac dysrhythmias. However, the study suffers from uncertainty about the Khamisiyah plume model. In the secondary studies, deployed veterans were generally more likely to report hypertension and palpitations, but those reports were not confirmed with medical evaluations. Thus, it does not appear that there is a difference in the prevalence of cardiovascular disease or diabetes between deployed Gulf War veterans and nondeployed.

Many veterans are understandably concerned about the possibility of birth defects in their offspring. The Araneta et al. (2003) and Doyle et al. (2004) studies yielded some evidence of increased risk of birth defects among offspring of Gulf War veterans. However, the specific defects with increased prevalence were not consistent. Overall, the studies are difficult to interpret because of the relative rarity of specific birth defects, use of small samples, timing of exposure (before or after conception), and whether the mother or the father was exposed. There was no consistent pattern of one of more birth defects with a higher prevalence in the offspring of male or female Gulf War veterans. Only one set of defects—urinary tract abnormalities—has been found to be increased in more than one well-designed study.

With regard to other adverse reproductive outcomes, the results of the Araneta et al. (2004) study, which had hospital discharge data available, are suggestive of an increased risk of spontaneous abortions and ectopic pregnancies among female Gulf War veterans. Findings specifically related to fertility (Ishoy et al. 2001a; Maconochie et al. 2004) and sexual problems (Ishoy et al. 2001b) relied on self-reports where there is potential for recall bias. In one study, there was no evidence of differences in levels of male reproductive hormones between Gulf War-deployed veterans and nondeployed veterans.

Numerous studies in several countries examined respiratory outcomes related to deployment to the Gulf War Theater. Five studies (Eisen et al. 2005; Gray et al. 1999; Ishoy et al. 1999; Karlinsky et al. 2004; Kelsall et al. 2004) representing four distinct cohorts from three countries (the United States, Australia, and Denmark), examined associations of Gulf War deployment with pulmonary-function measures or respiratory disease diagnoses. In none of those studies were statistically significant associations found. The uniformity of the findings is striking, especially given that the same five studies found that Gulf War deployment status was significantly associated with self-reports of respiratory symptoms among three of the four cohorts. Indeed, the overwhelming majority of studies conducted among Gulf War veterans— whether from the United States (Doebbeling et al. 2000; Gray et al. 1999; Gray et al. 2002; Iowa Persian Gulf Study Group 1997; Kang et al. 2000; Karlinsky et al. 2004; Kroenke et al. 1998; Petruccelli et al. 1999; Steele 2000), the UK (Cherry et al. 2001; Nisenbaum et al. 2004;

Simmons et al. 2004; Simmons et al. 2004; Unwin et al. 1999; Unwin et al. 1999) Canada (Goss Gilroy Inc. 1998), Australia (Kelsall et al. 2004), or Denmark (Ishoy et al. 1999)—have found that several years after deployment those deployed report higher rates of respiratory symptoms and respiratory illnesses than nondeployed troops. Of particular interest is the UK cohort study reported in Nisenbaum et al. (2004) and Unwin et al. (1999), which found substantially higher prevalences of respiratory symptoms and self-reported respiratory disease among those deployed in the Gulf War than among those deployed in another war theater, Bosnia.

Several studies examined respiratory outcomes specifically associated with chemical exposures experienced by Gulf War veterans, whereas the studies discussed above examined respiratory outcomes associated with deployment. The study of Cowan et al. (2002), which used objective exposure measures and methods[2], found associations between oil-well fire smoke and doctor-assigned diagnosis of asthma in veterans. A limitation of the study is that the participants were self-selected. The other key Gulf War study of oil-well fire smoke, which was based on the Iowa cohort (Lange et al. 2002), found no relationship between the same objective exposures and respiratory health outcomes; it had the advantage of avoiding the potential selection biases of the Cowan et al. study. However, its definitions of respiratory diseases were based entirely on self-reports of symptoms and cannot be viewed as adequate. The study by Smith et al. (2002) found no statistically significant associations between the same objective measures of exposure to smoke from oil-well fires and later hospitalization for asthma, acute bronchitis, chronic bronchitis, or emphysema. However, the participants were all active-duty veterans; most young adults are seldom hospitalized for those diagnoses, so most cases would not be expected to be captured.

The study by Gray and collaborators (1999) found a small increase in postwar hospitalization for respiratory system disease associated with modeled exposure to nerve agents at Khamisiyah. Limitations of that study probably include substantial exposure misclassification based on DOD exposure estimates that were later revised, lack of control for tobacco-smoking, lack of a clear dose-response pattern, and limited biologic plausibility of effects on the respiratory system in a setting in which no effect on nervous system diseases was seen. Karlinsky et al. (2004) found no statistically significant associations between pulmonary-function measures and exposure to nerve agents at Khamisiyah on the basis of the revised DOD exposure estimates.

In conclusion, as is the case for a number of other organ systems, respiratory symptoms and self-reported health outcomes are strongly associated with Gulf War deployment in most studies addressing this question that use comparison groups of nondeployed veterans. However, studies with objective pulmonary-function measures find no statistically significant association between respiratory illness and disease and Gulf War deployment in the four cohorts in which this has been investigated; thus, the studies leave an uncertain clinical interpretation of the increased symptoms and self-reported diseases.

---

[2] Exposure to smoke from oil-well fires was estimated by linking troop locations with modeled oil-fire smoke exposure. National Oceanic and Atmospheric Administration researchers modeled exposure on the basis of meteorologic and ground-station air-monitoring data (Draxler et al. 1994; McQueen and Draxler 1994) with a spatial resolution of 15 km and a temporal resolution of 24 hours. DOD personnel records were used to ascertain each study subject's unit and dates of service. Only Army personnel were included in the study because their location data were more precise. Two exposure measures were used: cumulative smoke exposure (the sum of the estimated concentration on all days when each subject was in the Gulf War Theater) in milligram-days per cubic meter (mg-day/m^3), with referent exposure < 0.1 mg-day/m^3 per day; and number of days when the subject was exposed at 65 µg/m^3 or greater.

Among studies that examined pulmonary outcomes in association with specific exposures in the Gulf War Theater, the positive study by Cowan et al. (2002), which used objective measures of oil-well fire smoke and doctor-assigned respiratory diagnoses, is the strongest methodologically. With respect to nerve agents at Khamisiyah, no study using valid objective estimates of exposure has found statistically significant associations with pulmonary-function measures or physician-diagnosed respiratory disease.

RECOMMENDATIONS

The adequacy of the government's response has been both praised and criticized; VA and DOD have expended enormous effort and resources in attempts to address the numerous health issues related to the Gulf War veterans. The information obtained from those efforts, however, has not been sufficient to determine conclusively the origins, extent, and long-term implications of health problems potentially associated with veterans' participation in the Gulf War. The difficulty in obtaining meaningful answers, as noted by numerous past Institute of Medicine committees and with which the present committee agrees, is due largely to inadequate predeployment and postdeployment screening and medical examinations, and lack of monitoring of possible exposures of deployed personnel.

Predeployment and Postdeployment Screening

Predeployment and postdeployment data-gathering needs to include physician verification of data obtained from questionnaires so that one could have confidence in baseline and postdeployment health data. Collection and archiving of biologic samples might enable the diagnosis of specific medical conditions and provide a basis of later comparison. Meticulous records of all medications, whether used for treatment or prophylactically, would have improved the data and their interpretation in many of the studies reviewed.

Exposure Assessment

Environmental exposures were usually not assessed directly, and that critically hampers the assessment of the effects of specific exposures on specific health outcomes. There have been detailed and laudable efforts to simulate and model exposures, but those efforts have been hampered by lack of the input data required to link the exposure scenarios to specific people or even to specific units or job categories. Moving beyond the current state requires that more detailed information be gathered during future military deployments. Specifically, working toward the development of a job-task-unit-exposure matrix, in which information on people with specific jobs or tasks or attached to specific units (according to routinely available records) is linked to exposures by expert assessment or simulation studies, would enable quantitative assessment of the effects of specific exposures.

Surveillance for Adverse Outcomes

The committee noted that several health outcomes seemed to be appearing with higher incidence or prevalence in the Gulf War-deployed veterans. For those outcomes, the committee recommends continued surveillance to determine whether there is actually a higher risk in Gulf War veterans. Those outcomes are cancer (particularly brain and testicular), ALS, birth defects (including Goldenhar syndrome and urinary tract abnormalities) and other adverse pregnancy outcomes (such as spontaneous abortion and ectopic pregnancy), and postdeployment psychiatric conditions. The committee also recommends that cause-specific mortality in Gulf War veterans continue to be monitored. Although there was an increase in mortality in the first few years after the Gulf War, the deaths appear to have been related to transportation injuries.

REFERENCES

Araneta MR, Schlangen KM, Edmonds LD, Destiche DA, Merz RD, Hobbs CA, Flood TJ, Harris JA, Krishnamurti D, Gray GC. 2003. Prevalence of birth defects among infants of Gulf War veterans in Arkansas, Arizona, California, Georgia, Hawaii, and Iowa, 1989-1993. *Birth Defects Research* 67(4):246-260.

Araneta MR, Kamens DR, Zau AC, Gastanaga VM, Schlangen KM, Hiliopoulos KM, Gray GC. 2004. Conception and pregnancy during the Persian Gulf War: The risk to women veterans. *Annals of Epidemiology* 14(2):109-116.

Black DW, Doebbeling BN, Voelker MD, Clarke WR, Woolson RF, Barrett DH, Schwartz DA. 1999. Quality of life and health-services utilization in a population-based sample of military personnel reporting multiple chemical sensitivities. *Journal of Occupational and Environmental Medicine* 41(10):928-933.

Black DW, Doebbeling BN, Voelker MD, Clarke WR, Woolson RF, Barrett DH, Schwartz DA. 2000. Multiple chemical sensitivity syndrome: Symptom prevalence and risk factors in a military population. *Archives of Internal Medicine* 160(8):1169-1176.

Black DW, Carney CP, Peloso PM, Woolson RF, Schwartz DA, Voelker MD, Barrett DH, Doebbeling BN. 2004. Gulf War veterans with anxiety: Prevalence, comorbidity, and risk factors. *Epidemiology* 15(2):135-142.

Blanchard MS, Eisen SA, Alpern R, Karlinsky J, Toomey R, Reda DJ, Murphy FM, Jackson LW, Kang HK. 2006. Chronic Multisymptom Illness Complex in Gulf War I Veterans 10 Years Later. *American Journal of Epidemiology* 163(1):66-75.

Bourdette DN, McCauley LA, Barkhuizen A, Johnston W, Wynn M, Joos SK, Storzbach D, Shuell T, Sticker D. 2001. Symptom factor analysis, clinical findings, and functional status in a population-based case control study of Gulf War unexplained illness. *Journal of Occupational and Environmental Medicine* 43(12):1026-1040.

Bullman TA, Mahan CM, Kang HK, Page WF. 2005. Mortality in US Army Gulf War Veterans Exposed to 1991 Khamisiyah Chemical Munitions Destruction. *American Journal of Public Health* 95(8):1382-1388.

Cherry N, Creed F, Silman A, Dunn G, Baxter D, Smedley J, Taylor S, Macfarlane GJ. 2001. Health and exposures of United Kingdom Gulf war veterans. Part I: The pattern and extent of ill health. *Occupational and Environmental Medicine* 58(5):291-298.

Coffman CJ, Horner RD, Grambow SC, Lindquist J. 2005. Estimating the occurrence of amyotrophic lateral sclerosis among Gulf War (1990-1991) veterans using capture-recapture methods. *Neuroepidemiology* 24(3):141-150.

Cowan DN, Lange JL, Heller J, Kirkpatrick J, DeBakey S. 2002. A case-control study of asthma among U.S. Army Gulf War veterans and modeled exposure to oil well fire smoke. *Military Medicine* 167(9):777-782.

David AS, Farrin L, Hull L, Unwin C, Wessely S, Wykes T. 2002. Cognitive functioning and disturbances of mood in UK veterans of the Persian Gulf War: A comparative study. *Psychological Medicine* 32(8):1357-1370.

Davis LE, Eisen SA, Murphy FM, Alpern R, Parks BJ, Blanchard M, Reda DJ, King MK, Mithen FA, Kang HK. 2004. Clinical and laboratory assessment of distal peripheral nerves in Gulf War veterans and spouses. *Neurology* 63(6):1070-1077.

Doebbeling BN, Clarke WR, Watson D, Torner JC, Woolson RF, Voelker MD, Barrett DH, Schwartz DA. 2000. Is there a Persian Gulf War syndrome? Evidence from a large population-based survey of veterans and nondeployed controls. *American Journal of Medicine* 108(9):695-704.

Doyle P, Maconochie N, Davies G, Maconochie I, Pelerin M, Prior S, Lewis S. 2004. Miscarriage, stillbirth and congenital malformation in the offspring of UK veterans of the first Gulf war. *International Journal of Epidemiology* 33(1):74-86.

Draxler RR, McQueen JT, Stunder BJB. 1994. An evaluation of air pollutant exposures due to the 1991 Kuwait oil fires using a Lagrangian model. *Atmospheric Environment* 28(13):2197-2210.

Eisen SA, Kang HK, Murphy FM, Blanchard MS, Reda DJ, Henderson WG, Toomey R, Jackson LW, Alpern R, Parks BJ, Klimas N, Hall C, Pak HS, Hunter J, Karlinsky J, Battistone MJ, Lyons MJ. 2005. Gulf War veterans' health: Medical evaluation of a US cohort. *Annals of Internal Medicine* 142(11):881-890.

Everitt B, Ismail K, David AS, Wessely S. 2002. Searching for a Gulf War syndrome using cluster analysis. *Psychological Medicine* 32(8):1371-1378.

Forbes AB, McKenzie DP, Mackinnon AJ, Kelsall HL, McFarlane AC, Ikin JF, Glass DC, Sim MR. 2004. The health of Australian veterans of the 1991 Gulf War: Factor analysis of self-reported symptoms. *Occupational and Environmental Medicine* 61(12):1014-1020.

Fukuda K, Nisenbaum R, Stewart G, Thompson WW, Robin L, Washko RM, Noah DL, Barrett DH, Randall B, Herwaldt BL, Mawle AC, Reeves WC. 1998. Chronic multisymptom illness affecting Air Force veterans of the Gulf War. *Journal of the American Medical Association* 280(11):981-988.

Goss Gilroy Inc. 1998. *Health Study of Canadian Forces Personnel Involved in the 1991 Conflict in the Persian Gulf.* Ottawa, Canada: Goss Gilroy Inc. Department of National Defence.

Gray GC, Coate BD, Anderson CM, Kang HK, Berg SW, Wignall FS, Knoke JD, Barrett-Connor E. 1996. The postwar hospitalization experience of US veterans of the Persian Gulf War. *New England Journal of Medicine* 335(20):1505-1513.

Gray GC, Kaiser KS, Hawksworth AW, Hall FW, Barrett-Connor E. 1999. Increased postwar symptoms and psychological morbidity among US Navy Gulf War veterans. *American Journal of Tropical Medicine and Hygiene* 60(5):758-766.

Gray GC, Reed RJ, Kaiser KS, Smith TC, Gastanaga VM. 2002. Self-reported symptoms and medical conditions among 11,868 Gulf War-era veterans: The Seabee Health Study. *American Journal of Epidemiology* 155(11):1033-1044.

Haley RW. 2003. Excess incidence of ALS in young Gulf War veterans. *Neurology* 61(6):750-756.

Hom J, Haley RW, Kurt TL. 1997. Neuropsychological correlates of Gulf War syndrome. *Archives of Clinical Neuropsychology* 12(6):531-544.

Horner RD, Kamins KG, Feussner JR, Grambow SC, Hoff-Lindquist J, Harati Y, Mitsumoto H, Pascuzzi R, Spencer PS, Tim R, Howard D, Smith TC, Ryan MA, Coffman CJ, Kasarskis EJ. 2003. Occurrence of amyotrophic lateral sclerosis among Gulf War veterans. *Neurology* 61(6):742-749.

Iowa Persian Gulf Study Group. 1997. Self-reported illness and health status among Gulf War veterans: A population-based study. *Journal of the American Medical Association* 277(3):238-245.

Ishoy T, Suadicani P, Guldager B, Appleyard M, Hein HO, Gyntelberg F. 1999. State of health after deployment in the Persian Gulf. The Danish Gulf War Study. *Danish Medical Bulletin* 46(5):416-419.

Ishoy T, Andersson AM, Suadicani P, Guldager B, Appleyard M, Gyntelberg F, Skakkebaek NE, Danish Gulf War Study. 2001a. Major reproductive health characteristics in male Gulf War Veterans. The Danish Gulf War Study. *Danish Medical Bulletin* 48(1):29-32.

Ishoy T, Suadicani P, Andersson A-M, Guldager B, Appleyard M, Skakkebaek N, Gyntelberg F. 2001b. Prevalence of male sexual problems in the Danish Gulf War Study. *Scandinavian Journal of Sexology* 4(1):43-55.

Ismail K, Blatchley N, Hotopf M, Hull L, Palmer I, Unwin C, David A, Wessely S. 2000. Occupational risk factors for ill health in Gulf veterans of the United Kingdom. *Journal of Epidemiology and Community Health* 54(11):834-838.

Kang HK, Bullman TA. 1996. Mortality among US veterans of the Persian Gulf War. *New England Journal of Medicine* 335(20):1498-504.

Kang HK, Bullman TA. 2001. Mortality among US veterans of the Persian Gulf War: 7-year follow-up. *American Journal of Epidemiology* 154(5):399-405.

Kang HK, Mahan CM, Lee KY, Magee CA, Murphy FM. 2000. Illnesses among United States veterans of the Gulf War: A population-based survey of 30,000 veterans. *Journal of Occupational and Environmental Medicine* 42(5):491-501.

Kang HK, Mahan CM, Lee KY, Murphy FM, Simmens SJ, Young HA, Levine PH. 2002. Evidence for a deployment-related Gulf War syndrome by factor analysis. *Archives of Environmental Health* 57(1):61-68.

Karlinsky JB, Blanchard M, Alpern R, Eisen SA, Kang H, Murphy FM, Reda DJ. 2004. Late prevalence of respiratory symptoms and pulmonary function abnormalities in Gulf War I Veterans. *Archives of Internal Medicine* 164(22):2488-2491.

Kelsall HL, Sim MR, Forbes AB, McKenzie DP, Glass DC, Ikin JF, Ittak P, Abramson MJ. 2004. Respiratory health status of Australian veterans of the 1991 Gulf War and the effects of exposure to oil fire smoke and dust storms. *Thorax* 59(10):897-903.

Knoke JD, Gray GC, Garland FC. 1998. Testicular cancer and Persian Gulf War service. *Epidemiology* 9(6):648-653.

Kroenke K, Koslowe P, Roy M. 1998. Symptoms in 18,495 Persian Gulf War veterans. Latency of onset and lack of association with self-reported exposures. *Journal of Occupational and Environmental Medicine* 40(6):520-528.

Lange JL, Schwartz DA, Doebbeling BN, Heller JM, Thorne PS. 2002. Exposures to the Kuwait oil fires and their association with asthma and bronchitis among gulf war veterans. *Environmental Health Perspectives* 110(11):1141-1146.

Levine PH, Young HA, Simmens SJ, Rentz D, Kofie VE, Mahan CM, Kang HK. 2005. Is testicular cancer related to Gulf War deployment? Evidence from a pilot population-based study of Gulf War era veterans and cancer registries. *Military Medicine* 170(2):149-53.

Macfarlane GJ, Thomas E, Cherry N. 2000. Mortality among UK Gulf War veterans. *Lancet* 356(9223):17-21.

Macfarlane GJ, Biggs AM, Maconochie N, Hotopf M, Doyle P, Lunt M. 2003. Incidence of cancer among UK Gulf war veterans: Cohort study. *British Medical Journal* 327(7428):1373-1375.

Maconochie N, Doyle P, Carson C. 2004. Infertility among male UK veterans of the 1990-1 Gulf war: Reproductive cohort study. *British Medical Journal* 329(7459):196-201.

McQueen JT, Draxler RR. 1994. Evaluation of model back trajectories of the Kuwait oil fires smoke plume using digital satellite data. *Atmospheric Environment* 28(13):2159-2174.

Nisenbaum R, Ismail K, Wessely S, Unwin C, Hull L, Reeves WC. 2004. Dichotomous factor analysis of symptoms reported by UK and US veterans of the 1991 Gulf War. *Population Health Metrics* 2(1):8.

Petruccelli BP, Goldenbaum M, Scott B, Lachiver R, Kanjarpane D, Elliott E, Francis M, McDiarmid MA, Deeter D. 1999. Health effects of the 1991 Kuwait oil fires: A survey of US army troops. *Journal of Occupational and Environmental Medicine* 41(6):433-439.

Reid S, Hotopf M, Hull L, Ismail K, Unwin C, Wessely S. 2001. Multiple chemical sensitivity and chronic fatigue syndrome in British Gulf War veterans. *American Journal of Epidemiology* 153(6):604-609.

Simmons R, Maconochie N, Doyle P. 2004. Self-reported ill health in male UK Gulf War veterans: A retrospective cohort study. *BMC Public Health* 4(1):27.

Smith TC, Gray GC, Knoke JD. 2000. Is systemic lupus erythematosus, amyotrophic lateral sclerosis, or fibromyalgia associated with Persian Gulf War service? An examination of Department of Defense hospitalization data. *American Journal of Epidemiology* 151(11):1053-1059.

Smith TC, Heller JM, Hooper TI, Gackstetter GD, Gray GC. 2002. Are Gulf War veterans experiencing illness due to exposure to smoke from Kuwaiti oil well fires? Examination of Department of Defense hospitalization data. *American Journal of Epidemiology* 155(10):908-917.

Smith TC, Gray GC, Weir JC, Heller JM, Ryan MA. 2003. Gulf War veterans and Iraqi nerve agents at Khamisiyah: Postwar hospitalization data revisited. *American Journal of Epidemiology* 158(5):457-467.

Steele L. 2000. Prevalence and patterns of Gulf War illness in Kansas veterans: Association of symptoms with characteristics of person, place, and time of military service. *American Journal of Epidemiology* 152(10):992-1002.

Storzbach D, Campbell KA, Binder LM, McCauley L, Anger WK, Rohlman DS, Kovera CA. 2000. Psychological differences between veterans with and without Gulf War unexplained symptoms. *Psychosomatic Medicine* 62(5):726-735.

Unwin C, Blatchley N, Coker W, Ferry S, Hotopf M, Hull L, Ismail K, Palmer I, David A, Wessely S. 1999. Health of UK servicemen who served in Persian Gulf War. *Lancet* 353(9148):169-178.

White RF, Proctor SP, Heeren T, Wolfe J, Krengel M, Vasterling J, Lindem K, Heaton KJ, Sutker P, Ozonoff DM. 2001. Neuropsychological function in Gulf War veterans: Relationships to self-reported toxicant exposures. *American Journal of Industrial Medicine* 40(1):42-54.

Wolfe J, Proctor SP, Erickson DJ, Heeren T, Friedman MJ, Huang MT, Sutker PB, Vasterling JJ, White RF. 1999. Relationship of psychiatric status to Gulf War veterans' health problems. *Psychosomatic Medicine* 61(4):532-540.

INDEX

A

Abnormal clinical and laboratory findings. *See* Symptoms, signs, and abnormal clinical and laboratory findings
Abortion. *See* Spontaneous abortion
Acetylcholinesterase (AChE), 23
Adverse pregnancy outcomes, *See* Birth defects and adverse pregnancy outcomes
AEHA. *See* US Army Environmental Health Agency
AFQT. *See* Armed Forces Qualifying Test
Air Force Technical Applications Center (AFTAC), 28
Air Force Women Study, 76-77
Air National Guard, 207-208
Alcohol abuse, 59, 126, 220, 223
All-cause hospitalization studies, 223-226
    primary studies, 223-224
    summary and conclusion, 224
ALS. *See* Amyotrophic lateral sclerosis
American College of Rheumatology (ACR), case definition of fibromyalgia, 4, 62, 188, 249-250
American Thoracic Society questionnaire, 59n
Amyotrophic lateral sclerosis (ALS), 3, 5-6, 153-156, 248, 251
    continued surveillance for, 8, 254
Anthrax vaccination, 14, 22-23
Anxiety, 3, 65, 122, 124, 127, 205-206, 248, 250
Armed Forces Qualifying Test (AFQT), 132, 136
Arthralgias, 5, 250
Arthritis and arthralgia, 64, 185-187
    primary studies, 185
    secondary studies, 185-186
    summary and conclusion, 186-187
"Arthromyoneuropathy," 72
Asthma, 3, 59, 64, 170-173, 175, 224, 248, 253
Attributable risk, 46
Australian Veteran studies, 70, 84-85, 126, 167, 171, 206
    exposure-symptom relationships, 70
    symptom clustering, 70

B

Beck Depression Inventory (BDI), 125, 132, 136

Behavioral Risk Factor Surveillance Survey, 58n
Bias, 47, 52, 56-57, 158, 204
    selection, 56, 188
Biologic and chemical warfare
    agents for, 71
    threat of, 14-15
    vaccination against, 67
Biologic monitoring, 39-41
    depleted uranium, 39-40
    oil-well fire smoke, 40-41
Bipolar disorder, 122
Birth defects and adverse pregnancy outcomes (ICD-10 O00-Q99), 3, 6, 192-195, 248, 252
    adverse pregnancy outcomes, 194-195
    birth defects, 192-195
    continued surveillance for, 8, 254
    Goldenhar syndrome, 194
    male fertility problems and infertility, 196
    primary studies, 192-193
    secondary studies, 193-194
    spontaneous abortion, 252
    summary and conclusion, 194
Brain cancer, 5, 116, 118, 251
    continued surveillance for, 8, 254
Brief Symptom Inventory (BSI), 58n, 76, 125
Bronchitis, 64, 170-175
BSI. *See* Brief Symptom Inventory

C

CAGE questionnaire, 58n, 123
Canadian Veteran Study, 65, 125, 163, 171, 189, 194, 228, 250
    exposure-symptom relationships, 65
    symptom clustering, 65
Cancer (ICD-10 C00-D48), 5, 115-121
    all cancers, 117-118, 120-121
    brain cancer, 116, 119
    continued surveillance for, 8, 254
    primary and secondary studies, 116-118
    summary and conclusion, 118
    testicular cancer, 116-117, 119-120
CAPS. *See* Clinician-Administered PTSD Scale
Carbon monoxide (CO) exposure, 26
Cardiovascular disease, 6, 166-169, 252
    primary studies, 166-167
    secondary studies, 167-168
    summary and conclusion, 168

INDEX

Case-control studies, 47, 61, 64, 72, 75, 158
Causality, assignment of, 53
CCEP. *See* Comprehensive Clinical Evaluation Program
Centers for Disease Control and Prevention (CDC) Chronic Fatigue Syndrome Questionnaire, 59n
CFS. *See* Chronic fatigue syndrome
Chalder Fatigue Scale, 59n
Charcot's disease. *See* Amyotrophic lateral sclerosis
CHD. *See* Congenital heart disease
Chemical and biologic warfare, *See* Biologic and chemical warfare
Chronic fatigue syndrome (CFS), 3-4, 57, 64-65, 74-75, 124, 161-165, 247, 249
    case definition, 162
    difficulties validating, 57
    primary studies, 162
    secondary studies, 162-163
    summary and conclusion, 163-164
Chronic Fatigue Syndrome Questionnaire, *See* Centers for Disease Control and Prevention (CDC) Chronic Fatigue Syndrome Questionnaire
"Chronic multisymptom illness," 3, 66-67, 75, 205, 207, 247
Chronic pain, 124
CIA-DOD report, 27-29
CIDI. *See* Composite International Diagnostic Interview
Circulatory system diseases, 166-169
Clinical findings, abnormal. *See* Symptoms, signs, and abnormal clinical and laboratory findings
Clinician-Administered PTSD Scale (CAPS), 71, 124-125
Cluster analysis, 2, 50-51, 204, 212-213, 248
CO. *See* Carbon monoxide exposure
Cognitive dysfunction, 3-4, 59, 65, 141
Cohort studies, definition of, 45-47
    Air Force Women Study, 76-77
    Australian Veteran Studies, 70, 84-85
    Canadian Veteran Study, 65
    Connecticut National Guard, 77, 102
    Danish Peacekeeper Studies, 68-69, 93
    defined, 45-47
    Department of Veterans Affairs Study, 60-63, 79-80
    Ft. Devens and New Orleans cohort studies, 70-71, 89-92
    general limitations of Gulf War cohort studies and derivative studies, 56-57
    Hawaii and Pennsylvania Active Duty and Reserve Study, 76, 102
    the Iowa Study, 58-60, 78
    Kansas Veteran Study, 64-65, 89
    larger Seabee cohort studies, 73-74, 97
    London School of Hygiene and Tropical Medicine Veteran Study, 68
    military-unit-based studies, 70-75, 98-104
    New Orleans Reservist Studies, 76, 100-101
    Oregon and Washington Veteran Studies, 63-64, 86-88

Pennsylvania Air National Guard Study, 74-75, 96
population-based studies, 58-70
as prevalence studies, 25
prospective, definition of, 46
retrospective, definition of, 46
Seabee Reserve Battalion Studies, 71-73, 94-96
United Kingdom Veteran Studies, 65-68, 81-84
University of London Veteran Studies, 65-67
University of Manchester Veteran Study, 67-68
Combat Exposure Scale, 25, 205
Expanded, 124
Committee on Gulf War and Health
approach to its charge, 1-2, 15
charge to, 1, 15
Complexities in resolving Gulf War and health issues, 16-17
individual variability, 17
limitations of exposure information, 16-17
multiple exposures and chemical interactions, 16
unexplained symptoms, 17
Composite International Diagnostic Interview (CIDI), 126
Comprehensive Clinical Evaluation Program (CCEP), 25, 189
Confounding variables, 46, 47, 52, 249
"Confusion-ataxia syndrome," 72
Congenital heart disease (CHD), 194
Connecticut National Guard, 77, 102
Connective tissue diseases, 185-187
Cox proportional-hazards modeling, 116, 167
Cross-sectional studies, see epidemiologic studies, cross-sectional studies
Cullen criteria for MCS, 227-228
Cyclosarin, exposure, 26-34, 35-36

D

Danish Peacekeeper Studies, 68-69, 93, 181
exposure-symptom relationships, 69
symptom clustering, 69
Department of Defense (DOD)
Comprehensive Clinical Evaluation Program (CCEP), 25
Gulf War Health Registry, 58, 115
Manpower Data Center, 62-63, 166, 180, 183, 185
Office of the Special Assistant for Gulf War Illnesses, 35, 40
vaccination records, 22
Department of Veterans Affairs Study, 60-63
exposure-symptom relationships, 61-62
medical evaluation findings (Phase III), 62-63
survey findings (Phases I and II), 60-62

symptom clustering, 60-61
Department of Veterans Affairs (VA), 1, 11, 57
    Gulf War Health Registry, 58, 115, 212-213
Depleted uranium (DU) exposures, 14, 24, 39-40
Deployment to the Persian Gulf, 11-13
Depression, 3, 5, 59, 64, 70-71, 73, 75-76, 122-127, 248, 250. *See also* Major depressive disorder
    major, 65
Dermatologic conditions, 4-5, 60, 62, 183, 250
Diagnostic medical tests, health outcomes with, 5-8, 251-254
Digestive system diseases, 180-182
Diseases of the circulatory system (ICD-10 I00-I99), 166-169
    primary studies, 166
    secondary studies, 167-168
    summary and conclusion, 168
Diseases of the digestive system (ICD-10 K00-K99), 180-182
    primary studies, 180
    secondary studies, 181
    summary and conclusion, 181
Diseases of the musculoskeletal system and connective tissue (ICD-10 M00-M99), 185-191
    arthritis and arthralgia, 185-187
    fibromyalgia, 188-191
Diseases of the nervous system (ICD-10 G00-G99), 153-160
    amoyotrophic lateral sclerosis, 153-156
    other neurological outcomes, 158-159
    peripheral neuropathy, 157-158
    primary studies, 154
    secondary studies, 154-155
    summary and conclusion, 159
Diseases of the respiratory system (ICD-10 J00-J99), 170-179
    Primary studies, 170-171
    Secondary studies, 183
    Summary and conclusion, 183
Diseases of the skin and subcutaneous tissue (ICD-10 L00-L99), 183-184
    primary studies, 183
    secondary studies, 183
    summary and conclusion, 183
Dispersion models, of the Khamisiyah demolition, 27-36
Distal symmetric polyneuropathy, 62-63, 157-158
DU. *See* Depleted uranium
Dyspepsia, 4, 62, 180-181, 205, 250

E

Ectopic pregnancy, 194-195, 252
    continued surveillance for, 8, 255
Eigenvalues, 50, 204

Environmental fate, for specific exposures, 37-38
Epidemiologic studies, description of, 48
    case-control studies, 47
    chance, 52
    cohort studies, 45-47
    confounding, 52
    cross-sectional studies, 47-48
    defining a new syndrome, 48-49
    identifying and evaluating, 45-53
    inclusion criteria, 51
    limitations of Gulf War veteran studies, 53
    misclassification, of outcomes, 57
    multiple comparisons, 52
    nested case-control studies, 47
    statistical techniques used to develop a case definition, 49-51
    summary, 53
    types of epidemiologic studies, 45-48
Expanded Combat Exposure Scale, 25, 124
Exposure (in the Persian Gulf), 21-41
    to depleted uranium, 24
    to nerve agents, 6-7, 174
    to oil-well fire smoke, 6, 22, 40-41, 172-174
    pyridostigmine bromide tablets for, 14, 23-24
    studies using biologic monitoring for specific exposures, 39-41
    studies using environmental fate and transport models for specific exposures, 37-38
    studies using simulation to assess the potential magnitude of exposures, 26-36
    summary and conclusions, 41
Exposure assessment, 21-41
    in epidemiologic studies, 21
    recommendations for, 7-8, 254
Exposure assessment with questionnaires, 21-24
    exposure to depleted uranium, 24
    exposure to oil-well fire smoke, 22
    exposure to pyridostigmine bromide, 23-24
    exposure to vaccination, 22-23
Exposure information, limitations of, 16-17
Exposure misclassification, 36
Exposure-symptom relationships
    in Australian Veteran Studies, 70
    in Canadian Veteran Study, 65
    in Danish Peacekeeper Studies, 69
    in Department of Veterans Affairs Study, 61-62
    in Ft. Devens and New Orleans cohort studies, 71
    in the Iowa Study, 59
    in Seabee cohort studies, 74
    in Oregon and Washington Veteran Studies, 64

INDEX 267

    in the Pennsylvania Air National Guard Study, 75
    in Seabee Reserve Battalion Studies, 72-73
    in University of London Veteran Studies, 67
    in University of Manchester Veteran Study, 68

F

Factor analysis, 2, 49-50, 59, 213-214, 248
Factor-analysis derived syndromes, 203-214
    Air National Guard, 207-208
    Australian Gulf War studies, 206
    cluster analysis, 212-213
    Department of Veterans Affairs, 204-205
    Department of Veterans Affairs Gulf War Health Registry, 212, 213
    Guy's, King's, St. Thomas's Schools of Medicine, 206, 212-213
    the Iowa study, 205-206
    Portland area veterans, 210-212
    primary studies, 203-206
    Seabee cohort, 208
    Seabee cohort and validation study, 208-210
    secondary studies, 207-212
    summary and conclusion, 213-214
    University of Manchester, 203-204, 212
Fatigue. *See* Chronic fatigue syndrome
Fibromyalgia, 3-4, 59, 62, 188-191, 247, 249
    difficulties validating, 57
    primary studies, 188-189
    secondary studies, 189
    summary and conclusion, 190
Fort Devens and New Orleans cohort studies, 70-71, 89-92, 124-125, 163, 183, 228
    exposure-symptom relationships, 71
    symptom clustering, 71

G

Gastrointestinal symptoms, 4, 60, 64, 250
Germany cohort, 125, 228, 248
GHQ-12. *See* Nonpsychotic psychologic illness testing (GHQ-12)
Goldenhar syndrome, 194
Gulf War
    deployment, 12-13
    environmental and chemical exposures, 13-14
    living conditions in, 13
    the setting, 12-15
    threat of chemical and biologic warfare, 14-15

"Gulf War syndrome" (GWS), 3, 17, 202, 209
GWS. *See* "Gulf War syndrome"

H

Hawaii and Pennsylvania Active Duty and Reserve Study, 76, 102
Health outcomes, 115-231, 247-255
    all-cause hospitalization studies of, 223-226
    based primarily on symptoms and self-reports, 3-5, 248-250
    birth defects and adverse pregnancy outcomes (ICD-10 O00-Q99), 192
    cancer (ICD-10 C00-D48), 115-121
    cardiovascular disease, 166-169
    chronic fatigue syndrome, 161-165
    diseases of the circulatory system (ICD-10 I00-I99), 166-169
    diseases of the digestive system (ICD-10 K00-K99), 180-182
    diseases of the musculoskeletal system and connective tissue (ICD-10 M00-M99), 185-187
    diseases of the nervous system (ICD-10 G00-G99), 153-160
    diseases of the respiratory system (ICD-10 J00-J99), 170-179
    diseases of the skin and subcutaneous tissue (ICD-10 L00-L99), 183-184
    injury and external causes of morbidity and mortality (ICD-10 S00-Y98), 219-222
    mental and behavioral disorders (ICD-10 F00-F99), 122-130
    multiple chemical sensitivity, 227-231
    neurobehavioral and neurocognitive outcomes (ICD-10 F00-F99), 131-152
    symptoms, signs, and abnormal clinical and laboratory findings (ICD-10 R00-R99), 202-218
Health Symptom Checklist (HSC), 124-125
Healthy-warrior effect, 56, 227, 251
Heaters, tent, 26
Hormones. *See* Male reproductive hormones
Hospitalizations
    all-cause, 223-226
    for unexplained illness, 202-203
HSC. *See* Health Symptom Checklist
Hypertension, 166, 167, 168, 252

I

ICD. *See* International Classification of Diseases and Related Health Problems
Impact of Event Scale, 76
Inclusion criteria, 15-16, 51
Infertility, and male fertility problems, 196-201
Injury and external causes of morbidity and mortality (ICD-10 S00-Y98), 219-222
    primary studies, 219-220
    secondary studies, 220
    summary and conclusion, 220-221

International Classification of Diseases and Related Health Problems, 9th Revision, Clinical Modification (ICD-9-CM), 126
International Classification of Diseases and Related Health Problems, 10th Revision (ICD-10), 115n
Intestinal conditions. *See* Gastrointestinal symptoms
The Iowa study, 58-60, 78-79, 163, 189, 205-206, 228, 250
    exposure-symptom relationships, 59
    symptom clustering, 59
    women's health, 59-60
Irritable bowel syndrome, 69, 74

J

Job-task-unit-exposure matrix, 21

K

Kansas Veteran Study, 64, 89, 167
Khamisiyah demolition
    chronology of, 29-30
    dispersion models, 29, 35-36
    maps of, 31-34
    potential exposure to sarin and cyclosarin, 6-7, 26-36, 166-168, 174-176

L

Laboratory findings, abnormal. *See* Symptoms, signs, and abnormal clinical and laboratory findings
Lagrangian modeling, 37-38
Laufer Combat Scale, 125
Leishmaniasis, 74
Limitations of Gulf War veteran studies, 2, 53
    of cohort studies and derivative studies, 56-57
London School of Hygiene and Tropical Medicine Veteran Study, 68
Lou Gehrig's disease. *See* Amyotrophic lateral sclerosis
Louisiana. *See* Fort Devens and New Orleans cohort studies

M

Major depressive disorder (MDD), 122-123
Male fertility problems and infertility, 196-201
    primary studies, 196

selected studies, 200-201
summary and conclusion, 196
Male reproductive hormones, levels of, 252
Manpower Data Center, 62-63, 78, 103, 162, 166, 180, 183, 185
Maps of the Khamisiyah demolition, *See* Khamisiyah demolition
MCS. *See* Multiple chemical sensitivity
MDD. *See* Major depressive disorder
Medical tests. *See* Diagnostic medical tests or objective measures
Mental and behavioral disorders (ICD-10 F00-F99), 122-130
- primary studies, 123-127
- secondary studies, 127
- summary and conclusion, 127

Metals, monitoring for, 40-41
Migraine headaches, 64
Military service, adverse health effects of, 15-16
Military-unit-based studies, 70-75, 98-104
- Ft. Devens and New Orleans cohort studies, 70-71, 89-92
- larger Seabee cohort studies, 73-74, 97
- Pennsylvania Air National Guard Study, 74-75
- Seabee Reserve Battalion Studies, 71-73

Ministry of Defence for UK Gulf War, Gulf War health registry, 58
Mississippi Scale for Combat-Related PTSD, 125, 127
Mood-cognition-fatigue, 207
Mood-cognition symptoms, 75
Motor neuron disease. *See* Amyotrophic lateral sclerosis
Motor-vehicle accidents, 219
Multiple chemical sensitivity (MCS), 3-4, 64-65, 227-231, 247-249
- Cullen criteria for, 227-228
- difficulties validating, 57
- primary studies, 227-228
- secondary studies, 228-229
- summary and conclusion, 229

Multisymptom-based medical conditions, 3-4, 203-205, 247-249, *See also* "Chronic multisymptom illness"
Multivariate analysis, 64, 69, 74, 228, 230
Musculoskeletal symptoms, 71, 75, 207, 213
Musculoskeletal system diseases, 185-191

N

National Adult Reading Test (NART), 132, 136, 142
National Comorbidity Survey Replication, 122
National Health and Nutrition Examination Surveys, 39
National Health Interview Survey, 58n, 60
National Health Service Central Register, 117
National Medical Expenditures Survey, 58n

Nerve agent exposure, 6-7, 174. *See also* Khamisiyah demolition
Nervous system diseases, 153-160
Neurobehavioral and neurocognitive outcomes (ICD-10 F00-F99), 131-152
    neurobehavioral deficits, 248
    neurobehavioral tests and confounding factors, 131-132
    primary studies, 142-143
    related findings of malingering and association of symptoms with objective test results, 140
    secondary studies, 144-152
    summary and conclusion, 140-141
Neurological outcomes, *See* Nervous system diseases
Neuropathy. *See* Peripheral neuropathy
Neuropsychologic symptoms, 3, 69, 71, 140, 213
New Orleans cohort. *See* Fort Devens and New Orleans cohort studies
New Orleans Reservist Studies, 76, 100-101
Nuclear Industry Family Study, 195

O

Obsessive-compulsive disorders, 73, 126, 208, 216
ODTP. *See* Oregon Dual Task Procedure
Oil-well fire smoke exposures, 6, 11, 13, 22, 37-38, 40-41, 172-174, 178-179
    modeling, 37-38
Operation Desert Shield, 12-13, 16
Operation Desert Storm, 13, 16, 24
Oregon and Washington Veteran Studies, 63-64, 86-88, 189
    exposure-symptom relationships, 64
    symptom clustering, 63-64
Oregon Dual Task Procedure (ODTP), 136, 140
Overview of health outcomes, summary, 247-254
    outcomes based primarily on symptoms and self-reports, 247-250
    outcomes with objective measures or diagnostic medical tests, 251-254

P

PAH. *See* Polycyclic aromatic hydrocarbons
PCL-M. *See* Posttraumatic Stress Disorder Checklist-Military
Pennsylvania Air National Guard Study, 74-75, 168
    exposure-symptom relationships, 75
    medical evaluation, 75
    symptom clustering, 75
Peripheral neuropathy, 157-160, 251-252
    primary studies, 157-158
    secondary studies, 158
Persian Gulf War. *See* Gulf War
PFTs. *See* Pulmonary-function tests

Physical and mental health screening questionnaire (SF-12), 70, 126
PIR. *See* Proportional incidence rate
PMRs. *See* Proportional morbidity ratios
Polycyclic aromatic hydrocarbons (PAH), 37, 40-41
Population-based studies, 58-70
    Australian Veteran Studies, 70, 84-85
    Canadian Veteran Study, 65
    Danish Peacekeeper Studies, 68-69, 93
    Department of Veterans Affairs Study, 60-63
    Iowa Study, 58-60, 65, 78-79
    Kansas Veteran Study, 64, 89
    London School of Hygiene and Tropical Medicine Veteran Study, 68
    Oregon and Washington Veteran Studies, 63-64, 86-88
    United Kingdom Veteran Studies, 65-68, 81-84
    University of London Veteran Studies, 65-67
    University of Manchester Veteran Study, 67-68
Portland area veterans, 210-212
Postdeployment and predeployment screening
    recommendations for, 7, 254
Posttraumatic stress disorder (PTSD), 3, 5, 59, 64-65, 70-71, 74, 76-77, 122-130, 248, 250
Posttraumatic Stress Disorder Checklist-Military (PCL-M), 123, 125-126, 128-129
Predeployment and postdeployment screening, *See* Postdeployment and predeployment screening
Pregnancy. *See* Ectopic pregnancy
Primary Care Evaluation of Mental Disorders (PRIME-MD),123-125, 128-130
Proportional incidence rate (PIR), 117, 119
Proportional morbidity ratios (PMRs), 180, 182, 203, 220, 222-224
Prospective cohort studies, 46
Psychiatric outcomes, 122-130
    primary studies, 123-127
    secondary studies, 127
    summary and conclusion, 127
PTSD. *See* Posttraumatic stress disorder
PTSD Checklist-Military, see Posttraumatic Stress Disorder Checklist-Military
Pulmonary-function tests (PFTs), 63, 170, 177
Pyridostigmine bromide (PB) use, 14, 23-24

R

Recommendations, 7-8, 10, 254-255
    exposure assessment, 7-8, 254
    predeployment and postdeployment screening, 7, 254
    surveillance for adverse outcomes, 8, 254-255
Respiratory outcomes, 6-7, 170-179, 252-253
    exposure to nerve agents, 174
    exposure to oil-well fire smoke, 6, 172-174

other exposures, 174
primary studies, 170-171
secondary studies, 171-172
Respiratory system diseases, 65, 170-179, 253

S

Sarin, exposure to, 26-34, 73
SCID. *See* Structured Clinical Interview for DSM-IV
SCL-90 test, 140, 146-149
Screening, recommendations for predeployment and postdeployment, 7
Seabee cohort studies, 73-74, 97, 167, 171, 185, 208, 229
    exposure-symptom relationships, 74
    larger, 73-74, 97
    symptom clustering, 73-74
    and validation, 208-210
Seabee Reserve Battalion Studies, 71-73
    exposure-symptom relationships, 72-73
    symptom clustering, 72
SF-12. *See* Physical and mental health screening questionnaire
SF-36. *See* 36-Item Short-Form Health Survey
Sickness Impact Profile, 59n
Signs. *See* Symptoms, signs, and abnormal clinical and laboratory findings
Simulation assessment of potential exposure magnitude, 26-36
    epidemiologic studies using fate and transport models to assess exposure to sarin and cyclosarin, 35-36
    Khamisiyah demolition and potential exposure to sarin and cyclosarin, 26-34
    tent heaters, 26
Sinusitis, 60
Skin diseases, 183-184. *See also* Dermatologic conditions
SNAP (Schedule for Nonadaptive and Adaptive Personality), 124
Somatization, 73, 208
Spontaneous abortion, 194-195, 252
    continued surveillance for, 8, 255
Stomach conditions. *See* Gastrointestinal symptoms
Structured Clinical Interview for DSM-IV (SCID), 124, 228
Subcutaneous tissue diseases, 183-184
Substance abuse, 3, 70-71, 248, 250
Surveillance for adverse outcomes
    recommendations for, 8, 254-255
Survey findings (Phases I and II), in Department of Veterans Affairs Study, 60-62
Symptom clustering, 202
    in Australian Veteran Studies, 70
    in Canadian Veteran Study, 65
    in Danish Peacekeeper Studies, 69
    in Department of Veterans Affairs Study, 60-61

in Ft. Devens and New Orleans cohort studies, 71
in the Iowa Study, 59
in larger Seabee cohort studies, 73-74
in Oregon and Washington Veteran Studies, 63-64
in the Pennsylvania Air National Guard Study, 75
in Seabee Reserve Battalion Studies, 72
in University of London Veteran Studies, 66-67
in University of Manchester Veteran Study, 67

Symptoms, signs, and abnormal clinical and laboratory findings (ICD-10 R00-R99), 202-218
    factor-analysis derived syndromes, 203-214
    hospitalizations for unexplained illness, 202-203
    unexplained illness, 202

Syndromes, defining a new, 48-49, 66-67
    factor-analysis derived, 203

T

Tent heaters, 13, 25, 26
Test of Memory Malingering (TOMM), 140
Testicular cancer, 5, 116-117, 119-120
    continued surveillance for, 8, 254
36-Item Short-Form Health Survey (SF-36), 59, 124
Threats, of chemical and biologic warfare, 14-15
Thyroid conditions, 64
TOMM. *See* Test of Memory Malingering
Transport models, for specific exposures, 37-38
Transportation accidents, 219-221

U

Unexplained illness, 17, 202
    hospitalizations for, 202-203
Unexplained symptoms, 17
United Kingdom Veteran Studies, 65-68, 81-84. *See also* University of London Veteran Studies; University of Manchester Veteran Study
    London School of Hygiene and Tropical Medicine Veteran Study, 68
University of London Veteran Studies, 65-67
    exposure-symptom relationships, 67
    symptom clustering, 66-67
University of Manchester Veteran Study, 67-68, 203-204, 212
    exposure-symptom relationships, 67-68
    symptom clustering, 67
Uranium. *See* Depleted uranium exposures
Urinary tract abnormalities, 6, 194, 252
    continued surveillance for, 8, 255

Urinary-uranium testing, 39-40
US Armed Services Center for Unit Records Research, 38
US Army Center for Health Promotion, 38
US Army Environmental Health Agency (AEHA), 37, 40

V

Vaccinations, 21-22
    against biologic and chemical warfare agents, 67
Veterans Programs Enhancement Act, 1, 11
Volatile organic compounds (VOCs), monitoring for, 40-41

W

WAIS. *See* Wechsler Adult Intelligence Scale
Warfare, threat of chemical and biologic, 14-15, 23, 65, 71
Washington State, 154. *See also* Oregon and Washington Veteran Studies
Wechsler Adult Intelligence Scale (WAIS), 132
Whiteley Index, 124
Women's health, in the Iowa Study, 59-60